ALLIS-CHALMERS
FARM EQUIPMENT
1914-1985

NORM SWINFORD

PUBLISHED BY THE
AMERICAN SOCIETY OF AGRICULTURAL ENGINEERS
2950 NILES ROAD, ST. JOSEPH, MICHIGAN

About ASAE — The Society for engineering in agricultural, food, and biological systems

ASAE is a technical and professional organization of members committed to improving agriculture through engineering. Many of our 10,000 members in the United States, Canada and more than t00 other countries are engineering professionals actively involved in designing the farm equipment that continues to help the world's farmers feed the growing population. We're proud of the triumphs of the agriculture and equipment industry. ASAE is dedicated to preserving the record of this progress for others. This book joins many other popular ASAE titles in recording the exciting developments in agricultural equipment history.

Allis-Chalmers
Farm Equipment 1914-1985

Editor and Book Design: Melissa Carpenter

Library of Congress Card Number (LCCN) 94-072224
International Standard Book Number (ISBN) 0-929355-54-7

Printed in the U.S.A.

Acknowledgements

The writing of *Allis-Chalmers Farm Equipment 1914-1985* has been especially challenging since the company archives were decimated during the late 1980s when A-C was in the throes of bankruptcy. The writer's personal tractor files accumulated over 30 years were reasonably complete but were by no means adequate on the rest of the line. This gave the writer an opportunity to renew old friendships by tapping the memories and personal files of company colleagues who loaned photographs and brochures to help fill in the blanks. Except as noted, the information in this book came from Allis-Chalmers printed materials and records as well as vivid memories of all those who were personally involved with various products and projects.

Thanks and sincere appreciation to all these helpful persons: Ray Dague and Bob Tweedy loaned materials and reviewed the entire manuscript for accuracy and to keep the writer honest. Bill Bryant and W. L. Voegeli checked the tractor sections and provided valuable facts. Alan Jones made his extensive library of A-C literature available and supplied many unusual photographs as well as verifying the facts on the green tractors. Ray Doggett, Dick Moe, and Keith Wheeler provided information on the implement line and reviewed those chapters. John Buzzell loaned a file on the Oxnard products. Craig Cannon provided many photographs and reviewed the chapter on industrial tractors. Paul Harrer provided a wealth of facts and photographs and checked the material on the LaPorte products. Ross Yohn, keeper of the files at the Gleaner plant, supplied volumes of combine material and reviewed the Gleaner history. Pat Murphy loaned material on the Fred E. Cooper products. Igor Kamlukin provided copies of his patents. J. T. Whitworth checked the lawn and garden section. Walt Buescher, Paul Freedy, Debbie Grede, Larry Gruenberger, Swede Muehlhausen, and Darrold Pries contributed important photographs.

Betty Hartwig opened the files of the West Allis Historical Society to reveal many good pictures of nearly forgotten products. Several interesting photographs were found in the LaPorte County Historical Society's museum at LaPorte, Indiana.

An Industrial Heritage by Walter F. Peterson, published by the Milwaukee County Historical Society in 1976, was the source of several references. This book is recommended to anyone who wants to study a very complete history of Allis-Chalmers as a company with insights beyond the multitude of products.

The Inside Story of the Innovative Years in Allis-Chalmers Tractor History was written by Harry G. Hoffman about 1979, but is unpublished. This is Mr. Hoffman's memoirs of his years from 1924 to 1931 when he worked for Harry Merritt as the advertising manager.

At ASAE Donna Hull, Melissa Carpenter, Suzanne Howard, and Bill Thompson took two fair-sized boxes of paper and photographs and made this book out of them. No small task. Their craftsmanship, dedication, and patience must be experienced to be fully appreciated.

To Dave, Ann, Mark, and Paul (our kids) for their interest and encouragement. And last, but far from least, to Betty for her support, patience, and understanding when the book came first.

This book is dedicated to the thousands of hard-working, conscientious, and dedicated people who made Allis-Chalmers the great company it was. Only the cruel realities of economics kept their efforts from having an entirely different outcome. Under even slightly more favorable conditions "Allis-Chalmers Farm Equipment 1914-1985" could have had a much later ending date.

Preface

This book focuses on the products and related events of Allis-Chalmers farm equipment up to the formation of Deutz-Allis. Future historians will carry on after 1985 and recount the continuation of the ALLIS name.

Allis-Chalmers is a name that holds many different meanings and also has led to several different ongoing businesses. Construction equipment became Fiat-Allis. Electrical equipment became Siemens-Allis. Agricultural equipment became Deutz-Allis in 1985 and became AGCO (Allis Gleaner Corporation) on June 22, 1990. By 1993, AGCO was the North American manufacturer and/or distributor of farm equipment branded AGCO Allis, White Tractor, Hesston, GLEANER, Deutz-Allis, SAME, and Massey Ferguson.

The author writes this A-C farm equipment history from the vantage point of an insider. Norm Swinford was a career marketing man at Allis-Chalmers during the glory years and kept his collection of A-C publications after he retired. He also kept in close contact with numerous A-C associates, including myself, and borrowed their folders of product information and data to add to his own photos and memories. This book is a uniquely accurate perspective of Allis-Chalmers farm equipment products during the twentieth century. It is rich with data that compliments a vast spread of carefully selected photographs and illustrations. It reveals some inside events and decisions at A-C not known to many others.

The author includes a broad coverage of all significant products well beyond the popular tractor scope. This includes engines, industrial equipment, implements, planters, lawn and garden products, combines, and harvesting equipment. Agricultural Engineering professionals were highly involved in these products. It is fitting that ASAE was selected as the publisher.

Robert H. Tweedy
Fellow and Past President
American Society of Agricultural Engineers

Table of Contents

♦ *Denotes a table.*

Allis-Chalmers Farm Equipment 1914-1985

Norm Swinford

The Rise of an Industrial Giant

From a small French burr millstone maker in 1847 to a $2 billion corporation in 1979, Allis-Chalmers became one of the more important machinery manufacturers in the United States. The depth and breadth of its product lines would have filled an encyclopedia of mechanical devices. Technology, as simple as those first millstones to as complex as nuclear physics and fuel-cell chemistry, was equal to any major company in the country during the 140 years of its active manufacturing history.

Few persons outside the company itself have a complete vision and understanding of this extremely complex and multi-faceted corporation. It could be said that Allis-Chalmers was a conglomerate long before the term became popular on the business pages.

It is important to know the background of the Allis-Chalmers Corporation to appreciate the contributions of the company and its people to the mechanization of agriculture and feeding the population of the world. During the "Land of Power" show, which was presented to about 7,200 customers in 1962, the following story of the corporation's importance to food production was given:

The headquarters and main plant of Allis-Chalmers in West Allis, Wisconsin. It covered over 160 acres, had 21 miles of railroad track, and the power plant could supply enough power for a city of 50,000 people.

Edward P. Allis was born in Cazenovia, New York, in 1824 and died on April 1, 1889.

"Allis-Chalmers is a major contributor to putting a loaf of bread on your table. Allis-Chalmers tractors and implements prepare the seedbed and plant the seed. A-C combines harvest the wheat. A-C motor graders maintain the farm-to-market roads. A-C electric motors power the grain elevators. Then the grain goes to the mill . . . probably over a concrete highway, and A-C could very well have provided the crushers and screens that prepared the aggregate for the concrete. Allis-Chalmers engines drive loaders and screens and A-C built the huge rotary kilns and grinding mills that produced the cement. A-C construction machinery could have built the roadbed. At the flour mills, it is quite likely that A-C equipment milled the grain into flour. Nearly 90 percent of the flour and flaked cereals consumed in the U.S. is milled on A-C mills. All along the line, electricity was used and Allis-Chalmers plays a major role in supplying the "juice", producing turbines, generators, transformers, and other electrical equipment. We are proud here at Allis-Chalmers to have had such a big hand in putting that loaf of bread on your table. By the way, when you pop a slice of bread into your toaster in the morning, be sure it's plugged into AC."

During the mid-1840s, Milwaukee, which got its name from an Indian word or phrase which meant "gathering place by the river," was a busy and growing young city. The local folks were proud of their industries which included breweries, brick factories, tanneries, blacksmith shops, wagon and boat shops, saw mills, grist and flour mills, all along the banks of the Milwaukee River. One of the growing industries was flour milling. One small mill in 1840 had grown to five by 1847, and they not only supplied flour to the entire region, but also exported over 34,000 barrels.

Charles Decker and James Seville established their Reliance Works in 1847 to make French burr millstones. It was acquired by Edward P. Allis in 1861 at a sheriff's sale.

The flour-milling industry needed equipment of all kinds, especially millstones. Two young men from Ohio, Charles Decker and James Seville, moved to Milwaukee in 1847 to cash in on the growth. They built a small building and hung out their shingle: "Decker & Seville - Manufacturers of French Burr Millstones, Grist and Sawmill Supplies." They prospered and were soon able to build a new shop which they named "The Reliance Works." By 1853, Decker and Seville employed 50 men and enjoyed a fine reputation. In 1857, 75 men were on the payroll and the firm was the largest iron business in the city. Their list of products included water wheels, shafting, gearing and millwork, castings, French burr millstones, hoisting screws, lighter screws, damsons, millricks, proof staffs, bolting cloths, and portable mills of all sizes. This growing enterprise also produced stoves and did general blacksmithing work.

The Panic of 1857 and its depression was hard on Decker and Seville. Heavy debt led to bankruptcy and a sheriff's sale on May 11, 1861. One account reported that the Reliance Works was valued at $6,000 and that Edward P. Allis acquired it for $22.72 — money was apparently scarce around Milwaukee that year.

E. P. Allis was born in Cazenovia, New York, in 1824. He graduated from Union College in Schenectady, New York, with a law degree, and shortly after made his way west to Milwaukee. He practiced law for a short time, but soon found that the business world was more attractive to him. He was involved in several different business enterprises before he made his big investment in the Reliance Works. From that inauspicious beginning he built Edward P. Allis & Company into a viable and thriving firm by surrounding himself with capable people and letting them do their jobs. He understood market

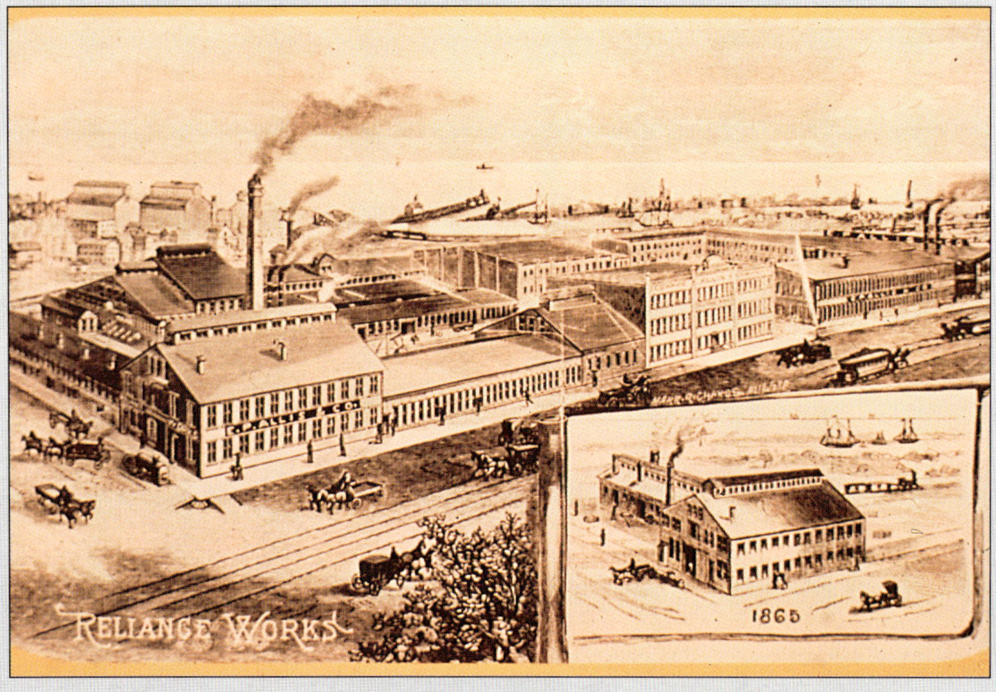

The Reliance Works grew over the years, but the company eventually outgrew it. After Allis Chalmers was formed in 1901, the company moved to West Allis.

Allis-Chalmers acquired the Monarch Tractor Corp. in 1928, bringing crawler tractors in three horsepower sizes.

needs and responded quickly with products to meet those needs. By 1865 sales were up to $100,000, and in 1867 business jumped to $150,000.

Allis expanded into steam power, and on New Year's Eve, 1869, the shriek of a steam whistle announced the arrival of the first Allis steam engine. To gain foundry capacity he purchased the Bay State Iron Works that same year, and by 1870 he employed 200 people and had sales of $350,000. He added steam pumps and built the largest centrifugal pump in America in 1884 and the first triple expansion pumping engine two years later. Following E. P. Allis' death on April 1, 1889, his family and a few trusted employees continued to operate the company. By that time the payroll had grown to 1,100 persons.

Allis-Chalmers Is Formed

In 1901, Allis-Chalmers was formed by the merger of the Edward P. Allis Company, Fraser and Chalmers Company and Gates Iron Works, both of Chicago, and Dickson Manufacturing Company, of Scranton, Pennsylvania. The corporate headquarters was first located in Chicago, later in New York City, but returned to Milwaukee in 1906.

Shortly after the merger in 1901, a large tract of land was purchased in what became West Allis, Wisconsin, and upon which the new factory was built. It would eventually grow into a great industrial complex and employ some 17,000 people in its prime. Expanding into the electrical equipment business, the company acquired the Bullock Electrical Manufacturing Company of Cincinnati, Ohio, in 1904. This became the Norwood plant, and its major products were large electrical generators. A-C built its first steam turbine-generator in 1906.

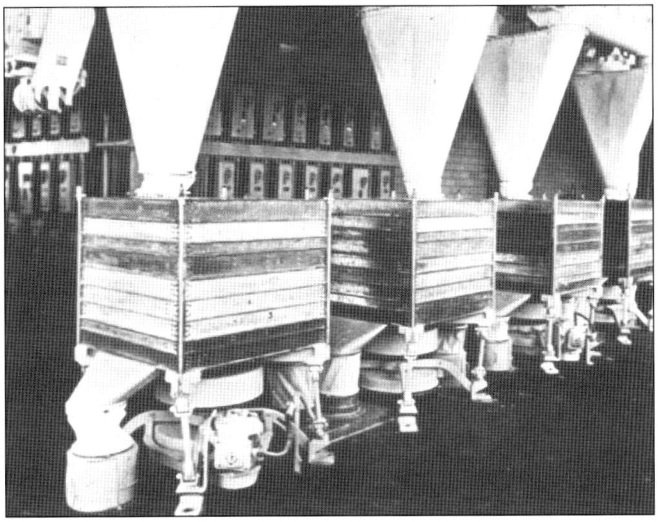

The Nordyke-Marmon Co. of Indianapolis, Ind., brought a line of flour mills in 1926. It was said that A-C mills processed 90 percent of the flour and flaked cereals in the United States.

The LaCrosse Plow Co. was purchased in 1929. It was already 69-years old, dating back to 1860 as the Barclay and Bantam Implement Works.

Searching for diversification, Allis-Chalmers built its first farm tractor in 1914. As the tractor business grew, the Tractor Division, with its own distribution system, was formed in 1926. The Monarch Tractor Corporation of Springfield, Illinois, was purchased in 1928, adding a line of crawler tractors. A year later, A-C acquired the LaCrosse Plow Company, of LaCrosse, Wisconsin, with their extensive line of plows, disk harrows, and other tillage implements. The General Machinery Division added the Nordyke-Marmon Flour Milling Company of Indianapolis, Indiana, in 1926. The Pittsburgh Transformer Company was acquired in 1927. Part of the Brown-Boveri Electric Corporation, which included the Condit Electrical Corporation, became the Boston plant in 1931.

The acquisition of Advance-Rumely, of LaPorte, Indiana, in 1931 not only brought a long line of threshers and combines, but also a complete distribution system with branches throughout the United States and Canada, and access to a large dealer organization. In 1938 the Brenneis Manufacturing Company, of Oxnard, California, contributed a line of deep tillage equipment needed to complement the crawler tractors which were popular on the West Coast.

After World War II, A-C continued its growth. The Gadsden, Alabama, plant was purchased in 1948. It built feeder voltage regulators, later cotton pickers, and the Model G tractor. To expand Tractor Division sales in the European market, Allis-Chalmers Great Britain was formed and the

The huge LaPorte plant was headquarters for Advance-Rumely when it became part of Allis-Chalmers in 1931. It was the largest manufacturer of threshers and steam engines in the United States.

Essendine, Lincolnshire, England, plant was purchased in 1950. A subsidiary called Canadian Allis-Chalmers Ltd. was formed in 1951 and it operated a plant at Lachine, Quebec (Lachine is a suburb of Montreal). The LaPlant-Choate Manufacturing Company of Cedar Rapids, Iowa, joined the company in 1952, adding pull-type and motor scrapers to the road machinery line. The Buda Company of Harvey, Illinois, became the Engine Division in 1953, and brought a needed line of diesel engines and lift trucks.

The acquisition of the Gleaner Harvester Corporation of Independence, Missouri, in 1955, had a profound impact on Allis-Chalmers farm equipment sales with their line of self-propelled combines. The Baker Company of Springfield and Beardstown, Illinois, builders of bulldozers and snowplows, was also added in 1955. A-C had been using Baker dozers for some years before the acquisition. In 1957, Allis-Chalmers purchased the assets of Thomas C. Pollard Pty. Ltd. of Newcastle, Australia, which had been building motor graders under license from A-C. The new operation became Allis-Chalmers Australia Ltd. and continued to build motor graders and other capital goods. Also in 1957, the company, with Radio Corporation of America, was selected by Princeton University to design and build a facility for advanced research into controlled thermonuclear reactions.

The LaPlant-Choate Mfg. Co. of Cedar Rapids, Iowa, produced motor scrapers and motor wagons. They joined A-C in 1952.

The Gleaner Harvester Corp., of Independence, Missouri, came aboard in 1955. Their self-propelled combines became a major part of the farm equipment line.

In 1958, the company purchased its Mexican subsidiary, Industrial Dufermex, S.A., in Mexico City, to build transformers. Also in 1958, certain assets related to diesel engines of Micromatic Hone Corporation of Detroit were purchased and added to the Harvey engine plant.

The Buda Co. of Harvey, Illinois, was acquired in 1953, bringing a line of diesel engines and lift trucks.

The S. Morgan Smith Company of York, Pennsylvania, became the York plant in 1959, bringing a line of hydraulic turbines, valves, water handling equipment, and other special equipment. The Tractomotive Corporation of Deerfield, Illinois, builder of wheel loaders, also joined the corporation in 1959. The same year Allis-Chalmers Italiana, S.p.A., Cusano, Italy, builder of crawler tractors, and Valley Iron Works Corporation of Appleton, Wisconsin, a wholly-owned subsidiary manufacturing papermaking machinery, became members of the corporation.

In 1960 the company acquired controlling interest in Establissements de Constructions Mecaniques de Vendeuvre, S.A., a French manufacturer of air-cooled diesel engines, engine-generator sets, and lift trucks at a plant in Dieppe. A wholly-owned subsidiary, it became known as Allis-Chalmers Material Handling Europe.

Acquisition of Schwager-Wood Company of Portland, Oregon, an electrical switching equipment manufacturer, was completed in 1963. The same year, the company's Canadian subsidiary, Allis-Chalmers Rumely, Ltd., opened a new plant in Guelph, Ontario, to build lift trucks and wheel loaders. Also in 1963, the company replaced the Mexico City plant with a new plant at San Luis Potosi, Mexico, to produce lift trucks. This plant also produced Gleaner combines for a short time, most of which were exported to Australia.

A-C was well-known for its hydraulic turbines and generators. Units like these are running in Hoover Dam in Arizona, Grand Coulee Dam in Washington, and Aswan High Dam in Egypt.

In 1965 the Simplicity Manufacturing Company, of Port Washington, Wisconsin, became a wholly-owned subsidiary. They built both Simplicity and Allis-Chalmers lawn and garden products. In 1968, A-C acquired the Henry Manufacturing Company, of Topeka, Kansas, which became the Industrial Tractor Division. The production of industrial tractors at the West Allis tractor plant was transferred to Topeka shortly after that acquisition.

In 1969, Allis-Chalmers acquired Standard Steel Corporation and its facilities in Los Angeles, California, and Decatur, Illinois. They built asphalt plants, fish meal plants, small kilns, and rotary driers and coolers. Also during 1969 the Company built an electric motor plant near Little Rock, Arkansas, and a materials handling equipment plant at Matteson, Illinois.

During 1970, license agreements with Siemens AG and Transformatoren Union AG, both of Germany, gave the company access to new electrical products. The Company and Kraftwerk Union AG, also of Germany, formed an affiliate, Allis-Chalmers Power Systems, Inc., to market steam turbine generators for

Steam turbines and generators were primary products of the electrical equipment group and hundreds of them are still running all over the world.

A-C built enormous rotary kilns for the production of cement, pelletizing iron ore, and processing fertilizer.

both conventional and nuclear power plants. Allis-Chalmers built an industrial controls plant in Wichita Falls, Texas, in 1972.

In 1972, Fabrica de ACO Paulista, of San Paulo and Sorocaba, Brazil, a maker of crushing and screening machinery, was purchased. The same year Allis-Chalmers bought a minority interest in Svedala-Arbra AB, Svedala, Sweden, that included an agreement covering the manufacture of mining and screening equipment. Later a majority interest was obtained. New electrical product facilities were built at Jackson, Mississippi; Sanford, North Carolina; and New Orleans, Louisiana.

A joint venture of the Allis-Chalmers Construction Machinery Division and Fiat S.p.A. of Italy was formed in 1974 to build and market the products of both companies. In the new worldwide operation, in which Fiat owned 65 percent and A-C owned 35 percent, there were two holding companies, Fiat-Allis B.V., a Netherlands corporation, and Fiat-Allis Inc., a U.S. corporation. The operating company was known as Fiat-Allis Construction Machinery, Inc., and headquartered at Carol Stream, Illinois. It included the Springfield and Deerfield plants formerly a part of A-C.

During 1974, Stephens-Adamson, Inc., became a wholly-owned subsidiary. They produced bulk material handling equipment and systems. Headquartered at Aurora, Illinois, they also had facilities at Clarksdale, Mississippi, and in Canada and Australia.

In 1975, Allis-Chalmers-Canada, Ltd., and National Steel Corporation of Canada, Ltd., entered into a partnership under the name of Sudbury Metals Company and leased the Falconbridge nickel-iron pellet refinery which was modified to the Allis-Chalmers direct-reduction process for converting iron oxide pellets into metallized iron pellets. The Sudbury Metals Company plant, Sudbury, Ontario, processed about 1,200 tons of oxide pellets per day.

Allis-Chalmers acquired capability in the processing of gases through the acquisition in October 1978, of American Air Filter, Inc., of Louisville, Kentucky. Its broad line of precipitators, scrubbers, air filters, and dust control systems were all designed, manufactured, and marketed for environmental air quality control.

In 1979, Allis-Chalmers acquired Barron Industries, Inc., of Birmingham and Leeds, Alabama. The company manufactured heavy-duty fans, mechanical dust collectors, hot gas dampers, expansion joints, and heat recovery and recycling equipment.

Allis-Chalmers had 71 plants worldwide producing a vast variety of machines in 1980. About 29,000 employees staffed these plants and over 100 branches, sales offices, and warehouses.

A major factor in automatic storage and retrieval systems, Hartman Metal Fabricators, Inc., of Victor, New York, became a member of the Allis-Chalmers Corporation in 1980. Hartman had the capability to design, manufacture, and install the special requirements of any computer-oriented automated storage and retrieval system. Products included computer-controlled storage cranes, high-rise rack structures, conveyor systems, electronic controls, and supporting equipment.

By 1980 the Allis-Chalmers Corporation was a company of approximately 29,000 employees. The Company had 71 plants worldwide and sales of $1.973 billion in 1979. It had more than 100 sales offices, branches, and warehouses in principal cities of the world. Nearly 5,000 dealers sold and serviced its products, which served the processing, minerals, coal, material handling, outdoor products, agricultural, and environmental air markets.

In the brief period from 1980 to 1985 when all its markets were seriously depleted and huge discounts were required to reduce inventories, the once-great industrial giant faded into bankruptcy. Its fine products were scattered, some among buyers who would continue their existence and others which would be lost completely. Like the products, the excellent workforce was scattered; some to surviving companies, others to a desolate job market. Though only a fraction of its former self, Allis-Chalmers lives on. Those of us who were a part of this great company hope that, like the fabled phoenix, it will rise again to its former greatness.

The bridge over South 70th Street in West Allis is still there but the name is gone. It is sadly missed.

Allis-Chalmers Enters the Tractor Business

The Search Begins

General Otto Falk, president of Allis-Chalmers from 1913 to 1932, was the driving force that put A-C in the tractor business. He had close ties to agriculture, owning a farm in Waukesha County, just west of West Allis. General Falk realized that, like the business he headed, agriculture must be mechanized if its productivity were to flourish. He also recognized that Allis-Chalmers must have greater diversification to offset the market fluctuations of the industrial equipment lines already in production.

He directed his staff in a search for products which could utilize the company's broad manufacturing resources. Three products were selected for further investigation: the rotary tiller, the tractor-truck, and the farm tractor.

The rotary tiller could be equipped with "stepper tracks" for work in soft soils.

General Falk sent J. F. Max Patitz, chief consulting engineer, to Switzerland to investigate the Motoculture "Motor-Driven Rotary Plow" in 1913. Patitz brought one back along with an engineer to operate it. Allis-Chalmers entered into an agreement to manufacture and market the rotary plow in the United States for $10,000 and a 6-percent royalty on all rotary plows sold (Petersen, 1976).

The Motoculture Rotary Plow from Switzerland was the pattern for A-C's rotary tiller. The company paid $10,000 plus a 6-percent royalty for manufacturing rights.

The Allis-Chalmers rotary tiller No. 1. Although it was based on the Swiss design there were many basic changes in the prototype, including the longitudinal engine arrangement.

A new machine on demonstration always draws a crowd. The A-C rotary tiller at work in 1915.

A-C made several major changes in the design before starting production. The wheel design of the Allis-Chalmers rotary tiller No. 1 was entirely different from the European machine. The Swiss version had a transverse engine, while the A-C engine was mounted longitudinally and drove the wheels and tiller through bevel gears.

The tractor had a single front wheel for steering and two drive wheels each 52 inches in diameter and 11 inches wide. The drive wheels had a center rim which allowed the machine to roll smoothly on the road while lugs on each side of the rim provided traction as the center rim sank into soft soil. The tiller weighed 4,500 pounds and was powered by a four-cylinder, 30-horsepower engine. There were three speeds of 0.8, 1.3, and 2.0 mph for tilling and one of 3.6 mph for traction. The tractor had individual rear wheel brakes for sharp turns. The tiller could be disconnected to permit the use of the tractor for other types of work, including belt work. No production numbers have been found, but it appears that only a few were built and sold (Petersen, 1976). As rotary tillage has since enjoyed some success, it was an idea whose time had yet to come.

An experimental tractor-tiller at work near Dallas, Tex., circa 1918. This big rig was rated at 25-45 hp. No records exist to indicate that this unit went beyond the experimental stage.

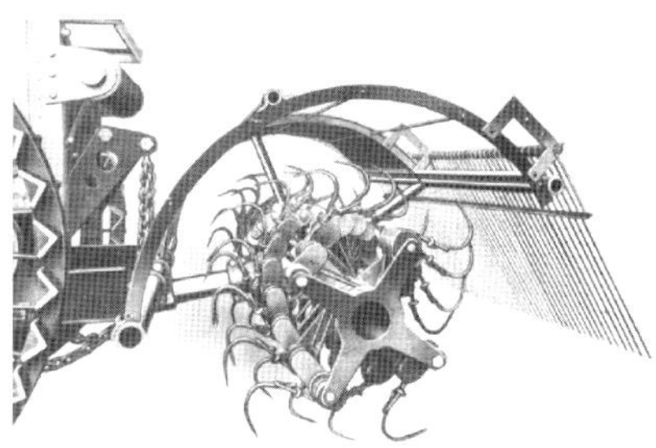

The tiller was 5 ft wide and could work as deep as 12 in. The claws were made of hardened steel and the claw holders were made of high-quality spring steel.

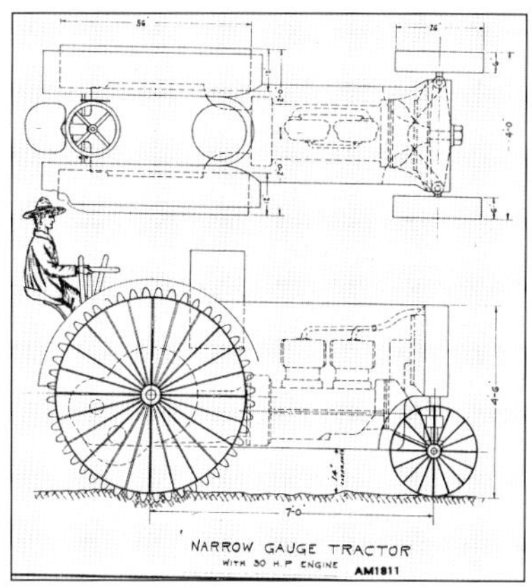

This engineering drawing was found in the West Allis tractor engineering files. There is no other information but it appears to be related to the 25-45 tiller.

The Tractor-Truck — 1915

Concurrent with the rotary tiller and tractor studies, work was begun on the Tractor-truck and by 1915 a prototype was running. The 5-ton machine was priced at $5,000, and looked like a natural for the unimproved roads of rural America (Petersen, 1976). A small number were built and some were even sold to Russia. (A. Solzhenitzen refers to them in his book "The Red Wheel." Also see Springfield crawlers.)

Two other companies built vehicles similar to the Tractor-truck during the same period. One was the Centiped Tractor built by the Phoenix Manufacturing Company of Eau Claire, Wisconsin, and the other was the Lombard Tractor Truck, built by the Lombard Auto Tractor Truck Corporation of New York City. Litigation over patent rights was initiated by Lombard but the outcome is not known.

The Tractor-Truck of 1915 looked like a natural for farm work and rural roads, but its $5,000 price tag may have been too high. Its successors joined the military and became half-tracks. Some were sold to Russia in World War I.

The Bull tractor was considered for a joint venture, but General Falk preferred that the company design its own.

The Bull Tractor

The third alternative proved to be the right one, although success was still well into the future. The true farm tractor was exactly what was needed to realize General Falk's objectives.

One approach was to enter into a joint venture with an established tractor manufacturer. The firm of Lyons, Knoll, and Hartsough of Minneapolis, Minnesota, offered their Bull tractor. General Falk declined this opportunity, however, preferring to

An early version, perhaps a prototype, of the Model 10-18. This is the earliest photograph found in the files. Note the bolt-type rear wheel spokes.

With only a single drive wheel, the Bull would have been easily stalled in poor traction conditions.

build and market an Allis-Chalmers design (Petersen, 1976). With only one drive wheel, operating the Bull must have been a real challenge in many conditions.

The Green Tractors
The Model 10-18 Tractor

After the false starts of the rotary tiller, tractor-truck, and Bull tractor, the company developed the Model 10-18. The prototype was finished in 1914 and production began the following year. The horizontally-opposed two-cylinder engine started on gasoline, then switched over to kerosene when it was warm enough to vaporize the heavier fuel. With a 5.25-inch bore and 7-inch stroke, the displacement was a healthy 303 cubic inches, and it ran a conservative 720 rpm. It was rated at 10-drawbar and 18-belt horsepower, but was never tested at Nebraska since only a few were built after the test law took effect in 1919. The 10-18 had one speed forward and one reverse, which was typical of early tractors. The gears were enclosed and ran in oil, which was unlike many early tractors. The 10-18 weighed 4,800 pounds.

From a sales brochure: "*The simple, durable, practical construction of the Allis-Chalmers Farm Tractor assures you unusual service. The Allis-Chalmers 10-18 horsepower Farm Tractor is built for results . . . It is of sturdy construction and has ample strength to meet all demands placed on it, yet is one of the light-*

This 10-18 is a later version and has the welded wheel spokes as shown in most photographs and in the parts list.

A Model 10-18 pulling a LaCrosse three 14-in. moldboard plow in North Dakota.

A prototype of the General Purpose tractor with a grain binder. It looks like the original drawings, especially the engine.

est weight farm tractors made. The materials used in it are approved by practical engineers and machinists as the best for their respective purposes. In its design all superfluous parts have been eliminated — it is a model of effective simplicity and has a remarkable capacity for hard work. Operating costs are low for it is economical of fuel. The frame is a one-piece steel casting — no joints, no rivets, no weak spots. The motor is built expressly for the tractor and is of the most efficient type — a great engine that more than makes good on its rating. In plowing it takes the place of six horses and will operate either two or three plows according to the condition of the soil plowed."

By 1918, the Model 10-18 was becoming obsolete with the emergence of the 15-30 and 18-30. With only half-hearted sales promotion, the company must have recognized that the days of the three-wheel tractor were gone. Although the Model 10-18 was not a great success, A-C was in the tractor business. This was just a small indication of things to come.

The Allis-Chalmers General Purpose Tractor

All three of these photographs (above) are captioned as the Allis-Chalmers General Purpose Tractor, the forerunner of the Model 6-12. No dates are available, but they must fall in the period of 1917 or early 1918.

Note the evolution of the prototype tractors. Radiators, fuel tanks, drive wheel supports, and steering mechanisms all change as the design progresses. Based on the photo numbers, the tractor pulling the binder is the earliest version, followed by the cultiva-

Ten Reasons Why You Should Buy an Allis-Chalmers Tractor

1. BUILT by a company that is world famous as manufacturers of better machinery.
2. EIGHT YEARS of experimental work under actual farm conditions, covering every possible design, has proved conclusively that each feature embodied is the best.
3. PRICE is based upon actual cost of manufacture in quantity by the most modern and efficient methods. Our enormous purchasing power assures us lowest prices on raw material.
4. COMPLETE stock of repair parts available at distributing points at prices in proportion to the price of complete machines, less the cost of assembling.
5. LOWEST plowing cost per acre.
6. HIGHEST drawbar horsepower per gallon of fuel. Our double-reduction drive transmits power from the engine to the drawbar with minimum loss.
7. CORRECT distribution of weight permits easier turning and better handling, and lessens soil packing.
8. LARGE water jackets and oversize radiation keep the motor cool even at high temperatures under peak loads.
9. LOW depreciation by reason of superior construction. Our tractors are built, not assembled.
10. PERFORMANCE is guaranteed. A sale is complete when the tractor has given full measure of service in the purchaser's hands.

From "The Nose Bag," (a company publication) dated January 1922.

This appears to be a production model of the General Purpose tractor, later called the Model 6-12. The single-wheel sulky was replaced by a two-wheel unit on later tractors.

The Ford trademark is easily seen on the engine block of this prototype, but Ford engines were not used in production.

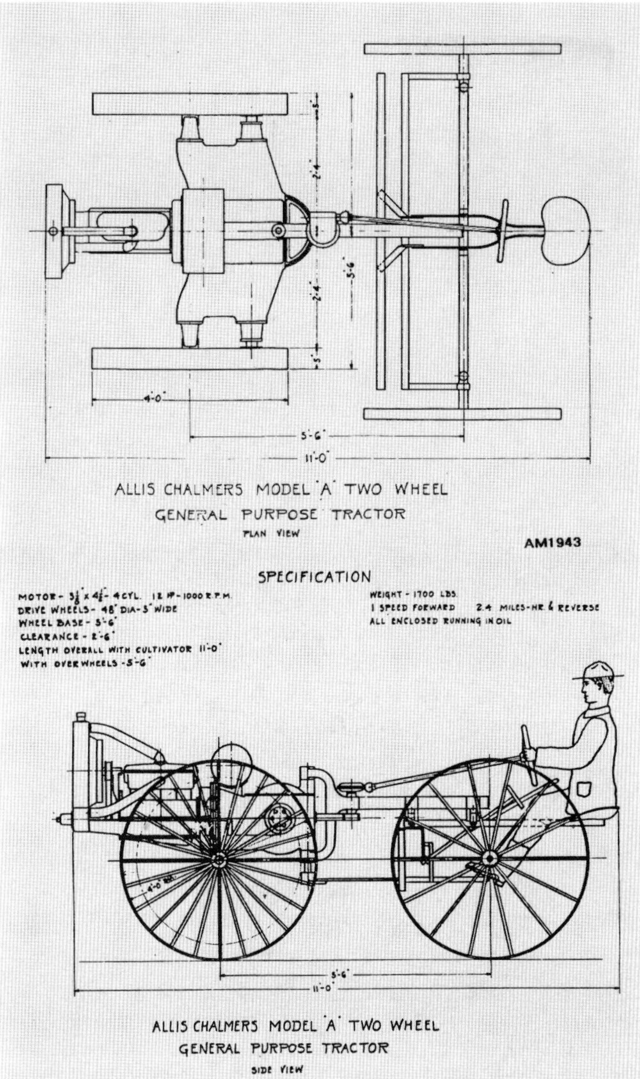

These drawings of the Model A General Purpose, dated December 1917, are similar to the prototypes.

tor. Close examination of the original photograph reveals "Ford" in its classic trademark script cast into the block of the engine. The last unit seems to be a production version.

The Model 6-12, B, and Duplex

Even though the Model 10-18 production was just beginning in early 1915, the General Purpose, later known as the Model 6-12, was already in development. Like other two-wheel designs of this period, it was intended to utilize horse-drawn implements with minimum modification. When hitched to an implement, the operator rode on the implement seat,

The General Purpose tractor with a combination implement truck pulling two-10 in. bottoms. One 16- or 18-in. bottom was usually recommended.

The 6-12 Model B. The external final-drive gears have disappeared and the two-wheel sulky was added. No production quantities have been found.

The Model B with orchard fenders working among low-hanging branches. The Model B was apparently designed especially for orchard operation.

The Model B disking in an apple orchard near Sturgeon Bay, Wisc.

enabling him to handle any implement controls. When the tractor was not hitched to an implement, the operator rode on the seat of a single- (later two-) wheel sulky. Almost all of the tractor's weight was concentrated over the drive wheels, resulting in good traction. The Model 6-12 was hitched to a wide variety of plows, harrows, mowers, rakes, cultivators, both corn and grain binders, and even a wagon and a manure spreader, all with the operator on the implement seat. This concept allowed a farmer to use both draft horse and tractor power without the investment in both types of implements, and may have hastened the acceptance of tractors by farmers who wanted to ease into this revolutionary age.

As near as can be determined, the 6-12 started out as the Model A General Purpose tractor in 1917, with production beginning in 1918 or early 1919. Production records began with 1919, so it is not known how many were built prior to that time. A total of 1,426 tractors were built from 1919 through 1926. In addition there were 45 "Cane" models built from 1920 through 1923. Just what the designation "Cane" refers to with regard to the 6-12 is unknown.

The 6-12 was powered by a four-cylinder L-head LeRoi engine with a 3.125-inch bore and 4.5-inch stroke (138 in.^{3}), rated at 1,200 rpm. It had one forward speed of 2.5 mph and weighed 2,500 pounds.

Two types of tractors were built. The better known Model 6-12 was called the General Purpose and had a high-mounted engine. The final drive consisted of pinion gears engaging large diameter ring gears which were mounted around the inner circumference of the drive wheels.

There was also the Model B version. With its low-mounted engine, it was well adapted to orchards. The large wheel-mounted final-drive ring gears disappeared, so it apparently had a pinion-internal ring-gear final drive similar to the 18-30. No mention of

The Model 6-12 final test area, probably about 1919.

the Model B appeared, however, in the production records. If any significant number were built they may have been included in the Model 6-12 quantities.

The price of the Model 6-12 started out at $850, but dropped to as low as $295 in 1923 (Petersen, 1976). No doubt the inconvenience of hitching it to horse-drawn implements, plus the rapid customer acceptance of the conventional four-wheel tractor design, hastened the abandonment of the two-wheel models throughout the industry.

In Nebraska Test No. 54, the test board, in commenting on the 6-12 advertising literature, stated: "None of the claims are excessive or unreasonable except the following: 'Today it stands forth . . . as the most efficient small power unit available'. 'There is no loss of power, no dead weight, no lost motion in the Allis-Chalmers general purpose tractor'." It seems that over the years the people responsible for writing tractor advertising copy haven't changed much.

The Duplex

The idea of having two smaller tractors for planting, cultivating, and other light work, and then combining them for heavy tillage has long had some appeal. Tandem tractors, mostly homemade, have been tried many times, although none very successfully. Tandem tractor hitches were tested extensively following World War II, but the difficulties of hitching and unhitching the tractors, synchronizing engine speeds, and controlling the rig in the field seemed to offset the advantages. Few if any ever reached even limited production.

Max Patitz, Allis-Chalmers chief consulting engineer, originated the concept of combining two 6-12 tractors for heavy tillage power. His patent for the "Duplex Hitch" was the subject of an infringement suit in 1919. The outcome of Patitz versus Mattson is not known, but probably wasn't relevant because the Duplex was not commercially successful. Patitz made a good case for the Duplex in the book "Allis-

The Duplex could pull a three-bottom plow in its one gear at 2.5 mph.

This exhibit was used in the patent rights case of Patitz vs. Mattson. No record of the outcome has been found.

Chalmers Farm Tractors Dealer Cooperative Sales Helps" which appears later in this chapter.

No separate records for the Duplex were available since production would obviously list only the 6-12 tractors. There is also no indication that the Duplex ever found any significant acceptance. Someone observed that the reason the Duplex was unsuccessful was because the operator couldn't get both engines running at the same time, or if he did, he didn't know which way the tractor would go.

The Duplex was two Model 6-12 tractors hitched back-to-back to provide power for larger implements. There is no indication that the concept was commercially successful.

With Four Wheels A-C Could Really Roll — Models 15-30, 18-30, 20-35, 25-40

In the mid-1910s it became apparent that farmers wanted four-wheel tractors. Two- and three-wheel tractors might be all right for some limited markets, but if a builder wanted to sell tractors in any quantity, a modern tractor must have four wheels. Allis-Chalmers' first conventional four-wheel tractor was launched in 1918. Quite a feat for a new tractor manufacturer, A-C brought three unique models to market in only four years.

The Model 15-30 was a three-plow tractor rated at 30-belt and 15-drawbar horsepower. Five were built in 1918. In early tests it produced more power than expected and was immediately renamed the 18-30. Nebraska Test No. 55 (1920) gave it 33.41-belt and 20.55-drawbar horsepower.

The 15-30 was A-C's first four-wheeler. Early 15-30 tractors had metal side sheets and narrow fenders.

This cross section could apply to either the 18-30 or the 20-35 tractors. The change gears for the two-forward, one-reverse transmission were arranged crosswise.

This 20-35 is a 1922 model. The 20-35 had this appearance until the redesign of 1927.

The engine was an Allis-Chalmers four-cylinder vertical valve-in-head design with a 4.75-inch bore and 6.5-inch stroke that displaced 461 cubic inches and was rated at 830 rpm. With two forward speeds of 2.31 and 2.82 mph, there was a choice of speeds, but not much with only 0.5 mph difference. The change-gear shafts were transverse in the transmission. This hefty unit weighed in at 6,000 pounds.

During 1921, the 18-30 was again tested (No. 83) burning gasoline and rated at 930 rpm. Still called the 18-30, it was rated at 22-38 on gasoline. Maximum results were 43.73-belt and 25.45-drawbar horsepower, more than enough to justify renaming the tractor the 20-35. Due to the higher rated rpm, speeds were up to 2.58 and 3.16 mph. The tractor weight also increased to 6,640 pounds.

Tractor sales had been very slow when George Gardner, manager of the newly-formed Tractor Division, recommended to General Falk that the company get out of the tractor business. General Falk didn't agree, so in 1926 Mr. Gardner's resignation was accepted and Harry Merritt was appointed Tractor Division manager. Mr. Merritt had been the assistant manager since 1923 when he came into the home office from the Kansas City branch (Hoffman, 1979).

The 15-30 pulling a four-bottom plow. When most farmers had walking plows or gang plows, this was big field capacity.

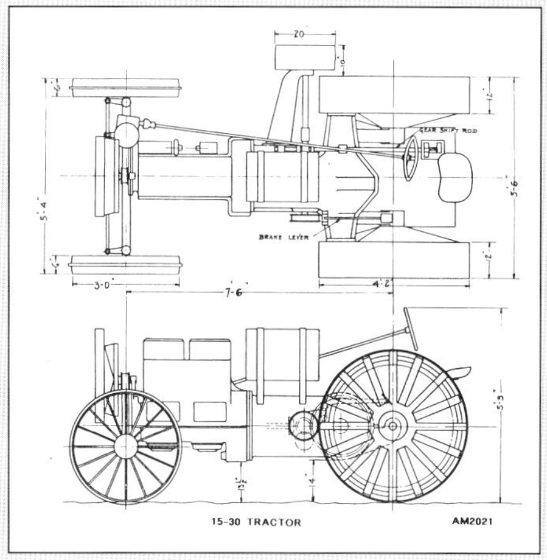

An early engineering layout for the 15-30 had an out-rigger belt pulley and steering gear located on the front axle.

Late 15-30 and the early 18-30 tractors had roll-down hood curtains and wider fenders.

The 18-30 could be equipped with hard rubber pads on the rear wheels for travel hard surfaced roads.

A 1929 Model 20-35 as it appeared after the 1927 redesign. The redesign changed the fenders, platform, and steering and supposedly reduced factory costs so that list price could be lowered substantially.

During 1926 Harry Merritt and Harry Hoffman, advertising manager, stripped a 20-35 down to the chassis and built a mock-up of hood, fenders, and platform which improved the tractor and could be built at less cost. Max Patitz, chief consulting engineer, was furious that the sales department was encroaching on his territory and also built a mock-up based on his own ideas. At a showdown with General Falk and the board, the Merritt design was selected for production. The 1927 complete redesign of the 20-35 reduced costs enough that a price of $1,495 still produced a profit for the company.

It would appear that Test No. 151 in 1928 was a result of that redesign. Although the bore, stroke, and rpm remained unchanged, the new E delivered 44.29-belt and 33.20-drawbar horses. The increased drawbar horsepower was due to greatly increased maximum pull, this in turn due to increased tractor weight, now 7,095 pounds.

It must have been the right plan, because production jumped from 682 in 1926; to 1,814 in 1927; to 4,867 in 1928. About the time the Model U appeared in 1929, the 20-35 became known as the E20-35. In 1931 the distillate-burning Model EK, with a 5.0-inch bore, and turning 1,000 rpm, developed 47.00-belt horsepower (Test No. 193).

The Model 25-40

Most well-designed machines grow in their capabilities throughout their production lives. The farm tractor is an outstanding example of that fact and the

The 18-30 or 20-35 running a thresher. This photograph has been found attributed to both tractors.

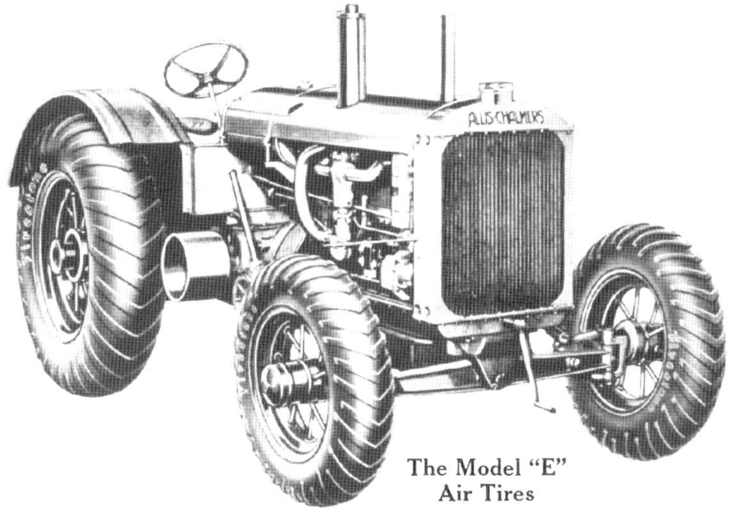

The Model E on air tires and with a 5.25-in. bore had plenty of power for a thresher but road gear was only 3.25 mph. List price was $1,450.

The 1935 Model E on steel with the standard 5.0-in. bore for field work was priced at $1,000.

Allis-Chalmers Model E was certainly no exception. Starting out at a modest 15-30 horsepower rating, the same basic tractor grew to a giant in its day of 33.82-drawbar and 47.00-belt horsepower. That's an 88 percent increase in drawbar and a 56 percent growth in belt power. It speaks well for the original engineers who saw fit to build in plenty of reserve.

Power growth came by increasing the bore from 4.75 to 5.0 inches (510.5 in.^{3}). The Model E with a 5.25-inch bore (563 in.^{3}) was available for belt work. Rated speed went from 830 to 930, then finally to 1,000 rpm. Modern for its time, the engine had removable sleeves, valve seat inserts, and air, oil, and fuel filters. It could burn either gasoline or kerosene, and was represented as being equally efficient on either fuel. The 5.00-inch bore was available in 4.0, 4.4, 5.2, and 5.8:1 compression ratios, and the 5.25-inch bore was offered with 4.0, 4.4, 5.2, and 6.0:1 ratios.

The Model E had only two forward speeds of 2.5 and 3.25 mph. This provided little choice in the field and was snail-slow on the road. With air tires and a good road gear to reduce time between sets, it could have been the answer to a thresherman's prayer. Although the new Model A eventually solved that problem when it came out in 1936.

The 1934 price list for the Model E shows four tractors:

1.	20-35	4.75-inch bore	930 rpm	$870
2.	25-40	4.75-inch bore	1,000 rpm	$970
3.	25-40	5.00-inch bore	1,000 rpm	$1,000
4.	25-40	5.25-inch bore	1,000 rpm	$1,025

All of these models have the same shipping weight, 6,280 pounds, so it is likely that differences are limited to bore and rpm.

The 25-40 rating seems to have coincided with the arrival of the new United-Model U with its 19-30 rating. Apparently more power was needed to help justify a higher price for the 25-40. The 25-40 began with Serial No. 24186 in 1930 and ended with Serial No. 25611 in 1936 when it was replaced by the Model A.

Throughout 19 years, 16,425 Models 15-30, 18-30, 20-35, and 25-40 tractors were built. That averaged only 864 per year, but it was enough to keep the company in the tractor business until the Models U and WC could change Allis-Chalmers' fortune completely.

The Rumely Six

The Rumely 6A appeared in catalogs and price lists of the early and mid-1930s, evidence that the depression of that period left high unsold inventories of all farm tractors.

"A four-plow tractor with the weight of a three" and "A six-cylinder engine for the price of a four" was the sales philosophy of the Rumely Six. It was capable of pulling a four 14-inch plow or driving a 32-inch thresher. In Test No. 185 it developed 33.57 horsepower on the drawbar and 48.37 on the belt. The Six was powered by a Waukesha 6-cylinder L-head engine with a 4.25-inch bore and 4.75-inch stroke with a 404-cubic inch displacement. It was rated at 1,200 and 1,360 rpm.

The Rumely Six had a three-forward, one-reverse transmission, but six-forward and two-reverse speeds were advertised, based on both the 1,200 and 1,360 rated rpm. The tractor had forward speeds from 2.5 to 4.75 mph and two reverse speeds of 3.0 and 3.5 mph.

The Rumely Six was comparable in power to the A-C Model E, but had a six-cylinder engine and three-speed transmission. It sold for $1,395 on rubber in 1935, but none were built after Allis-Chalmers acquired Advance-Rumely in 1931.

Electric starter and lights were available. When equipped with pneumatic tires, it had 7.50-18 front and 12.75-28 rear tires. The 6A was priced at $1,100 for steel wheels and $1,395 for rubber tires. Lights, starter, and distributor ignition were $100 extra. The serial numbers for 1930 were 501-502 and for 1931 were 503-1302. The 6A was discontinued when Allis-Chalmers purchased Advance-Rumely. Apparently the company re-identified some of the tractors since some of the early serial numbers had A-C serial tags while later serial numbers still carried the original Rumely identification tags.

The 12-20/15-25 Tractor (Model L)

By 1919, the new 18-30 was off to a good start and the 6-12 was two-thirds of the tractor production. The 10-18 was in its sunset years, and it was time to plan for its replacement. A new midsize tractor was needed and the success of the 18-30 dictated that the new model would be a conventional four-wheel design. The new tractor was the 12-20 (Model L). Records show that three were built in 1920, probably pilot models, followed by 334 in 1921.

Allis-Chalmers didn't have an engine the right size for a mid-range tractor, so it turned to Midwest of Indianapolis, Indiana, for a power source. The Midwest engine was a four-cylinder valve-in-head type with 4.125-inch bore and 5.25-inch stroke (280.6 in.^{3}) rated at 1,100 rpm.

The 12-20 was a three-plow tractor in most soils. It was powered by a Midwest four-cylinder engine of about 281 in.^{3}

After test No. 82, the 12-20 was renamed the 15-25. The 15-25 was discontinued when the 20-35 was redesigned and its price reduced in 1927. Only about 1,700 were built.

The transmission was a two-forward, one-reverse unit with transverse change-gear shafts similar to the 20-35. Low gear was 2.3 mph and high was 3.1 mph. Nebraska test weight was 4,550 pounds. In Test No. 82 (1921) the tractor was called the 12-20, rated at 14-27, and actually produced 21.42-drawbar and 33.18-belt horsepower. With that successful experience, the tractor was redesignated as the 15-25.

In the first three years of its production life, the 15-25 outnumbered the 20-35, but in the last three, the 20-35 numbers were well above the smaller unit. Clearly, the 15-25 didn't enjoy the popularity of the larger unit. In 1927, the same year that the 20-35 was redesigned for cost reduction, the 15-25 was discontinued altogether. The lower cost of the 20-35 allowed Allis-Chalmers to market a four-plow tractor for essentially a three-plow tractor price, and that left little room for the 15-25.

Allis-Chalmers Talks to Dealers and Customers

The following is a partial reproduction of a 63-page book by the title above. The original is a large book of 13 1/4 × 20 1/2 inches, printed in large print with many illustrations of the Allis-Chalmers tractor line, the West Allis plant, and various advertising materials that were available to A-C dealers.

The book was apparently developed for an A-C traveler to use with dealer prospects and presents the complete story of the Models 6-12, 10-18, 18-30, and Duplex. It was probably printed about 1920. Of the 63 pages, 12 were selected as being the most interesting to Allis-Chalmers enthusiasts.

Your Tractor

Your tractor should be regarded as a faithful friend and partner in your farming enterprise — not as a slave. Under your guidance and with your cooperation, it can be a decisive factor in making your farm a paying business. Given proper treatment, it will not only increase your profits but also shorten your working hours, lighten the physical strain on your body, and give you time for planning, relaxation, and a happier, more independent life. (From an Allis-Chalmers tractor operator's manual.)

Them new-fangle' tractors will be the ruination of the farmer 'cause they don't make no manure. — a non-believer

Introducing
Otto H. Falk
President
Allis-Chalmers Mfg. Co.

There is a Strong Company Back of Allis-Chalmers Farm Tractors

Every wide awake dealer looking to the sales of tractors in his territory as the foundation of a successful business knows one thing. It is folly to take on a line of tractors that is not adequately supported by a strong manufacturing organization---adequately financed---suitably fitted by manufacturing experience to build high-grade machinery---and suitably equipped through manufacturing facilities to build a complete product.

These are days when caution should govern the attitude of the dealer. With one of the largest sales fields in the entire power industry open—with a demand for tractors that has never been exceeded by the demand for any other power equipment---the tendency has been for a tremendous rush on the part of small organizations to attempt to build and sell tractor equipment.

The natural processes of selling must obtain, and those organizations which are weak at the start can bring only disappointment and loss. The very nature of the tractor service is such that only the right kind of a manufacturing organization can build a product which will be satisfactory, and from which the dealer can continue to realize a steadily growing profit, and a steadily increasing business.

To those familiar with the development of machinery, the name Allis-Chalmers needs no introduction. For 65 years the Allis-Chalmers Mfg. Company and its predecessor have been known, not only as among the largest manufacturers of machinery in the world, but as the manufacturer whose product, covering nearly every line in which machinery is used, has been universally superior both in achievement and in the quality of the product.

An Organization With Years Of Manufacturing Experience

Main Office Building—Allis-Chalmers Mfg. Company, Milwaukee, Wisconsin

Page One

I am leaving the specific facts regarding our tractor organization to the men most intimately in charge of its various phases, and shall attempt here to tell you only those facts which will best serve to satisfy you as to the kind of an organization you are tying up with, if you represent Allis-Chalmers tractors in your territory.

As to Size

Allis-Chalmers tractors will be manufactured in Milwaukee, at our own West Allis plant. The figures I will give you here apply to this plant alone.

The total floor area of the West Allis plant of the Allis-Chalmers Mfg. Company is 2,381,947 square feet.

The total ground area covered by this plant is 116 acres.

To furnish power the boiler H. P. is 10,700.

There are over 17 miles of railroad track in our West Allis plant alone—all operated by this plant for convenience in moving the product.

To give you an adequate idea of the productive capacity of this plant, the foundry alone can handle 350 tons a day, and individual castings have been produced weighing as high as 120 tons.

Other plants are located at East Norwood, Ohio, known as the Bullock Works, and in the city of Milwaukee, known as the Reliance Works.

In the West Allis plant approximately 250,000 square feet of floor space are devoted to the manufacture of Allis-Chalmers Tractors.

In this plant every part of the tractor is made, including the motors, transmissions, radiators and all with the exception of the carburetors, magnetos, spark plugs and such small parts as are standardized and best obtained from outside sources.

Our Pledge to Good Will

It has been our pledge to every field which we have served that nothing should ever be shipped from the factory that was not the best that our engineering and manufacturing facilities could produce.

Allis-Chalmers tractors have been in the experimental stage for four years. At any time during these four years we could have stepped into the open market with tractors fully as good, if not better, than others being sold. We were not ready. We were not certain in our own mind that we had reached the highest development possible at that time.

We have proceeded slowly, placing only a few tractors here and there, and these where we could directly control their operation.

$3,000,000.00 represents the actual investment which we have put into tractor manufacturing alone, to build the product right.

This is our investment in good will, from which our return can come only as it comes through you in sales made in your territory.

We have not considered the price in any way disproportionate, and yet it is far in excess of the total investment involved in many tractor factories that have already developed their selling organizations.

Now, the Allis-Chalmers tractor in all models is right. Into our new tractor plant, its equipment in machinery, into our sales organization and its work, into raw material, we are ready to put a much larger investment.

Our own investigation of the tractor manufacturing and sales field makes us believe that no less sums are adequate.

Our Policy

Your satisfaction—the satisfaction of the tractor buyers to whom you sell—the reputation of the Allis-Chalmers tractor as one of undoubted merit, unqualified efficiency and unquestioned value, is the backbone of our policy.

We shall serve the American farmer through the dealers in whom the farmer has confidence and from whom he makes his purchases.

To these dealers we offer the facts of unlimited financial strength—unexcelled manufacturing facilities—unrestricted production capacity and a good will second to none in the entire field of machinery manufacture and sales.

The men we have chosen to direct our tractor activities are men of ability, sincerity of purpose—all imbued with a willingness to work hand in hand with you to the ultimate development of a great tractor success built on a solid foundation.

It shall be my personal aim to keep so thoroughly in touch with every development in the tractor department of the Allis-Chalmers Mfg. Company that I shall know the individual problems which are presenting themselves to our sales representatives all over the country and wherever Allis-Chalmers tractors are sold.

The direct manufacturing responsibility will be in the hands of Mr. C. E. Search, General Works Manager of all Allis-Chalmers plants.

The designing of Allis-Chalmers tractors and all engineering developments will be in the hands of J. F. Max Patitz, Chief Consulting Engineer for all Allis-Chalmers products.

The control of the tractor division, including sales, will be under Mr. F. W. Kamm.

Advertising will be under the direction of Mr. B. M. Seymour.

These men will each tell you of their particular departments and how they will function to build the right kind of a product for you and assist you with the right kind of cooperation in developing the sales upon which rest the foundation of your own personal success.

We have entered the manufacturing of tractors with a firm conviction that it is one of the greatest contributions that we can make to the earning power of the country at large, and to those individuals who serve the country through agricultural pursuits.

We know that the tractor is an economic necessity.

We know that its development presents mechanical problems different from any which have ever before been faced by manufacturers of power machinery. We know that the tractor problem is an individual one which must be approached as an individual one by men with the proper realization of this one fact—a tractor cannot be simply a power plant with a method of transmitting power to driving wheels which will move an implement.

It must be a completely developed power plant in which conditions of load, conditions of operation and conditions of service are treated as they are presented under widely varying conditions of operation and under widely varied requirements of service.

Such a tractor is the Allis-Chalmers, presented in four models in order that no farm, however small or however large, may be without suitable tractor facilities.

Allis-Chalmers Club, Allis-Chalmers Mfg. Company, Milwaukee, Wisconsin.

Introducing
J. F. Max Patitz,
Chief Consulting
Engineer
Allis-Chalmers Mfg. Co.

Built to the Highest Engineering Standards

Five years ago the Allis-Chalmers Mfg. Company decided to manufacture tractors. To determine the most suitable type and size of tractor, a complete investigation of the various types of tractors being built at that time and a thorough inquiry into the conditions under which tractors must operate, was made.

Size of farm, kind of farm—whether dairy or grain, etc., crops grown, soil, weather, topography, altitude, all these are factors to be considered. Soil resistance, when plowing, for instance, varies as much as from 3 lbs. to 20 lbs. per square inch; it is a very important factor as plowing demands generally the greatest amount of work on a farm.

Temperature conditions must also receive serious consideration, as the temperature may range from 40° to 115-120° in the shade, and the tractor must work under a heavy load without overheating, surrounded by sandy and dust-laden air.

For best economy a tractor should work at about 80% of its full load; it is, therefore, heavily loaded all the time unlike other vehicles driven by internal combustion engines. In an automobile, only a small part of the power of the motor is necessary to propel it, after it is in motion; maximum load is put on the motor only under unusual conditions.

After the size of tractor has been settled, the question of the type of tractor to be made comes up—whether the tractor should be built with two wheels, the implements furnishing the balance of the rolling members, whether it should be built with three wheels with one steering wheel and two drive wheels, or with four wheels with two drive wheels and two steering wheels, or with four drive wheels or whether the creeper type would be the most practical one.

The considerations which led up to the design of the several types of tractors manufactured by the Allis-Chalmers Mfg. Company are briefly as follows:

Cultivating is comparatively light work, therefore, a small light tractor having high bottom clearance for corn, and cotton cultivation, is best for this work. A large tractor is not adapted to cultivating on account of its size, weight and bottom clearance, and cannot be used economically for this work.

The two wheel tractor with the implement furnishing the additional rolling members is best suited for cultivating. With this in view, the 6-12 Tractor was developed, but to extend its range of usefulness, it was designed to hitch up to all other implements used on a farm, so that it is what is called a General Purpose Tractor.

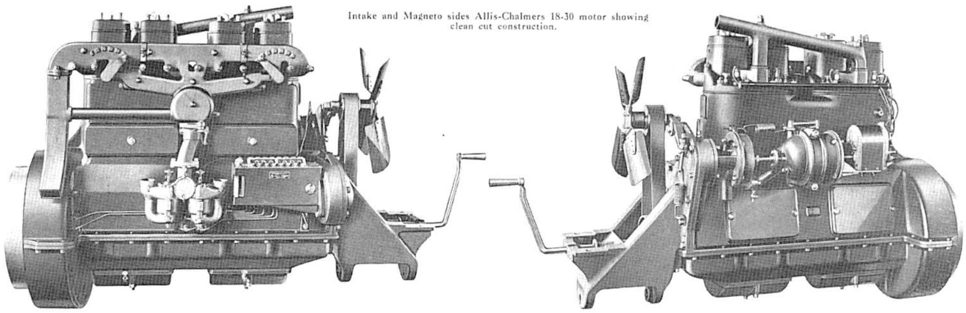

Intake and Magneto sides Allis-Chalmers 18-30 motor showing clean cut construction.

On farms of such size that two 6-12 Tractors would be needed for cultivating, either a large tractor would be required in addition for plowing and disking, or this work would have to be done by the two small tractors pulling one plow each and necessitating one man each to operate it. To meet these conditions the idea was conceived of coupling up two small tractors end to end, and using them as one four wheel drive tractor for the heavy farm work. A frame with platform was designed to accomplish this. This combination tractor is called the Duplex.

Allis-Chalmers double opposed 10-18 motor showing accessibility and sturdy construction.

The Duplex will pull three plows under favorable conditions; it is handled just like a single tractor, so one man can do the work of two. It takes only a short time to combine or separate the two tractors.

The advantages of the Duplex are; first, the power available when the two 6-12's are used together; second, the opportunity of having the most suitable power machinery when light work is being done, such as cultivating, seeding, mowing, raking, binding, etc., when the two 6-12's are used separately.

On farms of large acreage a tractor is desirable, sufficiently powerful to pull up to four moldboard or five disk plows and drive a good size separator. To meet these conditions, the 18-30 Tractor was developed. This is a size of machine which can be easily handled by one man, and, except for special conditions, is as large a tractor as is practical on any farm regardless of acreage.

Between the medium size farms and the large farms, comes the demand for a tractor of intermediate power, which has been met in the 10-18, a three wheel tractor, with two driving wheels and one steering wheel.

A description of the three general types of the Allis-Chalmers Tractors follows, giving the mechanical advantages as they have proved out in our experience.

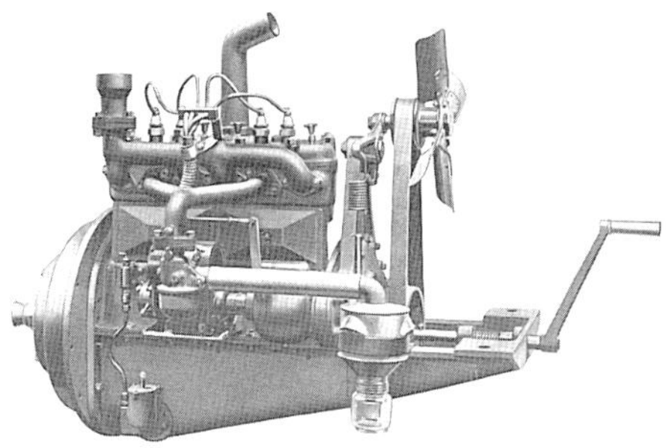

Allis-Chalmers 6-12 motor showing compact design.

Allis-Chalmers transmission for 18-30 made in Allis-Chalmers plant.

The Allis-Chalmers 6-12

This is the Allis-Chalmers tractor which we have designated as the ***General Purpose Tractor*** in that it may be utilized with every type of farm implement. In its operation we may say that this tractor is built primarily to give the utmost satisfaction in cultivating. It has a clearance of 26½ inches directly under the center of the tractor. In cultivating it may be driven directly over the corn row.

The Allis-Chalmers General Purpose Tractor delivers a normal draw bar pull of 1,000 pounds and 12 H. P. at the belt.

It is recommended for one 16 inch bottom. In many localities it is successfully used with two 10-inch or 12-inch bottoms, or two 24-inch disc plows.

The front wheel tread is 54 inches.

The tractor itself is of the two wheel type, supported in the rear by a truck which holds up the tractor when it is used with plows, disks, harrows and similar machinery, but which is removed in using such implements as furnish their own rear riding wheels.

When implements are used which require the use of the rear truck, the operator rides on the seat supported by the tractor, but on other implements he rides on the implement seat.

The draw bar is rigidly fastened to the tractor; it is adjustable lengthwise of tractor and may be swung sidewise to bring it in line with the pull of the implement.

The tractor itself, complete, weighs 2900 pounds.

Traction is by driving wheels, furnished with either lugs or angle iron cleats or extension rims.

Controls on 6-12 General Purpose Tractor showing steering wheel, clutch and transmission levers

The drive wheels have a diameter of 48 inches and a face of 6 inches.

The power plant consists of a four cylinder four cycle vertical type engine and is bolted to the front of the frame.

The frame is a casting. The axles for the driving wheel are pressed into it; the upper portion of it serves as fuel tank and also carries the hinge supports for the steering yoke and tubular beam to which the implements are connected.

The motor used in the 6-12 has a bore and stroke of 3⅛ x 4½ inches.

The motor block is made of the finest quality gray cylinder iron with cast iron pistons and the finest heat treated connecting rods, crank shaft and cam shaft.

Bearings are bronze backed babbitt shells, giving the highest quality construction throughout the working parts.

Cooling is by means of a tube and fin radiator, made in our shops, with an exceptionally ample cooling area and so constructed that the radiator will not clog up with chaff and dust. The water is circulated from the radiator through the motor by the thermo-syphon system with ample jackets for the water to circulate through.

Too much consideration cannot be paid to the effectiveness of the cooling system in a tractor.

In the Allis-Chalmers tractors every effort is bent to maintain the lowest possible range of temperature fluctuation between the water leaving the jackets and the water passing back around the motor from the radiator.

An efficient fan driven by a belt of ample width carries away the heat from radiator.

Lubrication is by the splash and circulating system, the oil being circulated by a positively driven plunger pump. Gasoline only is burned in the 6-12, on account of the small size of the motor; it is fed through a Kingston Carburetor, the air is cleaned through a Bennett Dry Air Cleaner.

Ignition is furnished by a High Tension Magneto.

The Allis-Chalmers 6-12 motor is supplied with a centrifugal governor, enclosed and dust-proof, which may be regulated to control motor speeds at any point between 600 to 1,000 R. P. M.

Bull ring gear in 6-12 General Purpose showing roller driving pinions and segment construction of ring gear

The driver can set the governor to the speed at which he wants the tractor to be driven, from his seat, and under conditions of varying load the governor automatically regulates and keeps the speed of the motor closely to the speed for which it has been set.

The drive in the Allis-Chalmers 6-12 is through a Borg & Beck dry plate clutch to the sliding gear transmission, and thence through the drive shaft to a roller pinion meshing in the bull ring gear mounted in the driving wheels.

The clutch is readily accessible and can be easily adjusted through an opening on the left side of the main frame member.

The transmission is made in our own plant, and is enclosed and dust-proof. We make our own differential and all other members of the drive system.

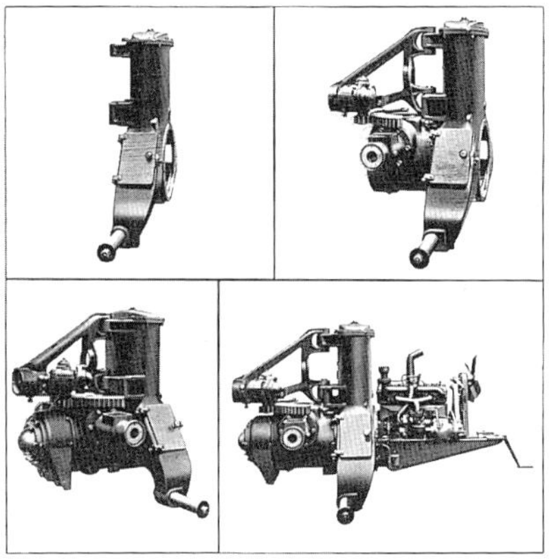

Progressive Assembly of Master Casting Allis-Chalmers 6-12.

The transmission is lubricated by a mechanically actuated splash system. When the tractor is moving, the transmission gears are constantly being sprayed with oil.

The roller pinions are lubricated through holes in the roller pins, which receive the oil from the transmission case.

The transmission is equipped with ball bearings and the wheels run on heavy duty plain bearings.

The steering gear is of the worm and gear type, working on a rack and pinion, the spur pinion on the worm gear shaft meshing with the cast iron rack bolted to the transmission case so that, as the steering wheel is turned, the front or driving wheels are directed in the proper course through the action of this pinion on the rack.

The steering wheel is located directly ahead of the driver and connected to the steering gears by a steering rod and universal joint.

The belt pulley is driven directly from the motor through the transmission shaft without any intermediate gearing and is located back of the transmission case. It is 10" diameter and has 5½" face.

The pulley is readily removed when the tractor is not being used for belt work.

The perfect governing of the motor speed obtained by the governor makes it possible to regulate the belt pulley speed at the right point to secure the best results from the implement being driven.

The 6-12 is so constructed that the rear truck allows for a plow power lift mechanism and also adjustments for other tools.

When a single 16-inch bottom plow is used, it may be raised by a positively operated dog clutch manipulated by a two way foot lever.

The cultivating feature of the Allis-Chalmers 6-12, its low operating cost, its availability for all types of tractor service, coupled with the quality of workmanship and material which we build into it, make it particularly desirable on small farms and indispensable as supplementary equipment on farms of big acreage.

The Allis-Chalmers 18-30

The Allis-Chalmers 18-30 is our largest tractor. It will pull three plows under all conditions—four under favorable conditions—will easily take the place of eight horses in plowing work, and will do all other work in proportion.

It will plow steadily at 2.8 miles an hour.

In it are incorporated features of design which will rank it far ahead of any other tractor of high power now on the market.

As the name implies, it is a tractor delivering 18 H. P. at the draw bar—30 H. P. at the belt. The normal draw bar pull is 3,000 pounds at slow speed, with a liberal surplus.

The tread is 66 inches.

The clearance at the lowest point is 14 inches.

Drive is through the two rear wheels, with the front wheels as guiding members.

The 18-30 is of the frameless type in that the frame is formed by the engine base, transmission case.

Over the front axle the frame is supported at one point, giving a three point suspension.

The front axle is a steel casting of hollow square section and the front axle is so pivoted on the frame as to allow maximum flexibility in accommodating the wheels to uneven ground.

The rear driving wheels are 50 inches in diameter and have a 12-inch face. The steering wheels are 36 inches in diameter and have a 6-inch face.

In the construction of the power plant of the Allis-Chalmers 18-30, there are many features of design that are outstanding. The motor is of the four-cylinder, four-cycle vertical type. The cylinders are cast en bloc and made of the finest gray iron.

The cylinder liners or cylinders proper are removable from the main motor block. These cylinders are made of the finest gray cylinder iron.

The fact that each cylinder liner may be removed individually is of the greatest importance in economical maintenance, since, should any one cylinder become in any way damaged, that one cylinder liner may be removed and replaced with a new one.

The cylinder dimensions are $4\frac{3}{4}$-inch bore, $6\frac{1}{2}$-inch stroke.

The normal motor speed is 830 R. P. M.

The valves are in the head, operated by rocker arms, which are in turn actuated through valve lifter rods by the cam shaft.

The crank shaft runs on three bearings of generous surface and the crank shaft, connecting rod and cam shaft are of the finest heat-treated steel.

Cooling is through a fin and tube radiator of our own make. The water is circulated through the cooling system by means of a high capacity pump forcing the water through the radiator and water jackets at a high rate of speed.

Worm's eye view of 18-30 Tractor

The nature of design in the Allis-Chalmers 18-30 motor makes it possible for us to secure cooling in which the temperature variations in the water from the jackets through the radiator to the jackets is low, not exceeding 10° in the entire cycle of cooling. By so constructing the cooling system, it is possible for us to obtain a motor which will not overheat under the hardest service in continuous operation.

A large fan driven by a wide belt carries away the heat from the radiator.

Lubrication in the motor is by force feed lubricator to the crank shaft and connecting rod bearings of the motor and to the gears driving the circulating pump and magneto shaft.

The oil, for the connecting rod bearings, is collected in a groove in the crank shaft web and driven by centrifugal force through holes in the crank shaft web to the connecting rod bearings, from which it is thrown off in a spray which lubricates the cylinder walls. Lubrication of the cylinders is further added to by the splashing of the connecting rods in the oil troughs.

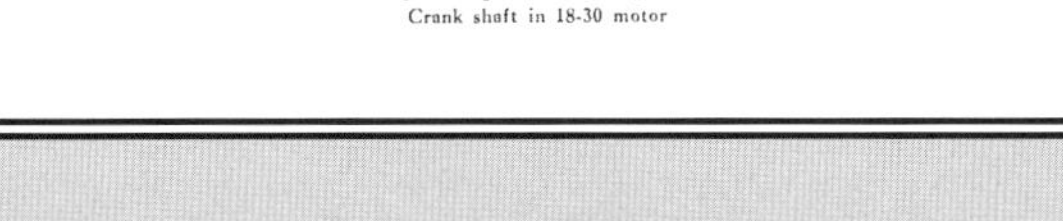

Crank shaft in 18-30 motor

The leads from the lubricator to the main bearings of the motor are located on the outside of the crank case, where they are readily accessible for inspection, and where any trouble may be quickly remedied without taking down the engine base.

Page Fifteen

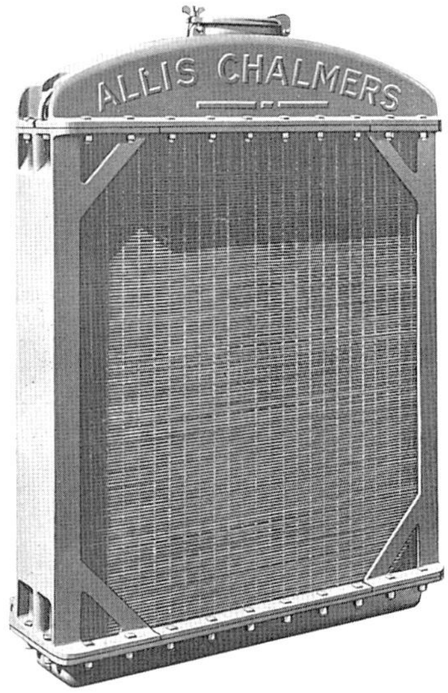

Radiator used on 18-30 built in Allis-Chalmers plant.

The valve rocker arms in the head are lubricated through oil cups, each individually filled and the overhead valve mechanism is enclosed in a dust proof housing.

Ignition is by high tension magneto with impulse starter.

For fuel, gasoline or kerosene is used, there being a three compartment tank, one holding gasoline for starting—the other holding kerosene for running the tractor, and a third compartment for water when running on kerosene.

Carburetor is the double-bowl Kingston Kerosene Burning Carburetor, which can be switched instantly from gasoline to kerosene or vice versa. Air is taken in through a Bennett Dry Air Cleaner.

The 18-30 motor is equipped with our own type of governor for automatically regulating the motor speed.

The power is transmitted from the motor to the drive wheels through an expanding shoe clutch of our own make, and through a sliding gear transmission to an enclosed gear drive in the driving wheels.

The feature of the expanding shoe type of clutch used in the 18-30 is its ease of adjustment, one simple adjusting screw controlling all the adjustments which may be necessary.

The transmission and drive systems of the 18-30 run on Hyatt roller bearings.

The transmission affords two forward speeds, 2.3 and 2.8 miles per hour.

Transmission, differential, rear axle and clutch are enclosed in a dust proof housing bolted to the engine flange.

The rear axle drive member carries a spur pinion meshing with a one piece internal ring gear in the drive wheels, the whole driving mechanism being enclosed in a dust proof housing and running in a bath of oil.

The drive wheels are supported on a live axle.

The draw bar of the 18-30 is swivelled and adjustable to any position which may be necessary to pull the implement properly.

Driver's seat on Allis-Chalmers 18-30 showing simplicity of control system.

The Allis-Chalmers 10-18

The Allis-Chalmers 10-18 is the only three wheel tractor we make. In power and pulling ability it is half way between the General Purpose Tractor and our big 18-30. It is the only Allis-Chalmers tractor in which the power plant is supported on a separate frame.

The Allis-Chalmers 10-18 is designed to operate on both kerosene and gasoline.

In plowing, it will pull two 14-inch bottoms in heavy soil.

The draw bar pull is 1650.

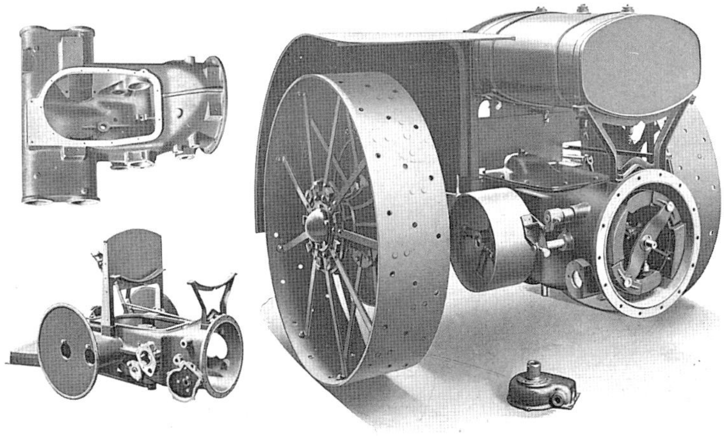

Progressive assembly of master casting Allis-Chalmers 18-30.

Fitting removable cylinder sleeve in Allis-Chalmers 18-30 motor built in Allis-Chalmers plant.

The belt pulley power is 18 H. P.

The total tread of the 10-18 at its maximum width is 70 inches.

The clearance at the lowest point, at the bottom of the engine base is 14 inches, and directly under the wheels is 25 inches.

It is equipped with an adjustable swivel draw bar.

Traction is on the two rear wheels and the one front wheel is used as the guide wheel, being turned by a worm and gear steering device.

The motor is our own make, of the two cylinder horizontal opposed type, L head construction.

Cylinder dimensions are 5¼" x 7" and the normal motor speed is 720 R. P. M.

Cooling, as on all Allis-Chalmers tractors, is through a fin and tube type radiator of our own make, the water being forced through the radiator and water jackets by a centrifugal pump.

Lubrication is by force feed lubricator to crankshaft and connecting rod bearings, and splash for cylinders, cam shaft, etc.

Ignition is by high tension magneto and fuel is fed to the motor by a double bowl carburetor, air being cleaned through a Bennett Dry Air Cleaner.

On the 10-18 the Allis-Chalmers type of centrifugal governor is used, giving the same governing results that are obtained on the other Allis-Chalmers models.

Clutch, transmission and drive system are our own construction, the expanding shoe type of clutch, similar to that used on the 18-30, being employed.

The 10-18 is equipped with one forward speed, giving a speed of 2.3 miles per hour.

All working parts are enclosed in a dust proof housing.

The 10-18 is admirably adapted to belt pulley work and is equipped with a pulley 14½" in diameter with a 6½" face.

Allis-Chalmers Duplex The desirability of coupling two General Purpose Tractors together under certain conditions has been mentioned.

The Allis-Chalmers Duplex is formed by joining two 6-12s together into one unit. The transmission controls of the two tractors are interlocked so that they operate as one.

To make this hitch it is necessary only to remove the tube which forms the support for the rear truck and the driver's seat—back the two tractors together, install a new pipe hitch, crank the two motors and start away.

This Allis-Chalmers Duplex Tractor with four driving wheels will pull three 16-inch bottoms, under ordinary conditions. On many farms it is the real solution of economical tractor power for heavy plowing work. The two 6-12s may be used as one tractor for plowing, disking, harrowing, and also belt work, and taken down in a few minutes into two separate tractors for cultivating, mowing, raking, etc.

Allis-Chalmers 18-30 Tractor Discing

The Allis-Chalmers 18-30 operating separator

Allis-Chalmers 18-30 plowing with four 16-inch bottoms

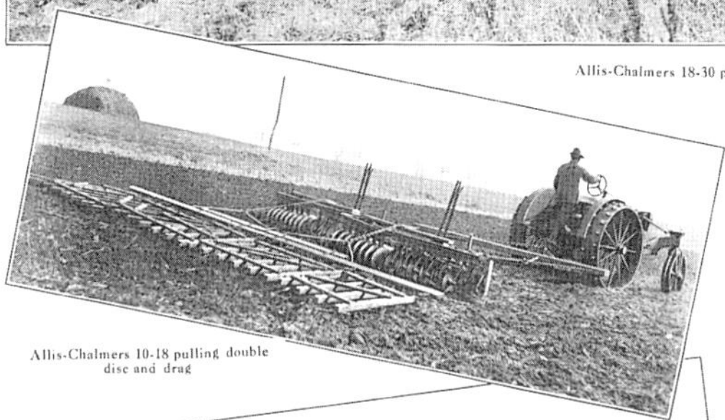

Allis-Chalmers 10-18 pulling double disc and drag

Allis-Chalmers 10-18 pulling four 24-inch discs

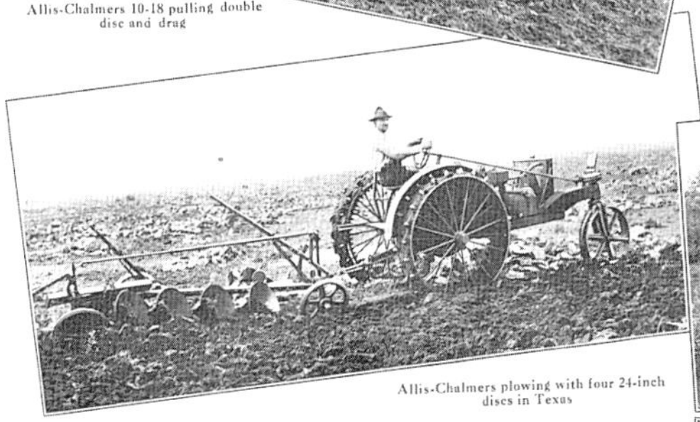

Allis-Chalmers plowing with four 24-inch discs in Texas

Allis-Chalmers Duplex plowing with three 16-inch bottoms

Allis-Chalmers 6-12 plowing with one 16-inch bottom

Allis-Chalmers 6-12 plowing with two 24-inch discs in heavy clay

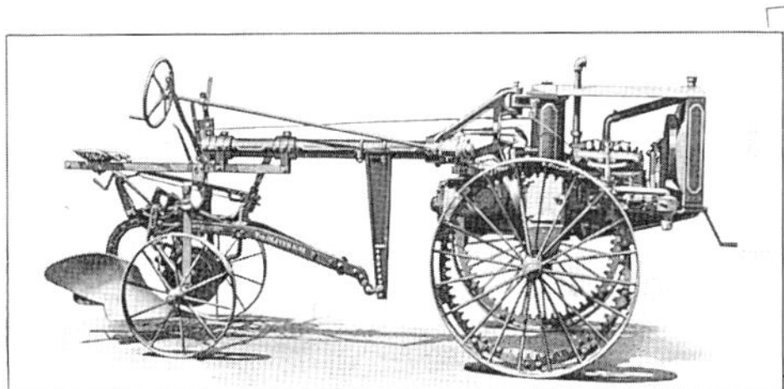

Allis-Chalmers 6-12 with one 16-inch sulky showing driver's seat

Allis-Chalmers 6-12 with double row cultivator showing driver's seat

Allis-Chalmers with a seeder showing driver riding on the implement

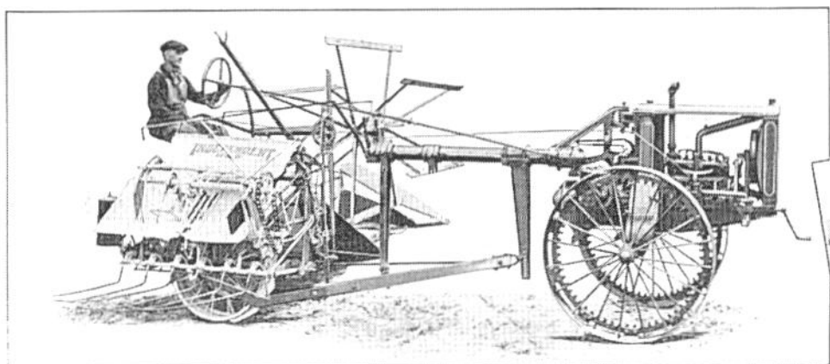

Allis-Chalmers 6-12 pulling binder showing extension steering rod.

Allis-Chalmers 6-12 pulling pea binder

Allis-Chalmers 6-12 with corn binder operated as a one-man outfit

Allis-Chalmers 6-12 pulling potato digger as a one-man outfit

Allis-Chalmers 6-12 mowing showing the adaptability of this tractor to all farm work

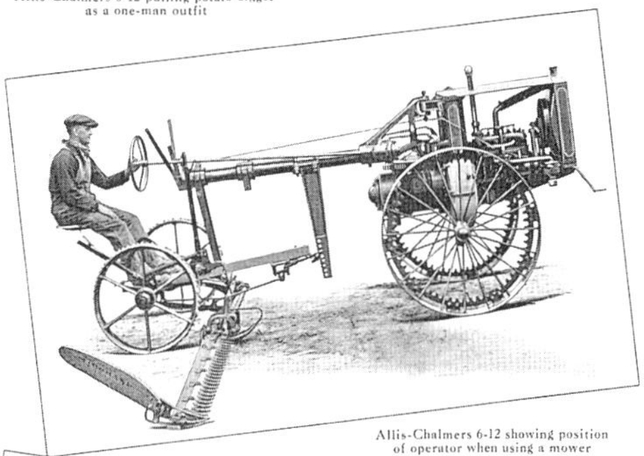

Allis-Chalmers 6-12 showing position of operator when using a mower

Allis-Chalmers 6-12 with rake, operator sits on the implement seat

Allis-Chalmers 6-12 pulling manure spreader, the drawbar construction allows hitch at convenient position

Allis-Chalmers 6-12 operating grinder showing adaptability to belt work

Allis-Chalmers 6-12 hitched to one row cultivator, 26½" clearance for cultivating

Allis-Chalmers pulling road grader, another proof of the General Purpose value of this implement

Every Tractor Dealer Will

Allis-Chalmers advertising will appear in all the prominent implement trade papers, telling the story of Allis-Chalmers development and what Allis-Chalmers means to the aggressive dealer in farm machinery.

ALLIS-CHALMERS FARM TRACTORS

The 18-30—the largest Allis-Chalmers model. A four plow tractor for large farms.

To the building of Allis-Chalmers Farm Tractors we have devoted the experience gained in over 60 years of manufacturing the highest quality machinery.

Experimental and development work have covered four years at an expense of $3,000,000.

All Allis-Chalmers models are built complete in our own plant from the motor to the final drive gears. Production facilities permit of the manufacturing of 50 complete tractors a day.

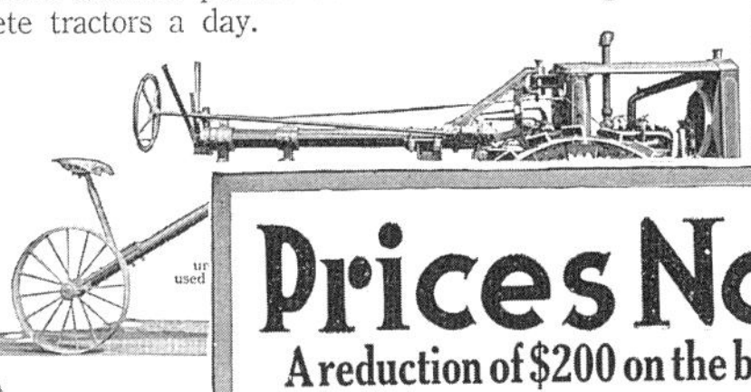

Ask to See This Book

We have stated our policy and our plans for co-operation in this book. Ask to see it. It is the most complete dealer sales help ever presented to you.

ur
used

Dealer

An examination
three Allis-Chalmer
dealership is the n

We are appoi
in selected territ
will bring a sale
yourself of the
we have to off

ALLIS-CHA
Tractor
Milwaukee,

Prices Now 18-30 $1785 6-12 $795

A reduction of $200 on the big Allis-Chalmers Farm Tractor

made in the face of general advances by other big manufacturers and on the Allis-Chalmers 18-30, which even at the old price, was accepted as the biggest tractor value in the industry.

ALLIS-CHALMERS-FARM-TRACTORS

Huge production in one of the largest plants in the world devoted to the manufacture of high grade machinery makes our remarkable reduction in price possible. We build Allis-Chalmers Farm Tractors complete, except carburetor and magneto, from radiator to drive wheels in our own plants.

Insure yourself the dealer supremacy that will come to you if you represent the manufacturer whose product is the dominant one. $3,000,000 spent in the past four years on experimental and development work insure superior design. Financial and manufacturing facilities insure production.

For Dealers Who Qualify—Our sales cooperation is the most comprehensive ever put back of any tractor dealer. Ask to see our Dealer's Cooperative Selling Helps.

Distributors and Dealers—Be sure that the tractor line you tie up with is going to bring you a permanent business. Allis-Chalmers Mfg. Co. and its predecessors have been known for 65 years as one of the largest manufacturers of high grade machinery in the world.

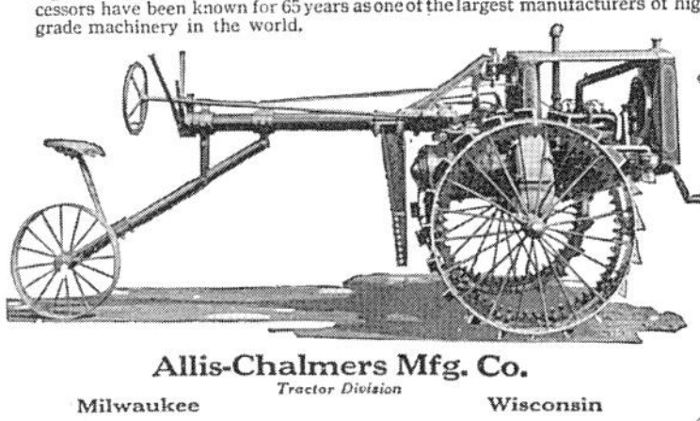

Allis-Chalmers Mfg. Co.
Tractor Div.,
Dept. 0000
Milwaukee, Wis.

Gentlemen:—Kindly send me full information regarding your proposition.

Name ____________

Street ____________

City ____________

State ____________

Allis-Chalmers Mfg. Co.
Tractor Division
Milwaukee Wisconsin

Page Thirty-Six

Read This Exclusive Story

Here are sample advertisements in our series for trade papers. They are direct, forceful and to the point.

The trade paper is the mouthpiece for the implement dealers throughout the country. We shall make Allis-Chalmers development an open book for the perusal of those dealers who are alive to the real opportunities offered them.

Model 6-12 Nebraska Test Results and Specifications

Model 6-12		Engine	
Test No.	54	Make	LeRoi
Year	1920	Cylinders	4
Serial No.	10633	Rated rpm	1,200
Fuel	Gas	Bore (in.)	3.125
Mfr. Rating	6-12	Stroke (in.)	4.50
Max. Belt hp	12.37	Displacement (in.^{3})	138
Max. Dbr. hp	6.27		
Weight (lb)	2,500		

Model 10-18 Production

1914	1915-1918	1919	1920	1921	1922	1923
1	(no info)	71	31	45	38	4

Model 6-12 Production*

	1919	1920	1921	1922	1923	1924	1925	1926
Model 6-12	303	447	66	243	198	99	62	8
6-12 Cane	---	---	20	6	6	13	---	---
Model B	INA†	INA†	INA†	INA†	INA†	INA†	INA†	INA†

* No serial number information available, but appears to have started at Serial No. 10001.

† Information not available.

Models 10-18, 6-12, 12-20, 18-30, and 20-35 Nebraska Test Results and Specifications

	10-18	6-12	12-20	18-30	18-30	20-35	EK 25-40
Test No.	None	54	82	55	83	151	193
Year	---	1920	1921	1920	1921	1928	1931
Serial No.	---	10633	20134	5729	5929	13620	24559
Fuel	Gas	Gas	Gas	Ker.	Gas	Gas	Dist.
Mfr. Rating	10-18	6-12	14-27	18-30	22-38	20-35	27-42*
Max. Belt hp	18	12.37	33.18	33.41	43.73	44.29	47.00
Max. Dbr. hp	10	6.27	21.42	20.55	25.45	33.20	33.82
Weight	4,800	2,500	4,550	6,000	6,640	7,095	7,200
Make	A-C	LeRoi	MW†	A-C	A-C	A-C	A-C
Cylinders	2	4	4	4	4	4	4
Rated rpm	720	1,200	1,100	830	930	930	1,000
Bore (in.)	5.25	3.125	4.125	4.75	4.75	4.75	5.00‡
Stroke (in.)	7.0	4.50	5.25	6.50	6.50	6.50	6.50
Displ'mt (in.^{3})	303	138	280.6	460.7	460.7	460.7	510‡
Transmission Speeds:							
1	2.3	2.5	2.3	2.31	2.58	2.5	2.5
2	---	---	3.1	2.82	3.16	3.25	3.25
Rev.	INA§	INA§	INA§	INA§	INA§	3.25	3.25

* Not rated, SAE/ASAE code shown.

† Midwest.

‡ 5.25-in. bore (563 in.^{3}) optional.

§ Information not available.

Model L 12-20 and 15-25 Tractors Serial Numbers and Production Totals

Year	Serial No. to Serial No.	Production Total
1920	---	3
1921	20001-20334	334
1922	20335-20497	163
1923	20498-20905	408
1924	20906-20995	90
1925	20996-21370	375
1926	21371-21681	311
1927	21682-21705	24
Total		1,708

A 10-18 discing in the central Experimental Farms in Ottawa, Canada.

Serial Numbers and Production Totals
Models 15-30, 18-30, 20-35, 25-40, and A

	15-30/18-30 (SN to SN)†	20-35 (SN to SN)†	E20-35 (SN to SN)†	E20-35 (SN to SN)†	E25-40 (SN to SN)†	Model A* (SN to SN)†	Production Totals‡
1918	5000-5005	---	---	---	---	---	5
1919	5006-5160	---	---	---	---	---	96
1920	5161-6014	---	---	---	---	---	694
1921	6015-6160	---	---	---	---	---	104
1922		— Records Conflict —		— Records Conflict —		— Records Conflict —	181
1923	---	6161-6396	---	---	---	---	334
1924	---	6397-6754	---	---	---	---	315
1925	---	6755-7368	---	---	---	---	635
1926	---	7369-8069	---	---	---	---	682
1927	---	8070-9869	---	---	---	---	1,814
1928	---	9870-10000	12001-16761	---	---	---	4,867
1929	---	---	16762-20250	22001-23251	---	---	4,339
1930	---	---	---	23252-24185	24186-24842	---	1,592
1931	---	---	---	---	24843-24971	---	128
1932	---	---	---	---	24972-25023	---	53
1933	---	---	---	---	25024-25061	---	38
1934	---	---	---	---	25062-25308	---	248
1935	---	---	---	---	25309-25581	---	272
1936	---	---	---	---	25582-25611	25701-25725	28/25
1937	---	---	---	---	---	25726-26304	578
1938	---	---	---	---	---	26305-26613	310
1939	---	---	---	---	---	26614-26781	168
1940	---	---	---	---	---	26782-26895	114
1941	---	---	---	---	---	26896-26914	19
1942	---	---	---	---	---	26915-26925	11

* Model A shown here since serial numbers continue from the E25-40.
† Serial number to serial number.
‡ Separate production records do not always agree with serial numbers.

Con Razor of Allen, Tex., hauling the wheat raised on 1,500 acres, 8 miles to market, total weight of load 17,800 lb. The tractor was doing the work of eight good mules.

Tractors of the Growing Thirties

Going Orange is Going Great

Going orange is going great was an advertising theme used in the late 1960s, about 40 years after A-C tractors "went orange." One of the many company legends a newcomer to the Farm Equipment Division would hear was that Harry Merritt, manager of the Tractor Division, while on a trip to California in 1929, saw fields of brilliant orange wild poppies. He was so impressed with their visibility that he brought some home and had the original Persian orange developed from their color. Thus the dark green tractors became the famous A-C orange. This was a great change and it wasn't long until other tractor manufacturers were changing to brighter colors. It sure made them easier to spot in the cornfield.

There's another story, as there always is, chronicling the A-C color change. Harry Hoffman, who worked for Harry Merritt as advertising manager for many years and was a part of these events, told it differently:

"In 1928, the United Tractor and Equipment Corporation of Chicago, Illinois, approached Allis-Chalmers to manufacture a tractor of about 4,000 lb to replace the Fordson they had been using. Henry Ford had found that he couldn't make money on the Fordson tractor and was discontinuing production in the United States, although he continued building it in Great Britain. The company entered into an agreement to build a tractor.

The success of the redesigned 20-35 established the direction for developing a high power-to-weight, maneuverable tractor and the United was the result.

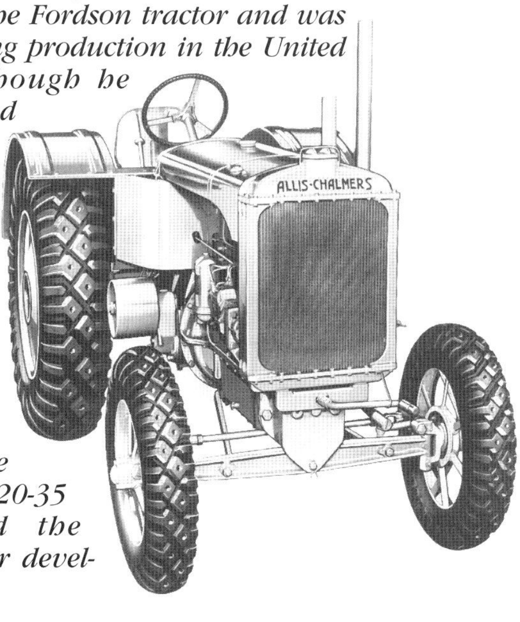

This ad in the September 1929, Country Gentleman announced the United tractor built by Allis-Chalmers. United was a distributor of a long line of agricultural and industrial equipment in Chicago, Ill., and needed a tractor to replace the Fordson which was no longer built in the United States.

For some period of time everything went well. However, due to problems within the United organization, the agreement was soon terminated. A-C then added the United tractor and its equipment to their line of products as it filled the vacancy left when the 15-25 was dropped. Since the United Tractor and its allied equipment were painted a Persian orange color, this was adopted as the standard color for all Allis-Chalmers farm equipment."

Two interesting and perhaps equally plausible stories — take your pick. Maybe the facts are some combination of the two accounts. One fact remains — Harry Hoffman was there.

At Serial No. 12001 the Model U got 28-in. rear tires, new fenders, a high back seat, and a 4.5-in. bore which increased the displacement to 318 in.^{3}

This interesting modification of a Model UI has Goodyear tracks. In the original photograph "Goodyear Tire and Rubber Company" and "McNeil Boiler Co." are discernible along the top of the rubber track. A request to Goodyear for information produced no results.

This nicely restored United with a Trackson conversion is an example of the many ways the United and Model U found their way into industrial products. For its time the Model U may have been the most widely-used farm tractor in non-farm applications.

Advertising of the air-tired Model U emphasized high-speed operation both on the road and in the field. Other companies were slow to accept rubber tires so A-C had an exclusive for some time.

The Model U with a Continental engine, prior to Serial No. 7405. This was one of the first production tractors with air tires in 1932. It had 6.50-16 front and Firestone 11.25-24 rear tires on French and Hecht wheels.

The United-Model U Tractors

The Model U industrial tractor with Continental engine and solid rubber tires for operation on hard surfaces. It was an early approach to adapting a farm tractor to industrial applications.

Designed to replace the Fordson, the United-U more than achieved that goal. With about 6 more belt horsepower than the Fordson, the U moved into the three-plow class. A-C needed a three-plow tractor to replace the 15-25 which had been discontinued in 1927. The Model E was upgraded to 25-40, becoming a four-plow tractor that was above the largest market segments, both in weight and price.

For a time the United-Model U apparently was sold by both the United and Allis-Chalmers dealer organizations, but the production record does not differentiate the number of units for either group. In Test No. 170 (1929) the tractor was called the Allis-Chalmers (United) Model U. Production records show that the United was built for three years: in 1929, 1,962 units were built; in 1930, 4,579 units; and in 1931, 707 units. After 1931, all production was designated as Model U tractors.

The United-Model U tractors through Serial No. 7404 were powered by a Continental engine. It was a four-cylinder L-head type with a 4.25-inch bore and 5.00-inch stroke (283.7 in.^{3}) rated at 1,200 rpm.

This early Model U industrial tractor is equipped with Goodyear Air-Wheels, 46×20 10 rear and 26×11-6 front. The Continental engine now had a muffler.

The transmission was a four-speed sliding-gear box with the fourth gear locked out on steel-wheeled tractors. This was done by relocating the shifter fork so it couldn't engage the shift collar. It didn't take long for owners and users to discover that they could easily install the shifter fork in the collar and have a four-speed tractor.

The Continental engine was replaced by the UM engine which was manufactured for A-C by the Waukesha Motor Company. They built over 30,000 of these engines. In late 1932, it was installed in the Model U beginning at Serial No. 7405. The UM engine was a four-cylinder overhead-valve type with a 4.375-inch bore and 5.0-inch stroke (300.7 in.^{3}) and rated at 1,200 rpm.

In 1936 the bore of the UM engine was increased to 4.5 inches (318 in.^{3}) beginning at Serial No. 12001. The Model U was not retested at Nebraska with the larger bore. At the same time other major changes were made. Steel rear wheels were increased from 42 to 45 inches in diameter, and rear tires went from 24 to 28 inches. The padded seat from the UI was installed, the fenders and platform were redesigned, and some steering gear changes were made.

The latest version of the Model U was capable of pulling three 14-inch bottoms at 3.5 mph on steel and 5 mph on rubber. That 43 percent increase could amount to another 4 to 5 acres per day and was ample justification for the $210 price differential for

The Model U with UM engine and 24-in. rear tires, Serial No. 7405-12000. The engine had a 4.375-in. bore and 300 in.^{3}

rubber tires. Road speed was up to 11.75 mph, fast enough to really reduce road time. Practically all of the company advertising emphasized speed of operation, both in the field and on the road.

Tractors Would Never Be the Same — Low-Pressure Pneumatic Tires

The Model U had the distinction of being the first farm tractor to be equipped with pneumatic tires by a tractor manufacturer. Of all the major advances made in farm tractors since their inception, the substitution of air tires for steel-lugged wheels is no doubt the most profound. Solid rubber tires and high pressure truck tires had been tried by others and met with some success for industrial and highway applications, but they failed in field work.

From *Farm Implement News*, dated October 13, 1932:

Low Pressure Air Tires on Farm Tractors

"The challenge comes from the low-pressure balloon-type heavy-duty rubber tire which, when mounted on a tractor in place of steel-lugged wheels, effects a revolution in working capacity, puts the riding comfort of the automobile into the tractor seat, and extends the tractor's usefulness to hauling on concrete.

No one knows who started this tractor farming on air. Two Iowa dealers, one a tire dealer and the other an IHC dealer, started installing truck tires on Farmall tractors equipped with mounted corn pickers and found they could pick corn in conditions that left their steel-wheeled counterparts stalled. That turned into quite a conversion business.

The original air-tired Model U is on permanent display at the Wisconsin State Historical Society museum at Stonefield Village near Cassville.

Harry Merritt was the general manager of the Tractor Division from 1926 to 1941. His vision and entrepreneurship were directly responsible for the air-tired tractor, the All-Crop harvester, and other innovations which led to Allis-Chalmers' position of leadership in that era. He was awarded the McCormick medal in 1941 by the American Society of Agricultural Engineers for his outstanding contributions to the mechanization of agriculture. In 1993 the air-tired tractor and the All-Crop harvester were recognized by the Equipment Manufacturers' Institute as two of the most important advancements in American agriculture from 1893 to 1993. Harry Merritt was the driving force of these two extremely significant contributions at Allis-Chalmers.

An Allis-Chalmers representative who traveled that territory got wind of the idea and wondered if rubber tires would enable a Model U which already had a high speed fourth gear that was locked out on steel-wheeled tractors to use that gear for light field and road work."

This information must have filtered back to Harry Merritt, who was always receptive to new ideas. Harry G. Hoffman, who was the advertising manager for Harry Merritt at the time, wrote about the beginnings of the air-tired tractor in his memoirs.

"When Mr. Merritt conceived the idea for low-pressure tires, he sat down with C. E. Frudden, (Chief Engineer) and Walter Streblow (Assistant Chief Engineer) to share his idea. Merritt decided to call Firestone at Akron and see if a meeting could be arranged for a high level discussion. Firestone agreed and came to Milwaukee and listened to what they called Harry Merritt's hare-brained scheme. Several days later they called back and said they had a pair of airplane tires, but they had no tread on the face of the tire. Merritt said, 'Let's put them on a tractor and see what happens.'

They learned that when they dropped the pressure from the recommended 70 psi to 15 psi it was clear that it could work."

According to the *Farm Implement News*, the Firestone tires were a pair of 12×48 smooth airplane tires like those used on the Ford tri-motor and other large planes of that time. French and Hecht made suitable wheels and rims, and the stage was set to

A 1933 demonstration of air tires with Barney Oldfield in the seat and Bill Roberts on the fender. Bill Roberts was president of Allis-Chalmers from 1951 to 1955.

determine if this crazy idea would work. First tests with normal airplane inflation pressure were not very successful, but when the pressure was reduced to about 12 psi, good results were achieved.

C. E. Frudden, chief engineer of tractors, and his staff ran many tests with various implements in the field and with loaded trailers on the road. A friendly and helpful German farmer, near Waukesha, Wisconsin, allowed his farm to become the test facility. He was loaned the rubber-tired tractor to use in place of his own steel-wheeled U. At the end of one week he was convinced. He could pull his field cultivator twice as fast while the steel-wheeler couldn't start the load in third gear. He could plow just as well with rubber, but without the constant jolting of steel wheels. To spread manure, it was "load it and step on it." His comments on spreading manure were described as "more vivid than dainty." His steel-wheeled U was rarely used after the testing was completed.

Testing proved that at around 2 mph only 5 horsepower was required to move the tractor on rubber, but it took 10 horsepower for the steel-wheeled tractor. At 5 mph in the field those numbers rose to 9 and 18.5 horsepower, respectively. That left little to pull the implement. A-C had proven that low-pressure pneumatic tires would work. Now it was just a matter of convincing farmers of that fact — easier said than done.

Like most revolutionary ideas, air tires faced far more detractors than proponents, and the Company faced quite a challenge in gaining farmer acceptance. Ellzey Brown, sales promotion manager, had the job of promoting air tires. He came up with the idea of equipping Model U tractors with special high-speed fourth gears and holding tractor races. This led to hiring famous race car drivers such as Frank Brisko, Lou Meyer, Ab Jenkins, and Barney Oldfield to run the races which were held at state fairs and other events all over the country.

At the A.A.A. automobile races at the Wisconsin State Fair grounds in 1933, Albert Schroeder gave a plowing demonstration in the infield, then the tractor was unhitched and Frank Brisko set a new record for

Model U tractor races were held at fairs all over the country. Barney Oldfield, the famous race driver in the lead here, drove in many of them. He always won — it was in the contract.

a tractor at 35.4 mph. A cross-country run from Milwaukee to Chicago was another attention-getter. The tractor covered 88.1 miles in 5 hours and 1 minute. In September of the same year Barney Oldfield, the first man to drive an automobile over 60 mph, drove a Model U at 64.28 mph at Dallas. Finally Ab Jenkins set the record by driving a Model U 67.877 mph on the Utah salt flats.

Dealers even got into the act. A farmer who had ordered a steel-wheeled tractor might be delivered an air-tired tractor because the dealer was "temporarily out of steel-wheeled tractors," and promised steel wheels in a week or so — if he still wanted them. Few took the dealer up on his offer. The A-C air-tired tractors were off and rolling.

The All-Crop/Model UC

By the late 1920s, the tricycle row-crop tractors were becoming popular in corn and cotton-growing country, and A-C responded with the All-Crop, later called the Model UC. Production began in 1930 with 36 units being built. In 1930 and 1931 production records, the All-Crop was identified as the "Cultivator" before becoming the UC in 1932.

Like the Model U, the All-Crop tractor was powered by the Continental four-cylinder L-head engine. Starting at Serial No. 1269 in 1933, the UC received the UM overhead-valve engine with 300.7 cubic inches, also rated at 1,200 rpm.

The next upgrading occurred in 1936 at Serial No. 2282 when the bore was increased to 4.50 inches, raising the displacement to 318 cubic inches. At the same time there were other significant changes. Steel rear wheels went from 42 to 45 inches in diameter, and rear tires increased from 11.25-24 to 11.25-28.

The All-Crop tractor, later called the UC, had the Continental 283.7-in.^3 engine prior to Serial No. 1269. It was called the "Cultivator" in 1930 and 1931 production records.

Albert Schroeder (left) was introduced by the writer at the "Land of Power" show. He was always eager to relate his experiences with the first air-tired tractor and his appearance at each show was enjoyed by the audience.

The friendly German farmer mentioned in the account of the tire test was Mr. Albert Schroeder of Waukesha County, Wisconsin. He continued to use the Model U for many years, and was very proud of his association with this important project. The writer had numerous visits with him, and he never tired of relating his experiences. During the "Buy a Tractor and Take a Trip" Land of Power show is 1962, he appeared at each performance and talked to many of the over 7,000 visitors about the beginning of air tires and the performance of the Model U. The tractor, with its original tires, is on permanent display at the Wisconsin State Agricultural Museum in Stonefield Village near Cassville, Wisconsin.

Albert Schroeder used the first air-tired Model U from 1932 to 1946. He was proud of his participation in the project and was quick to tell everyone about rubber tires.

The Model UC with the UM engine had 300 in.^{3}, full fenders, and 24-in. rear tires. Serial No. 1269 through 2281.

Cross section of the Model U. Transmission had longitudinal change gear shafts and differential inside a single bull gear.

In 1936, effective Serial No. 2282, the Model UC received the 4.5-in. bore (318 in.^{3}), three-quarter fenders, a new seat, and 28-in. tires.

One of several cane tractors built by Thomson Machinery, Thibodaux, La., using UC components. Thomson purchased UC parts until 1960, long after the UC tractor was out of production.

The Model UC was first A-C tractor with offset final drives, a design which continued through 1981 with the Model 185. It provided extra clearance for late cultivation.

Rear fenders were reduced from full to three-quarter fenders, and a padded seat similar to the UI was installed.

It shared the four-speed transmission with fourth disabled in steel wheeled tractors. The Model UC had offset final drives, a design that remained a hallmark of Allis-Chalmers tractors for nearly 50 years. It also provided extra crop clearance which was a desirable feature to the farmer who insisted on cultivating his corn until it was 4-feet tall.

The UC was built from 1930 through 1941, with production curtailed by World War II. A cane model was produced from 1937 through 1953, most of them sold to Thomson Machinery in Thibodaux, Louisiana. Model UC components continued to be supplied to Thomson through 1960 for use in tractors of their own design.

The 1936 Model A had the Model E engine with 460.7 in.^{3} (gas) or 510.5 in.^{3} (distillate) with a power train similar to that of the Model U. It was designed to operate threshers.

The Model A Tractor

The Model A replaced the Model E in 1936. Strangely, within the company it was treated almost as an extension of the earlier tractor, although about all it had in common was the engine and the serial number sequence. The E stopped at Serial No. 25611 and the A resumed with Serial No. 25701.

The Model A gasoline engine had a 4.75-inch bore and 6.5-inch stroke (460.7 in.^{3}) with a compression ratio of 4.4:1, while the distillate version had a 5.0-inch bore (510.5 in.^{3}) with a 4.0:1 compression ratio. The tractor was rated at 1,000 rpm. The Model A was not tested at Nebraska, but carried a manufacturer's rating of 33-drawbar and 44-belt horsepower which was later increased to 39.7 and 51.2 horsepower.

The transmission of the A was similar to that of the Model U, with longitudinal main and countershafts and four forward speeds, which were 2.5, 3.75, 5.00, and 9.00 mph. This would have been a significant improvement over the Model E which had only two forward speeds.

The shift pattern was changed at Serial No. 26526:
Prior to Serial No. 26526:

```
      R   3
  ____|___|
 |    |   |
 4    1   2
```

Serial No. 26526 and up:

```
      1   3
  ____|___|
 |    |   |
 4    2   R
```

The steel-wheeled version of the Model A tractor in 1937. From 1936 through 1942, 1,225 Model A tractors were built.

The Model A was likely the result of a continuing need for a thresherman's tractor in a more modern form. The Advance-Rumely threshers were still being produced in 1936 (their last year), but Rumely OilPulls had been discontinued in 1930 and the Rumely 6A was dropped in 1931.

With its 9-mph road gear, an A on rubber could move quickly from set to set and gain valuable threshing time. Since its primary mission was belt work, its big engine was the Model A's most important asset. As threshing declined rapidly in the late 1930s, when small combines such as the All-Crop harvester took over, the Model A and other large belt-power tractors lost their market. By 1943 it was gone, probably hastened by World War II and its shortage of materials. It would be 10 years before Allis-Chalmers again offered a 44-horsepower tractor, and then it would be an entirely different kind of tractor.

The WC with a No. 20 two-row cultivator, one of six different models for corn and cotton cultivation. It was a drive-in design, attached with only two pins plus the power-lift arm.

A 1934 Model WC tractor with the A-C W201 engine. The first 25 WC's had a Waukesha four-cylinder 186-in.^{3} engine.

The Amazing Model WC

Until the Model WC, the Allis-Chalmers tractor line was best described as something less than a roaring success. The WC certainly changed all that. By 1933, two-plow row-crop tractors had a firm grip on the Corn Belt tractor market. The Great Depression only briefly impeded farmers' rush to retire their horses to the bluegrass pasture. A-C's timing seemed almost perfect when they launched the WC in late 1933, geared up production in 1934, and pulled out all the production stops in 1935, just as farmers were recovering from that economic disaster.

Called the Model W during development, it was changed to Model W All-Crop as the tractor was announced in late 1933, then changed again to Model WC in early 1934. From an early brochure: "Full two plow power; lighter weight; sturdy construction; extreme simplicity; ready accessibility; maximum flexibility; easy manipulation; effortless application of power; satisfying performance; greater speed — that, in a nutshell, summarizes this new tractor achievement by Allis-Chalmers engineers — the Model WC All-Crop farm tractor."

The new Model W cultivating corn on the Brookhill Farm near Genesee, Wisc., 1933. This tractor was the first to be designed for rubber tires.

Obviously a lot of American farmers agreed; from 1933 to 1948 they bought over 178,000 Model WC tractors. The peak annual production was 29,208 in 1937. By comparison, A-C built only slightly over 28,000 total tractors between 1919 and 1933.

Promotion of the WC was tied to the family farm, of which there were millions. With the All-Crop harvester and matched equipment, the owner of a WC tractor could have complete control (if you don't count weather and market prices!) of his farming operation with very little hired help. It brought new meaning to the word independence. Farmers everywhere were quick to understand and turned to Allis-Chalmers in numbers that made the company one of the leading tractor manufacturers.

This trailer was matched to the WC for farm hauling. Capacity was 3 tons and vacuum brakes were optional. Only 25 were built at LaPorte in 1936.

Cross section of the Model WC tractor. Transmission and differential assemblies were purchased from outside suppliers.

The first 25 WC's built in 1933 were powered by a Waukesha engine with a 3.625-inch bore and 4.5-inch stroke (186 in.^{3}). In early 1934, A-C began building the famous, reliable, and economical "W" 4- × 4-inch, 201-cubic inch engine. It delivered 21.48-belt horsepower at Nebraska.

Allis-Chalmers developed air tires on the Model U, then followed with the WC which was designed for rubber tires. On rubber it could pull two-14's at 4.75 mph in average conditions, and do 20 to 25 percent more work per day on considerably less fuel. Tests in many different conditions substantiated a fuel saving of 20 to 30 percent compared to a steel-wheeled tractor of similar power. Field operation had changed immensely — now it was possible to pull a rotary hoe fast enough for it to be much more effective, and do it without jarring the tractor operator's teeth. One of the biggest joys to the owner of a rubber-tired WC was the ability to go from farm to farm over hard-surfaced roads. The next thing the farmer bought was a rubber-tired wagon so he could haul grain to the elevator. Rubber tires were $150 extra which sounds reasonable enough, but the tractor price went from $675 to $825, a very substantial 22 percent.

The WC was the first tractor tested at Nebraska on rubber (Test No. 223). The same tractor was also tested on steel wheels and the drawbar horsepower results were astounding. Maximum drawbar horsepower on steel was 14.36, while on rubber it was 33.5 percent greater at 19.17. It must be recognized that the rubber-tired unit developed its maximum drawbar horsepower at 9.0 mph. Thus began the move to high-speed farming.

During 1938 the WC was again subjected to testing by the Cornhuskers (Test Nos. 303 and 304). Again both steel and rubber performance were measured along with both distillate and gasoline fuels. Maximum belt horsepower on gasoline was now up to 29.93, or 39 percent more than the original results. The fuel was 70 octane gasoline so apparently the engine was the high compression version — not bad growth for only four years.

The sliding-gear type transmission had four forward speeds of 2.5, 3.5, 4.75, and 9 mph at 1,300-engine rpm. Individual rear wheel brakes were hand operated which was acceptable for turning. However, stopping required removing both hands from the steering wheel, a feature that met with some criticism by operators.

Wide-front axle and single-front wheel versions were also offered. Specialty crops in narrow rows, grown on beds, required the single-front wheel. The wide-front axle found only a limited market and few were built. A WC Cane model was offered in 1938, but production was low.

An optional mechanical power lift took the hard work out of lifting mounted equipment and at $20 (1934) it would have been a real bargain. The power lift was driven by a worm gear on the main drive shaft from the engine clutch to the transmission, therefore the drive to the plow-clutch type lift ran whenever the engine clutch was engaged. The power-lift shaft

The WC with single-front wheel and special rear-wheel equipment for California in 1936. Rear tires were a narrow 7.50-36 or 9.00-36.

The WC was adapted to southern agriculture with the No. 251 bedder and No. 250 center-beam attachment, shown here sweeping out cotton middles.

Hugo Liebert, tractor plant superintendent, and Walter Streblow, chief engineer, tractors, with the last WC tractor, Serial No. 178202, in April 1948.

drove the WC lister planters so a heavy-duty power lift was available for this application.

Front-mounted cultivators were the drive-in type and required only two pins plus the power lift arm to attach. The drive-in cultivator with quick hitching was an A-C exclusive at that time. A full line of two-row corn and cotton cultivators, listers, bedders, and bedder planters equipped the WC for row-crop farming anywhere in the country.

Rear-mounted equipment such as listers and bedders were hitched at a single hitch point located below the flywheel housing, well ahead of the rear axle. Only three pins were required to hitch the implement. With the WC the ease of hitching single hitch-point implements was emphasized. Later with the WD, WD45, and CA the superior implement performance would also be highlighted.

Time for a New Look
The Styled WC

There was big news in late 1938, when the WC was restyled to match the Model B. A new rounded hood, radiator shell, fuel tank, and shell fenders made everyone forget that what was underneath hadn't really changed much. Perhaps the most important change was the new electrical system with starter and lights. Everybody was glad to see the crank stored on the fender and seldom removed from that spot. Another welcome change was increasing the rear tire size to 11-28, which improved performance in poor traction conditions. New rear wheels had four tread settings although minimum and maximum treads remained at 65 and 76 inches.

Farmers liked what they saw in the restyled WC and production increased almost 12 percent in 1939. Even with that increase, the WC production never approached the over 29,000 built in 1937. That didn't affect the company all that much, however, since the factory was busy building the new Model B as fast as it could.

Mechanical specifications remained similar to those of the tractors tested in Tests 303 and 304, so

The WC and 60 All-Crop brought harvesting independence to thousands of farms. This pair made Allis-Chalmers a leader in farm equipment.

The WC and experimental two-row corn harvester at the LaPorte plant in June 1940. The corn harvester went into production in 1941 and nearly 44,000 were built.

The WC with wide front axle had limited acceptance due to lack of clearance and tread adjustment.

A single-front wheel WC and side-mounted mower in western hay. Good maneuverability and excellent visibility provided safe, convenient operation.

the styled model was not retested. Several versions were offered, which included the conventional dual-front, single-front wheel, wide-front axle, and cane. A WC adjustable-front axle appeared in the 1947 and 1948 production records. In 15 years, a total of 178,202 WC tractors were produced in West Allis. The success of the Model WC firmly established Allis-Chalmers as a major tractor manufacturer.

The WC Speed Maintainer and W Speed Patrol

The WC Speed Maintainer and the W Speed Patrol were based on the WC tractor. A total of 411 single-front wheel maintainers were assembled at LaPorte plant from 1938 through 1941. The W Speed Patrol replaced the WC Maintainer and from 1940 through 1942, 678 units were produced at LaPorte. After

A 1938 WC Speed Maintainer grooms a country road. Earliest maintainers were built on flat radiator WC tractors.

The W Speed Patrol with a rotary broom was a versatile highway maintenance outfit. It was built at the LaPorte and Springfield plants from 1940 to 1950.

World War II, Speed Patrol production resumed at the Springfield plant in 1945 and continued through 1950 with 3,073 units built there. The Model D grader was introduced in 1949 and replaced the WC-based graders. There was no Speed Patrol based on the Model WD tractor.

The Model WF

After the row-crop tractors were widely accepted, the conventional four-wheel tractor became known as the standard tread or wheatland tractor, depending on its size. Like their predecessors, they had fixed front and rear treads, generally 50 to 56 inches. Sometimes the wheels could be reversed for a minimal adjustment, but this was not a major consideration. Most of these tractors had full fenders and lower operator's stations.

Hay and small grain country and orchards had no need for a cultivator and actually preferred the standard tractor. It was easier to drive and handle, particularly when plowing in the furrow, although they couldn't turn as short. Ultimately, adjustable front axles eliminated both the dual-front and standard models, but that didn't happen until the late 1960s.

This preference explained the need in 1938 to build a standard tread version of the WC, called the Model WF. The engine and power train were similar to the WC, but shorter rear axle sleeves reduced the

A 1937 prototype of the Model WF equipped with skeleton rear wheels. WF tractors Serial No. 4 through 1903 were unstyled from 1937 to 1940.

A hybrid Model WF with new style grill and hood, but with old style full crown fenders. According to parts books, none were produced like this.

The WF orchard tractor had special fenders and a lowered operator's station. It was not differentiated so production quantities are not known.

The styled Model WF started in very early 1940 and was built until early 1951, nearly three years after the WD replaced the WC.

width and rear tread. The final drive housings were rotated forward which shortened the wheelbase and lowered the tractor. Its narrow width lent the WF to modification as an orchard tractor, much like the Model U. With the operator's station relocated low behind the axle and special fenders to protect tender branches and fruit, the WF was ready to head for the trees. Production records do not separate orchard models from the regular version. Offered with either a low octane fuel (4.2:1 compression ratio) or gasoline (5.0:1 compression ratio) engine, it had essentially the same power as the WC. The WF was not tested at Nebraska and was restyled to match the WC in 1940, starting at Serial No. 1904.

In 1949 and 1950, 150 tractors were sold to the Tractomotive Corporation in Deerfield, Illinois, to become the base units for reversed loaders. A total of 8,500 Model WF tractors were built from 1937 through 1951, continuing three years after the WC was replaced. As the WD tractor with adjustable front axle was accepted, there was little need for a standard model.

Goodbye Dobbin The Model B Tractor

The small farmers of the 1930s, '40s, and '50s were indeed fortunate that men like Harry Merritt, Bill Roberts, C. E. Frudden, and Walter Strehlow rose to executive ranks in Allis-Chalmers. They were men who understood the problems, needs, and motivation of farm families who extracted their livelihood from the soil. They saw that, in addition to the larger farmers who could easily afford and were well on their way to mechanization, there were a vast number of family-sized operations that needed and could afford tractor power if it were sized and priced right. The WC was exactly what many of these family farms wanted.

These same men also perceived an enormous need on the part of even smaller farmers, part-time farmers, and specialty crop growers who had one or two teams of horses, but could ill afford to commit a substantial part of their limited acreage to the production of horse feed. A-C estimated that there were perhaps as many as 4 million farms that needed mechanization but were still well below the size of operation that could justify a WC tractor and implements. Here was an opportunity for the company to step in and meet those needs and at the same time help improve farm life and profits.

The Model B was conceived to bring iron horsepower to the small farmer. Production of the Model B began in 1937 with 96 tractors, followed by its introduction in early 1938. As the "successor to the horse," the B was designed from its inception to ride

Allis-Chalmers calculated that there were 4 million farms which could utilize Model B power efficiently. Strong advertising presented this message to prospective dealers.

The Model B was the first of a new breed of small tractors that mechanized American small farms and retired the draft horse to the bluegrass pasture.

The tractor square dance was originated by Allis-Chalmers salespeople and it was a crowd-pleaser. It must have been a good idea — another major tractor company imitated it.

on pneumatic tires, although some were built during World War II with steel wheels due to the shortage of rubber. With a price tag of an unheard of $495 on rubber, the Model B immediately captured the attention of an eager and receptive market. Advertising and sales promotion concentrated on the high cost of

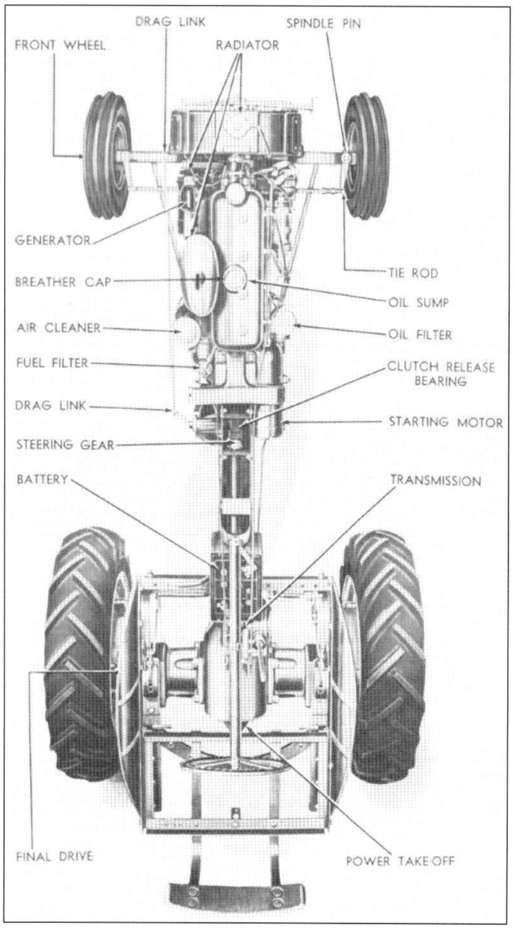

The Model B in its "birthday suit" reveals the narrow torque-tube design that let the operator see his work. Unlike offset designs, tractor stability was retained.

horse farming, emphasizing that it took 5 acres of cropland to feed one horse; 25 acres for a five horse operation. "Being chambermaid to five horses takes 270 hours per year — why not use that time and the 25 acres to raise profit-producing livestock?" The Models B and RC, and later, the Model C effectively retired ol' Dobbin to the bluegrass pasture, if he was fortunate enough to have an owner who could not face the alternative.

About two-thirds the power of the WC, the one-row, one-plow B was just the right size for thousands of small farms all over North America. And while some companies offered a one-plow tractor, the Model B just seemed better fitted to more needs. It enjoyed immediate and lasting acceptance. Other tractor manufacturers were soon scrambling to bring out new models to meet the threat of the versatile little Model B. With a complete line of mounted implements, planters, mowers, and even a power-matched Model 40 All-Crop Harvester, the Model B owner could be as fully-equipped and independent as the larger operators.

The Model B didn't look like the larger A-C tractors. The well-known industrial designer, Brooks Stevens, of Milwaukee, rounded the corners of the radiator shell, fuel tank, and fenders to produce a more modern appearance which would identify A-C tractors for the next 20 years. Its unusual wasp-waist torque tube design became an industry blueprint for one-row tractors, and was carried over into small two-row tractors because of commonality. Within a few months after the Model B appeared, a number of competitive units hit the market. Some were offset, that is, the engine and power train were on one side and the operator on the other. While this enhanced visibility somewhat, decreased tractor stability was

The adjustable front axle provided more tread adjustment range, adapting the B to more crops.

the undesirable side effect. A-C had a wide cushioned seat to allow the operator to move to either side for better visibility, yet maintained tractor stability with the power train components on the centerline.

The first approximately 96 tractors were powered by a four-cylinder Waukesha engine that had a 3.00-inch bore and 4.00-inch stroke, about 113 cubic inches. Beginning in 1938, a new A-C engine with a 3.25-inch bore and 3.50-inch stroke (116 in.^{3}) went into production. In Nebraska Test No. 302, the B developed 15.68-belt and 12.97-drawbar horsepower while burning distillate. In 1943, at Serial No. 64501, the bore was increased to 3.375 inches, expanding the displacement to 125 cubic inches, and the rated rpm was increased from 1,400 to 1,500. In 1950, the B was retested (Test No. 439) on gasoline and horsepower rose to 22.25 belt and 19.51 drawbar.

One of the most important reasons for the Model B's popularity was its extreme versatility. It was adapted to a wide variety of crops and row spacings. Special models such as the Asparagus Special and Potato Special were offered. The adjustable front axle appeared in 1945. The adjustable front axle plus rear wheels with greater offset provided a much greater range of wheel treads, both front and rear. A forerunner of this configuration was built in 1940 and was designated the Model BC. Essentially this was a prototype Model C. The B found its niche if non-farm work everywhere. One adaptation was the Drott grader conversion, which at first glance, looks like a shrunken W Speed Patrol.

The Model IB found many jobs in commercial and industrial applications, and over 2,800 were built during its 20 year life span. With its low center of gravity, short turning radius, and low cost of operation, it

The 1940 Model B Potato Special had an adjustable front axle and 7.00-28 rear tires to adapt it to the narrow rows of bushy potatoes. Tread range was 36 to 56 in.

The Model IB prototype appears to have an inverted standard front axle. It was an original "low rider."

The Model B Asparagus Special of 1940 had a tall front axle, special drive gears, and 8.00-38 rear tires to cultivate asparagus after it had ferned out. The tractor had 29 in. of clearance under the front axle and 28 in. under the rear. The tread range was 41 to 52 in.

was a natural choice for such work as pulling gang mowers or powering a rotary broom. In 1949, 80 units were built for Tractomotive Corporation of Deerfield, Illinois. Tractomotive became a part of Allis-Chalmers in 1959 and continued to utilize farm tractors for small loaders for many years.

From 1947 to 1949 a total of 2,000 Model B skid units (engine-power train components) were exported to A-C International's Essendine plant in Great Britain, where assembly was completed with local wheels, tires, etc. In 1963 and 1964, the British would return the favor by sending their ED40 to the United States.

Like the Models UC and WC, the Model B had offset final drives. This design provided lots of room under the tractor and permitted the use of implements having a hitch point well ahead of the tractor's rear axle. Ground engaging tools, such as the Model B plow, benefited greatly. Implement depth was regulated by raising or lowering the hitch point and the implement would follow in equilibrium, requiring only light forces to keep it from tipping. It worked like the old walking plow, which when properly hitched, required a minimum of effort on the part of the farmer in the furrow.

In the case of the B, a lever raised or lowered the hitch point. Tractor hydraulics only raised the plow at the furrow end and held it in transport position. Ten years later, the Traction-Booster system would add the precision control of hydraulic weight transfer, but the operation of the implement remained unchanged.

A-C made an complete line of implements for the Model B. As the source of power and carrier, the B was the centerpiece of a system, with nearly every tool and attachment that an owner might need. If A-C didn't have some specialized implement, other suppliers were quick to make whatever was needed to keep the B fully employed.

Most equipment was mounted on the cultivator frame. To mount the cultivator, the tractor was backed astraddle the frame and the lift links attached. This frame not only carried the one-row cultivator, but also could be fitted with several types of planters and fertilizer attachments. This put everything up front for complete observation and efficient operation.

The side-mounted Model B mower let the operator finish the day without a stiff neck from looking back at a rear cutterbar. He could see everything in front of the cutterbar and avoid obstructions that would damage the sickle. The V-belt drive could slip to prevent sickle damage if something lodged in the guards. It was made for the B and C tractors with 5- or 6- foot cutterbars. A later version was called the No. 5 mower and available with 5-, 6-, or 7-foot cutterbars.

This conversion by Drott produced a small grader that looked like a little W Speed Patrol.

The production Model IB had a substantial frame and bumper to carry attachments. Over 2,800 units were built from 1939 to 1958.

With a broom or a blade the Model IB was well-equipped and ready to keep streets and drives clean and clear. This is a Hough broom.

This nicely restored English Model B has an American engine and power train. Two thousand skid units were shipped to the United Kingdom between 1947 and 1949.

The No. 66 four-row beet and bean cultivator had two 100-in. toolbars and could be equipped with a variety of tools for work in vegetables.

A Model B with an all-steel cab. This is a prototype, there was no production.

Raising or lowering the hitch-point controlled plow depth and the plow followed its line of draft in equilibrium like a walking plow.

Although the Model B was specifically designed of rubber tires, this 1942 model did not get an exemption from serving on steel wheels in World War II. As soon as tires became available most war tractors had their wheels cut down and rims welded on.

A Model B and 40 All-Crop harvester replaced binder, thresher, many teams with bundle wagons, and many days on the threshing rig.

Model B tractors starting with Serial No. 64501 could be retro-fitted with Snap-Coupler hitch to use CA rear-mounted implements. Tractors beginning at Serial No. 114527 could use independent front-rear gang cultivators.

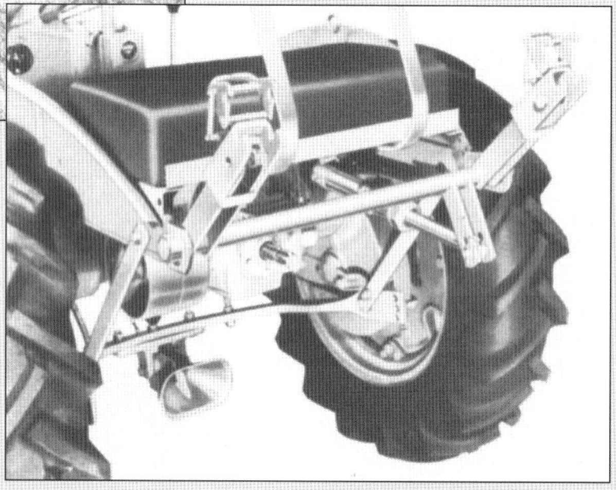

The side-mounted B-C mower put the driver's work in front of him for safer operation. It had V-belt drive to protect the sickle and cut 5 or 6 feet.

The Model B and its well-matched partner, the Model 40 All-Crop harvester, gave even the smallest farms complete freedom from the binder and thresher. From the start, the Model 40 All-Crop harvester was a wonder. Here was a little machine that could harvest many crops efficiently with the power of a small tractor, replacing the binder, thresher, many teams and bundle wagons, and a whole dining room full of hungry men. Everybody won!

In 1956 the Model B sold for $1,285. The Model B tractor was still listed in the price list in 1960, although production ended in 1957. There were still a few tractors in dealers' inventory at the price they carried when the last tractors were shipped in 1957. The last published price was $1,440 for a standard front axle tractor and $1,470 for an adjustable front axle model. Ten-inch tires added $40.80 and the rear PTO and belt pulley was $86.10. The hydraulic pump required the PTO and listed for $65. Compare the advertised price of $495 in 1938 to the last price in 1957. Of course, the last tractors had lights,

This prototype one-row mounted corn picker for the Model B was shown at Laporte in June 1940. It was never put in production. Operation would have been difficult since the operator could not see the gathering unit.

The Model RC could utilize WC front-mounted cultivators like this No. 21 two-row.

starter, other standard features, and many improvements over the '38 model. There was nearly a $1,000 increase, an average of $50 per year. Maybe that wasn't so bad after all — it was a great little tractor at any price.

The Model RC Tractor

The Model RC was Allis-Chalmers' first attempt at building a small two-row tractor. The approach, at least on the surface, made a lot of sense — just pop a B125 engine into a WC chassis and install smaller tires to help hold the price down. The only problem with down-sizing a tractor is that it usually turns out to be unprofitable because it carries so much of the larger unit's costs. This must have been the case since the RC didn't last very long in the line.

In Test No. 316, the RC developed 18.21-belt and 15.25-drawbar horses, burning distillate. It was powered by the same 125-cubic inch engine that drove the later Model B's and the forthcoming Model C. Rated engine speed was 1,500 rpm. It weighed 4,005 pounds as tested, only 530 pounds less than the test weight of the WC.

The RC had four forward speeds, about 20 percent slower than the WC. Since it was based on the WC power train, it had the same belt pulley, PTO, and mechanical power lift. One big advantage was its ability to utilize many of the WC two-row implements, which was a real benefit to customers and dealers as well as the company itself. About 5,500 Model RC's were built from 1939 to 1941.

The Model C

Since the Model C followed the RC by just a little over a year, the RC may have been planned as an interim model, providing time to complete design and test work on the C. The BC tractor of January 1940, appears to be a prototype Model C.

This 1940 experimental Model BC tractor had a 52- to 84-in. adjustable front axle. There was no front pedestal or gear box. The linkage-type steering for a two-row tractor would not show up in production until the Model D12 about 19 years later. The deep-offset rear wheels allowed a 52- to 84-in. tread range.

The Model RC pulling a No. 4 light-series two-furrow plow.

By upgrading the Model B into the larger Model C, Allis-Chalmers was able to market a light two-plow, two-row tractor at a competitive price. The company aggressively promoted the advantages of the Model C's front-mounted implements, emphasizing superior visibility and precision control. Rear-mounted implements were mostly interchangeable with the Model B, but a line of two-row front-mounted implements was developed especially for the C. Adjustable front axle and single-front wheel models were available.

In Test No. 364, the gasoline-powered C developed 23.30-belt and 18.43-drawbar horsepower. The distillate-powered version, in Test No. 363, produced 19.40 on the belt, 15.96 on the drawbar. The Model C had the same four-cylinder 125-cubic inch engine as in the RC, rated at 1,500 rpm. The gas burner had a 5.75:1 compression ratio, while the low-octane fuel engine had a 4.75:1 ratio. The three-speed transmission was common with the Model B. The B and C were the first A-C tractors to have the belt pulley mounted on the rear, driven by the power take-off. The Model C shared the same hydraulic lift system with the B, which was a lift-lower system with no draft control.

The Model C had a test weight of 3,205 pounds, which was 800 pounds less than the test weight of the RC, yet they were essentially the same horsepower. This supports the concept of upgrading a smaller unit rather than down-sizing a larger unit to design a new model. That 800 pounds would add a lot of cost to a tractor. In 1947 the Model C had a list price of $880 and a two-row cultivator was $104. The Model C proved to be a worthy companion to the B and WC. Over 84,000 were built during its 10-year life span, confirming its value and acceptance by farmers.

The Model C had its own line of front-mounted equipment. A-C emphasized the better work of front-mounted implements because of good visibility.

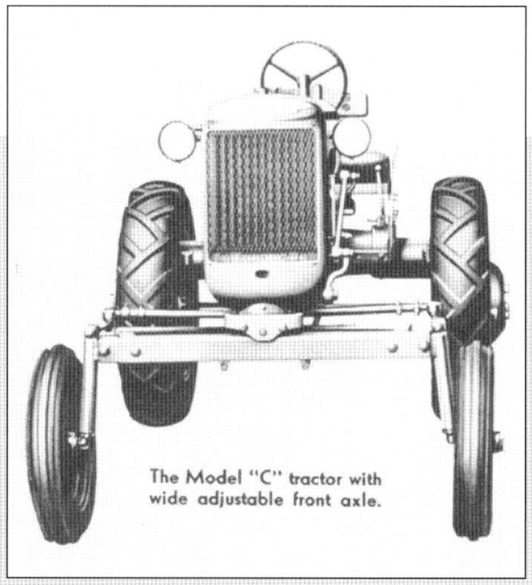

The Model "C" tractor with wide adjustable front axle.

An adjustable front axle with 52- to 82-in. treads was offered for the Model C but the conventional dual front was the most popular.

This experimental Model C had sliding-hub adjustable rear axles to provide up to an 84-in. rear tread. This was an alternative to the deep offset wheels of the BC and followed them by only seven months in August 1940. Power-Shift rear wheels were still over four years away.

The two-way plow for the Model C threw furrows all one way to eliminate dead furrows and help control erosion. Rear-mounted implements fit both B and C tractors.

Nebraska Test Results and Specifications United, Model U, All-Crop, Model UC

Model	United, U	All-Crop	Model U	Model UC
Test No.	170	189	237	238
Year	1929	1931	1935	1935
Serial No.	U946	UC217	U8263	UC1704
Max. belt hp	35.04	36.09	34.12	34.09
Max. dbr. hp				
Steel	25.63	24.98	28.78	24.17
Rubber	---	---	30.07	28.85
Weight (lb)				
Steel	4,821	5,965	5,030	5,710
Rubber	---	---	5,140	6,115
Engine	Cont.*	Cont.*	A-C	A-C
Bore (in.)	4.25	4.25	4.375†	4.375†
Stroke (in.)	5.0	5.0	5.0	5.0
Displ'mt (in.^{3})	283.7	283.7	300.7	300.7
Carburetor (in.)	1.25	1.25	1.25	1.25
Rpm	1,200	1,200	1,200	1,200
Trans. speeds:				
1	2.3	2.3	2.3	2.3
2	3.3	3.3	3.3	3.3
3	5.0	5.0	5.0	5.0
4	---	---	10.0	10.0
Rev.	2.7	2.7	2.7	2.7

* Continental.

† 4.50-inch bore effective Model U Serial No. U12001 and Model UC Serial No. U2282.

WC Speed Maintainer and W Speed Patrol Serial Numbers

Laporte Plant

Year	WC Speed Maintainer (Serial No. - Serial No.)	W Speed Patrol (Serial No. - Serial No.)
1938	IC1-IC150	---
1939	IC151-IC400	---
1940	IC401-IC407	IE1-IE466
1941	IC408-IC411	IE467-IE634
1942	---	IE635-IE678

Springfield Plant

Year	WC Speed Maintainer (Serial No. - Serial No.)	W Speed Patrol (Serial No. - Serial No.)
1945	---	IE681-IE819
1946	---	IF820-IF1437
1947	---	IF1438-IF2278
1948	---	IF2279-IF3515
1949	---	IF3516-IF3746
1950	---	IF3547-IF3753

Nebraska Test Results and Specifications for the Model WC

	W	WC		WC		WC	
Test No.	None	223		303		304	
Year	---	1934		1938		1938	
Serial No.	---	WC109		WC65623		WC65625	
Fuel	Gas/Dist.	Dist.		Dist		Gas	
Max. belt hp	---	21.48		25.45		29.93	
	---	Steel	Rubber	Steel	Rubber	Steel	Rubber
Max. dbr. hp	---	14.36	19.17	18.72	20.41	22.29	24.16
Weight (lb)	---	3,190	3,792	3,175	4,545	3,165	4,535
Ship. wt.*	2,700/3,300	2,700	3,300	2,700	3,300	2,700	3,300
Engine	Waukesha	A-C		A-C		A-C	
Cylinders	4	4		4		4	
Rated rpm	1,300	1,300		1,300		1,300	
Bore (in.)	3.625	4.00		4.00		4.00	
Stroke (in.)	4.50	4.00		4.00		4.00	
Displ'mt (in.^{3})	186	201		201		201	
Comp. ratio	INA†	4.2:1		4.2:1		5.0:1	
Carburetor (in.)	INA†	1.0		1.125		1.125	
Trans speeds:							
1st	2.5	2.5		2.5		2.5	
2nd	3.5	3.5		3.5		3.5	
3rd	4.75	4.75		4.75		4.75	
4th	9.00‡	9.25‡		9.00		9.00	
Rev.	2.0	2.0		2.0		2.0	
Tires:							
Front	5.25-17	5.50-16		5.50-16		5.50-16	
Rear	11.25-24	11.25-24		11.25-24		11.25-24	

Changes in specifications:

	Prior to Serial No. 74330 Flat Radiator	Serial No. 74330 and Up Styled
Tires:		
Front	5.25-17	5.50-16
Rear	11.25-24	11-28
Rear tread (in.)	65 or 76	65, 68, 73, 76

* Shipping weight from A-C specifications.

† Information not available.

‡ In some specifications, fourth gear was given as 9 to 18 mph.

Serial Numbers for the United, Model U, All-Crop, and Model UC

United, Model U

Year	United, U (Continental)	Model U (UM engine)	Model U (4.5-in. bore)	Model U	Model U	Model U
1929	U1-U1974	---	---	---	---	---
1930	U1975-U6553	---	---	---	---	---
1931	U6554-U7261	---	---	---	---	---
1932	U7262-U7404	U7405-U7418	---	---	---	---
1933	---	U7419-U7684	---	---	---	---
1934	---	U7685-U8062	---	---	---	---
1935	---	U8063-U9470	---	---	---	---
1936	---	U9471-U9988	U12001-U12821	---	---	---
1937	---	---	U12822-U14854	---	---	---
1938	---	---	U14855-U15586	---	---	---
1939	---	---	U15587-U16077	---	---	---
1940	---	---	U16078-U16721	---	---	---
1941	---	---	U16722-U17136	---	---	---
1942	---	---	U17137-U17469	---	---	---
1943	---	---	U17470-U17801	---	---	---
1944	---	---	U17802-U17819	---	---	---
1945	---	---	---	---	---	---
1946	---	---	---	---	---	---
1947	---	---	18309-18612	20174-20178	---	---
1948	---	---	18613-19421	20179-20300	21000-21035	---
1949	---	---	---	20301-20421	21036-21986	22018-22023
1950	---	---	---	20422-20512	---	22024-22547
1951	---	---	---	---	---	INA*
1952	---	---	---	---	---	INA*

All-Crop, UC, and UC Cane

Year	All-Crop, UC Continental	Model UC UM engine	Model UC (4.5-in. bore)	Year	Model UC Cane
1930	UC1-UC38	---	---	1944	5038-5067
1931	UC39-UC1099	---	---	1947	5068-5267
1932	UC1100-UC1231	---	---	1948	5268-5525
1933	UC1232-UC1268	UC1269-UC1293	---	1949	5526-5643
1934	---	UC1294-UC1551	---	1950	5644-5805
1935	---	UC1552-UC2000	---	1951	5806-5938
1936	---	UC2001-UC2281	UC2282-UC2770	1952	5939-6142
1937	---	---	UC2771-UC3756	1953	6143-6217
1938	---	---	UC3757-UC4546		
1939	---	---	UC4547-UC4769		
1940	---	---	UC4770-UC4971		
1941	---	---	UC4972-UC5037		

* Information not available.

Production of United, U, UI, All-Crop, UC, and UC Cane Tractors

Year	Model U	Model UC	UC Cane	Model UI	UC Thomson Parts*	U-Based OEM Units†	Components for UM42 UE54, and UKO54 Patrols‡
1929	1,962	---	---	---	---	---	---
1930	4,579	36§	---	---	---	---	---
1931	707	1,058§	---	---	---	---	---
1932	156	131	---	---	---	---	---
1933	267	63	---	---	---	---	INAll
1934	387	258	---	---	---	---	INAll
1935	1,474	423	---	---	---	---	INAll
1936	1,338	740	---	---	---	72	INAll
1937	2,032	875	172	62	---	226	464
1938	738	693	60	81	---	40	301
1939	329	126	63	40	---	38	189
1940	453	195	35	92	---	49	52
1941	271	49	31	80	---	43	---
1942	265	---	0	135	---	125	---
1943	332	---	0	2	---	---	---
1944	18	---	30	0	---	30	---
1945	0	---	0	1	---	---	---
1946	0	---	0	0	---	---	---
1947	307	---	208	1	---	60	---
1948	968	---	250	---	---	152	---
1949	1,078	---	118	---	---	51	---
1950	615	---	162	---	---	12	---
1951	337	---	133	---	---	26	---
1952	396	---	204	---	---	24	---
1953	0	---	75	---	---	13	---
1954	---	---	---	---	253	---	---
1955	---	---	---	---	225	---	---
1956	---	---	---	---	80	12	---
1957	---	---	---	---	45	---	---
1958	---	---	---	---	200	---	---
1959	---	---	---	---	0	---	---
1960	---	---	---	---	125	---	---
Total	19,009	4,647	1,541	494			

* UC Thomson parts: Engine and power train components for Thomson Machinery, Thibodaux, La., for their cane tractors and equipment.

† Various U-based OEM components: These include skid winches and other industrial units, mostly for Fred Cooper, Inc., Tulsa, Okla.

‡ Components for UM42, UE54, AND UKO54 were supplied to LaPorte plant which built these units from 1933 to 1940. Quantities for 1933 through 1936 are not shown in West Allis production.

§ All-crop and UC called "Cultivator" in 1930-31 production records.

ll Information not available.

Serial Numbers for Models WC and WF

	Model WC Flat Radiator	Model WC Styled	Model WF Flat Radiator	Model WF Styled
1933	WC1-WC28	---	---	---
1934	WC29-WC3126	---	---	---
1935	WC3127-WC13869	---	---	---
1936	WC13870-WC31783	---	---	---
1937	WC31784-WC60789	---	WF4-288	---
1938	WC60790-WC74329	WC74330-WC75215	289-1335	---
1939	---	75216-91533	1336-1891	---
1940	---	91534-103516	1892-1903	1904-2299
1941	---	103517-114533	---	2300-2703
1942	---	114534-123170	---	2704-3003
1943	---	123171-127641	---	---
1944	---	127642-134623	---	3004-3194
1945	---	134624-148090	---	3195-3509
1946	---	148091-152844	---	3510-3747
1947	---	152845-170173	---	3748-4110
1948	---	170174-178202	---	4111-5499
1949	---	---	---	5500-7317
1950	---	---	---	7318-8315
1951	---	---	---	8316-8353

Model WC Prices (1934 and 1947)

1934	
Model WC with steel wheels, belt pulley, fenders, and platform	$675
Model WC with 5.25-17 front, 11.25-24 rear tires, belt pulley, fenders, platform	$825
1947	
Model WC on rubber with lift and PTO	$1,341
Two-row cultivator for WC	$156

Models U and UC Prices (1934)

Description	List Price
Model U tractor with steel wheels	$840
Model UO Orchard tractor, steel wheels	$930
Model U Air-tires, 6.00-16, 11.25-24 tires	$1,050
Model UO Air-tires, 6.00-16, 11.25-24 tires	$1,140
Model UC All-Crop with steel wheels	$940
Model UC All-Crop, 6.00-16, 11.25-24 tires	$1,150

Starter and lights became available for the Model U in December 1939.

Model A Serial Numbers and Production Totals*

Year	Serial No. to Serial No.	Production Totals
1936	25701-25725	25
1937	25726-26304	578
1938	26305-26613	310
1939	26614-26781	168
1940	26782-26895	114
1941	26896-26914	19
1942	26915-26925	11
Total		1,225

* Separate production records do not always agree with serial number records.

Nebraska Test Results and Specifications for Models B, C, and RC						
	Model B first 96	**Model B <SN 64501**	**Model B SN64501-up**	**Model RC**	**Model C**	**Model C**
Test No.	None	302	439	316	363	364
Year	---	1938	1950	1939	1940	1940
Serial No.	---	B3712	B73430	RC3004	C78	C81
Fuel	---	Dist.	Gas	Dist.	Dist.	Gas
Max. belt hp	---	15.68	22.25	18.21	19.40	23.30
Max. dbr. hp	---	12.97	19.51	15.25	15.96	18.43
Weight (lb)						
Unballasted	---	---	2,199	---	---	---
Ballasted	---	2,620	4,193	4,005	3,205	3,205
Shipping wt.*	---	2,060	2,060	---	2,200	2,200
Engine	Waukesha	A-C	A-C	A-C	A-C	A-C
Cylinders	4	4	4	4	4	4
Rated rpm	1,400	1,400	1,500	1,500	1,500	1,500
Bore (in.)	3.00	3.25	3.375	3.375	3.375	3.375
Stroke (in.)	4.00	3.50	3.50	3.50	3.50	3.50
Displ'mt (in.^{3})	113	116.1	125.2	125.2	125.2	125.2
Comp. ratio	---	---	5.75:1	4.75:1	4.75:1	5.75:1
Carburetor (in.)	---	0.875	0.875	0.875	0.875	0.875
Speeds (mph):						
1st	2.5	2.5	2.5	2.0	2.5	2.5
2nd	4.0	4.0	4.0	2.8	3.75	3.75
3rd	7.75	7.75	7.5	3.75	7.5	7.5
4th	---	---	---	7.5	---	---
Rev.	2.0	2.0	3.0	1.75	2.75	2.75
Tires:						
Front	---	5.00-15	4.00-15	4.75-15	4.00-15	4.00-15
Rear	---	7.00-24	9.00-24	10.00-28	9.00-24	9.00-24
Optional	---	9.00-24	10.00-24	---	10.00-24	---

* Shipping weight from A-C specifications.

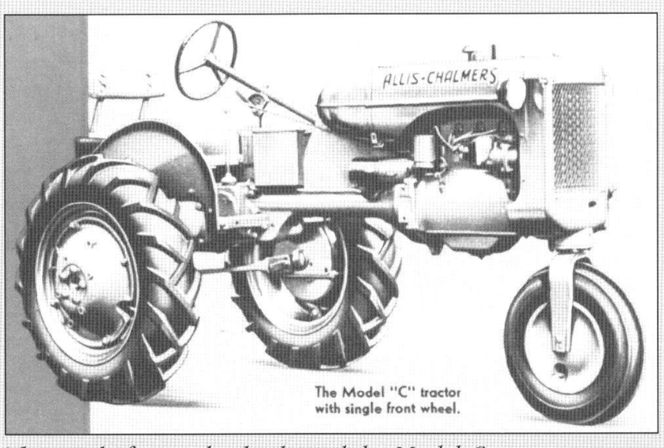

The single-front wheel adapted the Model C to narrow-row vegetable planting and cultivation. The front tire was a 6-12.

The Model B cultivator frame also carried planter and fertilizer attachments for many different crops. This is a 601A planter.

Serial Numbers for Models B, IB, RC, and C

	Model B	Model B	Model B	Model B	Model B	Model IB	Model RC	Model C
1937	1-96*	---	---	---	---	---	---	---
1938	97-11799	---	---	---	---	---	---	---
1939	11800-33501	---	---	---	---	---	4-4391	---
1940	33502-49720	---	---	---	---	---	4392-5416	1-111
1941	49721-56781	---	---	---	---	---	5417-5504	112-12388
1942	56782-61400	---	---	---	---	---	---	12389-18781
1943	---	64501-65501†	---	---	---	---	---	18782-23907
1944	---	65502-70209	---	---	---	---	---	23908-30694
1945	---	70210-72264	---	---	---	‡	---	30695-36377
1946	---	72265-73369	---	---	---	1001-1002	---	36378-39167
1947	---	73370-74079	75080-80555	---	---	1003-1009	---	39168-51514
1948	---	---	80556-85833	87834-92294	---	1010-1281	---	51515-68280
1949	---	---	---	92295-102392	---	1282-1555	---	68281-80517
1950	---	---	---	102393-103578	106579-114526	1556-1878	---	80518-84030
1951	---	---	---	---	114527-118673	1879-2218	---	---
1952	---	---	---	---	118674-122309	2219-2567	---	---
1953	---	---	---	---	112310-124200§	2570-2847§	---	---
1954	---	---	---	---	124201-124710	INAII	---	---
1955	---	---	---	---	124711-126496	INAII	---	---
1956	---	---	---	---	126497-127185	INAII	---	---
1957	---	---	---	---	127186-127461	INAII	---	---
1958	---	---	---	---	---	INAII	---	---

* Waukesha engines.
† Serial No. 64501 is the start of the 3.375-in. bore.
‡ 260 units built prior to 1946 have "B" serial numbers.
§ Distributor ignition effective: Model B Serial No. 123996, 1953. Model IB Serial No. 2836, 1953.
II Information not available.

This well-restored Model RC belongs to Tom O'Brien of Coggon, Iowa. Only about 5,500 were built from 1939 through 1941.

Production Records for Models B, IB, RC, and C

	Model B	Model B Cane	Model B Adj.	Model B (UK)*	Model IB	Model RC	Model C Std.	Model C Cane	Model C Adj.	Model C SFW
1937	96†	---	---	---	---	---	---	---	---	---
1938	11,699	---	---	---	---	3	---	---	---	---
1939	21,594	---	---	---	113	4,385	---	---	---	---
1940	16,119	---	---	---	70	1,040	104	---	---	---
1941	6,961	---	---	---	75	88	12,274	10	---	---
1942	4,619	---	---	---	---	---	6,374	10	---	---
1943	1,001	---	---	---	---	---	5,126	---	---	---
1944	4,678	30	---	---	---	---	6,787	---	---	---
1945	2,024	30	---	---	---	---	5,683	---	---	---
1946	1,107	---	---	---	2	---	2,790	---	---	---
1947	5,305	---	380	480	7	---	12,081	---	266	---
1948	7,608	---	2,629	1,339	272	---	14,512	---	1,969	285
1949	6,868	80‡	3,232	181	274	---	9,618	---	2,360	258
1950	7,186	10	1,939	---	323	---	2,838	---	554	121
1951	2,754	40	1,352	---	340	---	---	---	---	---
1952	1,906	205	1,525	---	351	---	---	---	---	---
1953	1,158	50	683	---	278	---	---	---	---	---
1954	314	196		---	225	---	---	---	---	---
1955	969	38	774	---	235	---	---	---	---	---
1956	380	10	304	---	100	---	---	---	---	---
1957	175	15	86	---	75	---	---	---	---	---
1958	---	---	---	---	110	---	---	---	---	---
Total	104,515	508	13,100	2,000	2,850	5,516	78,187	20	5,149	664

* Skid units exported to the United Kingdom.

† Production quantities do not agree with serial number.

‡ 80 Model B tractors shipped to Tractomotive.

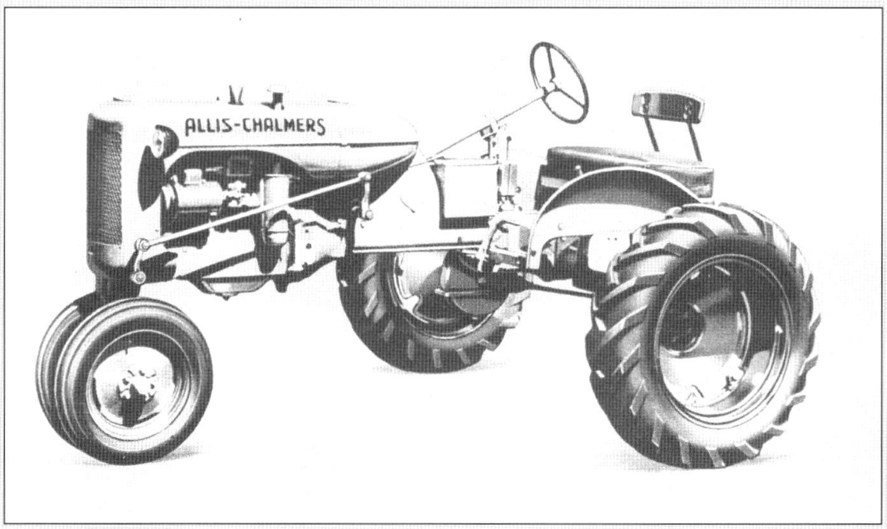

The Model C brought economical two-row power to the A-C line. Slightly over 84,000 units were built from 1940 through 1950.

It Wasn't All Farm Tractors

This vehicle was identified only as the V40 tank and was apparently intended as a personnel carrier.

Allis-Chalmers made a vast array of military machines and components during World War II, most of them by plants and divisions other than farm equipment operations. Those machines and products are not within the scope of this book. However tractor engineering was involved with a number of non-tractor machines and some of them are presented here.

The Fuel Cell Tractor

The fuel cell tractor created a wave of excitement when it was shown in 1959. An electric tractor — WOW! Everyone had visions of revolutionizing the tractor industry. The fuel cell was developed by the Allis-Chalmers Research Division and the tractor was built to demonstrate a practical application of the principle.

In the tractor, 1,008 fuel cells produced direct current when a mixture of fuel gases, primarily propane, and oxygen reacted in an electrolyte. The by-products of the process were pure water and carbon dioxide, completely non-polluting and non-toxic.

The tractor was a highly modified Model D12 with an electric motor driving the transmission shaft. Although it was slow, the tractor could develop 3,000 pounds of drawbar pull, more than enough to pull a two-bottom moldboard plow. The tractor was incredibly quiet; the only tractor that could slip up from behind and scare the daylights out of someone (we speak from experience).

The M7 snow tractor was designed for military use in the arctic. Front skis could be replaced with wheels for dry surface operation and it was powered by a Willys engine. The tractor plant built 291 units in 1944.

The 1942 mobile pump unit was designed as a fire-fighter. There was no production.

Nov. 16, 1948. I. A. KAMLOOKHINE 2,453,750

ENDLESS TRACK MECHANISM

Filed May 1, 1944 2 Sheets-Sheet 2

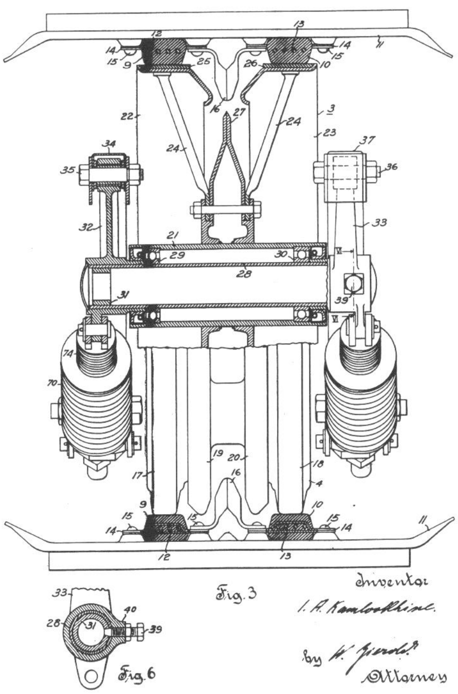

Nov. 16, 1948. I. A. KAMLOOKHINE 2,453,750

ENDLESS TRACK MECHANISM

Filed May 1, 1944 2 Sheets-Sheet 1

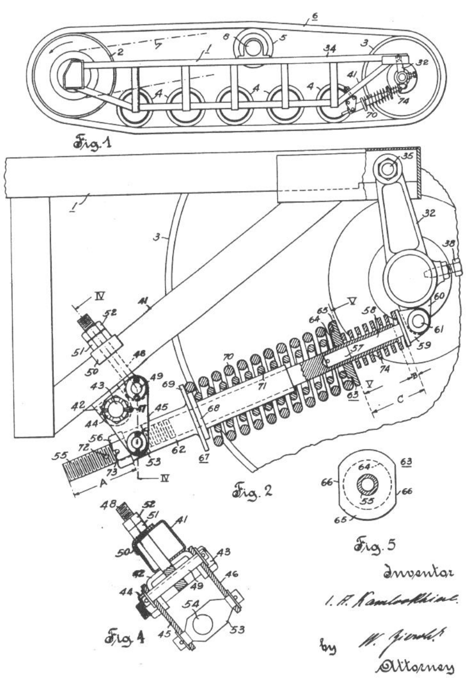

This patent covers the track mechanism of the snow tractor built in 1944. The track was designed by Mr. Igor Kamlukin of tractor engineering, later Simplicity Mfg. Co.

The fuel cell was very expensive and was soon ruled out for farm tractor use. It found its proper niche in the space program where cost was not a limiting factor and clean operation was a highly desirable feature. After the tractor had completed its mission as a demonstration vehicle, it was donated to the Smithsonian Institute.

The Fuel Cell Golf Cart

Another demonstration vehicle, the fuel cell golf cart, was powered by an even more advanced-design fuel cell than the tractor. It operated on a liquid solution of hydrazine and water. A gallon of hydrazine could produce as much electricity as 4 gallons of gasoline in a generator set. Fuel tank capacity could be reduced by 75 percent. Hydrazine smells like ammonia and is non-flammable, and the exhaust, like the tractor, was pure water. No wonder the fuel cell became so important to the space program. Even after 30 years the fuel is still expensive, although research continues on fuel cells as a alternate source of non-polluting power for automobiles.

The 1959 Allis-Chalmers fuel cell tractor demonstrated that the fuel cells could generate enough electricity to pull a two-furrow plow. It was extremely quiet and exhaust was pure water.

The Ontos ("thing" in Greek) was built for the Marine Corps in LaPorte. It was a high-speed M-50 tractor equipped with a 0.50 caliber spotting rifle and six 106-mm recoilless rifles for destroying tanks. It was powered by a General Motors engine and had an Allison power-shift transmission. The contract required a running prototype in six months from contract-letting. About three-fourths of the tractor engineering department worked around the clock to accomplish this enormous undertaking and spent most of two years on the Ontos project.

The Marine Diesel Engines

It was as big as a barn! During the early 1980s Allis-Chalmers built three gigantic marine diesel engines for Sulzer Brothers of Switzerland. These monsters were installed in large container-carrying vessels which were built in Louisiana.

The fuel-cell golf cart was demonstrated to over 7,200 visitors at the "Land of Power" show in 1962. The young lady in the seat is the writer's daughter, Ann.

The engine could develop 43,200 horsepower, turning only 126 rpm. It was a 12-cylinder power plant that weighed 1,160 tons. The bedplate, or engine base, was 80 feet long and weighed 80 tons. Each cylinder jacket weighed 15 tons. The engines were assembled in the West Allis plant, tested, then dismantled and shipped to New Orleans for installation.

American President Lines, the ships owners, chose the slow-speed marine diesels over conventional steam turbine power plants because of the diesels greatly reduced cost of operation. They anticipated saving 60,000 barrels of fuel, about 1 million dollars, per year per ship.

Though this huge engine was unrelated to farm equipment, it is presented as an example of the company's tremendous capabilities.

The silhouette of a man at the left end of the engine illustrates the incredible size of the Sulzer diesel. This cutaway drawing shows the works of an eight-cylinder engine. A-C built only 12-cylinder models.

The enormous crankshaft was lowered onto the bedplate by radio-controlled overhead cranes. The bedplate was 80 feet long and weighed 80 tons.

You Should Have Seen the Ones That Got Away . . .

The W-BC tractor had the W201 engine, BC power train, B front axle, and WC fenders. The tires were 11-26 rear and 5.50-16 front. Dated January 1940, it preceded Model C production by only a few months.

The 1940s must have been an interesting period for Allis-Chalmers tractor engineers. First, it was a transition period for farm tractors. Every major tractor manufacturer had recently introduced new models with modern styling and innovative features. Electric lights and starters were firmly established. Many lines had added another speed or two in the transmission, or replaced mechanical lifts with hydraulics. No one debated rubber tires versus steel wheels anymore; now the argument was about which tread pulled the best, or shed mud the best, or wore the longest.

During this period, A-C tractor engineering pursued a number of projects that were taken through the design and development stages, even into advanced prototypes, only to have them cancelled for one or another reason. This would seem to have been costly in terms of engineering time and money, but in many cases these features would appear in production several years later. Some of the abandoned projects, with as much detail as possible, follow:

- The W-BC tractor, January 1940: W engine, BC torque tube and transmission, BC final drives, WC wheel guards, 11-26 Goodyear tires rear, 5.50-16 Goodyear tires front. Weight was 2,775

The rear-engine tractor was based on Model B hardware but with no provision for rear-mounted implements.

By March 1940, the W-BC had a front pedestal with steering gear and dual front wheels, similar to the Model C.

The Model H four-wheel drive tractor had planetary steering which interrupted or reversed the drive to the inside wheels. The steering that increased its maneuverability also limited its ability to pull a load through turns.

pounds, including 514 pounds of solution.

- The W-BC tractor, March 1940: Front support installation per layout UL-1198-C. The Model RC was a small engine in a big frame, but the W-BC was the opposite — a large engine in a small frame. Dated early 1940, the W-BC tractor could have been a parallel project to the Model C. It could also have been an attempt to provide WC power at a lower price.
- The Rear Engine tractor: No details or date given. Apparently of Model B-C ancestry before rear mounted implements were widely accepted. From this concept to the Model G was a long, tedious search for the right solution.

This tractor appears to be a version of the Bonham tractor of Utah. Allis-Chalmers took over the development of this concept which led to the Model H.

The Model H Four-Wheel Drive Tractor

Albert Bonham of Clinton, Utah, began building four-wheel drive tractors in the mid-1930s. He called his tractor the Power Horse and it was equipped with an Allis-Chalmers engine. It was a lever-steered design and could be driven from the seat or with reins like a team of horses. The design was also built by the Eimco Corp., of Salt Lake City for a time. About 1,000 tractors had been built by World War II when further production was curtailed by war-time restrictions. Allis-Chalmers became interested in the project and continued development well into 1945 (Larsen, 1981).

Six prototype tractors were tested over 1,500 hours in various applications. The test report contained many good comments, but several areas needed work, including larger tires, more crop clearance, and more tread adjustment. The greatest problems involved the steering system. This system was a planetary gear and clutch. When a steering lever was pulled, power was interrupted to both wheels on that side. Pulling the lever farther back reversed the wheels, so that the tractor could spin on its vertical axis. When power to the inside wheels was interrupted in a turn, the Model H didn't have enough traction to pull its load around the corner. The system didn't function well on hillsides. The Model H was never produced, but it sure looked like a good idea.

The Models D and F Tractors

The Model D tractor was planned as a replacement for the time-proven Model WC. It was rated at 31.5-brake horsepower. The 40-horsepower Model F would have replaced the defunct Model UC. Although neither was produced, both were quite far along in development. The LaCrosse engineering department had designed and built a complete line of matched front- and rear-mounted implements for the Model D. Both tractor projects were active in 1944 and 1945.

The Model D with adjustable front axle was built by May 1945, but dual-front tractors were running by August 1944. Note the wide cushion seat like a B or C.

A mockup of the Model D was compared with a production Model WC in early 1944. The Model D had the W201 engine, six-forward speeds, and a transmission clutch like the WD. Grill was changed on prototype units.

The Model F was a 40-hp tractor with a four-cylinder 251- in.^{3} engine and six-speed transmission. It weighed 4,470 lb with liquid ballast in the tires.

Both the Model D and Model F tractors had a transmission clutch to provide a continuous power take-off similar to the WD. The six-speed transmission was covered by a patent issued to Igor Kamlukin.

Offset final drives were used, and the Model F final drive appears to be similar to the Model WD. Although equipped with a hitch for rear-mounted implements, it didn't seem to have anything in common with either B-C design or later WD-CA arrangements. There was no draft control, the plow depth being controlled by natural draft lines and a gauge wheel. These implements appear to be the forerunners of all the single-hitch-point rear-mounted tools to follow, but the Snap-Coupler was still almost a decade away.

The estimated cost of the factory tooling was so high that the company abandoned the Models D and F in favor of the highly successful WD and CA tractors. The only Model D and F tractors built were prototypes, total quantity unknown. There was no factory production of either tractor.

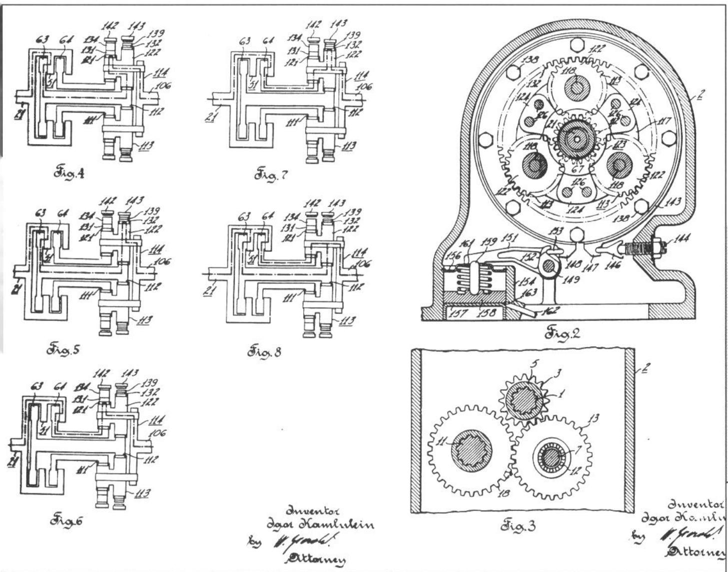

This five-speed power-shift transmission was invented by Igor Kamlukin of the tractor engineering department. One Model D and one Model F prototype tractor were built with this transmission and the units placed on test. This interesting transmission was abandoned with the rest of the Model D and F projects.

The D and F tractors had a six-forward, two-reverse transmission which had the minimum number of gears for those speeds. It was patented by Igor Kamlukin.

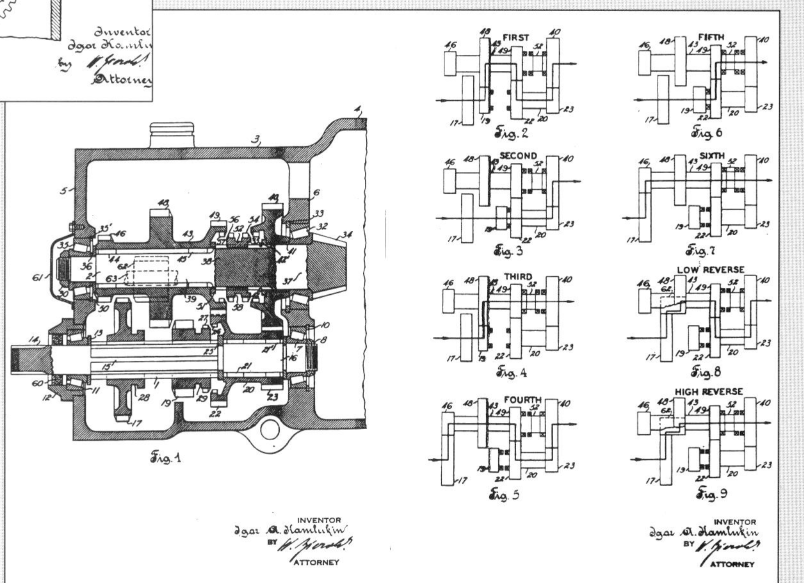

The Model F and 60 All-Crop harvesting rye on the Zinke farm at Elm Grove, Wis., in August 1947. An oil-bath transmission clutch provided a continuous PTO.

LaCrosse engineering developed a complete line of implements for the Model D. Besides this two-row cultivator, there was a pickup plow, front-mounted bedder, toolbar bedder, drill planter, independent-beam bedder, field cultivator, and semi-mounted mower.

Model H Specifications

Engine (Model C)	125 in.^{3}, 1,800 rpm, 27.5 hp		
Main clutch	Rockford 8.5-in. single dry plate		
Steering clutches	Two Rockford 8.34-in.		
Final drive	Roller chain and spur gear		
Tires	8-20		
Treads	40.5, 47, 53.5, and 60 in.		
Brakes	Contracting band on st. clutch		
Hydraulic pump	Three piston pump, 3,000 psi		
Power take-off	1.375-in. six spline, 545 rpm		
Steering	Two hand levers; also reverse		
Speeds (mph)	Gears	Forward	Reverse
	1	2.24	1.19
	2	3.54	1.88
	3	4.96	2.63
	4	8.12	4.31

This 1944 hammermill would have been a new tool in the A-C line. It was matched to Model D power but could be operated by a Model B or Model C at reduced capacity. A large diameter rotor was used to provide high hammer speed at relatively low rpm. It weighed 501 lb including the dust collector.

Model D and Model F Specifications

	Model D	Model F
Engine	D (WC)	F*
Bore and stroke (in.)	4.00 × 4.00	4.00 × 5.00
Displ'mt (in.^{3})	201	251
Rated rpm	1,300	1,450
Max. brake horsepower	31.5	40.0
Main clutch (in.)	10	11
Fuel tank capacity (gal)	16.5	21
Transmission clutch	2-7 in. plates	3-7 in. plates
Transmission speeds	(6 forward, 2 reverse)†	
Rear tires	11-28	12-30
Rear treads, power shift (in.)	56 to 90.25	57.25 to 90.25
Front tires	5.50-16	6.00-16
Front treads:		
Adjustable front axle	52 to 82.68 in.	Not available
Brakes	(8 in. contracting band)†	
Hydraulic pump pistons	2	3
Relief pressure (psi)	(3,200 to 3,700)†	
PTO shaft	(1.375 in. six spline)†	
Rated speed (rpm)	546	537
Speeds (mph)		
1	1.75	1.75
2	2.76	2.75
3	3.74	3.75
4	5.03	5.00
5	5.90	5.87
6	10.71	11.00
Rev. 1	2.34	2.25
Rev. 2	4.96	5.00
Wheelbase (in.)	89.87	91.50
Length (in.)	134.00	138.50
Row-crop clearance (in.)	25.25	30.50
Shipping weight‡ (dual)	3,820	4,470
Adjustable front axle	3,935	Not available

* The engine for the Model F was based on the W201 but with a 1-in. longer stroke.

† Applies to both models.

‡ With liquid ballast in rear tires.

This pickup plow was designed for the Model D tractor in 1944. Depth of 2 to 10 in. was controlled by the beaming screw which raised or lowered the hitch point and the gauge wheel. It had forged beams and weighed 478 lb.

Tractors — 1948-1985

The WD2, second prototype of the WD tractor in September 1947. It was close to the production tractor with power take-off, hydraulic lift with draft control, power-shift wheels, and transmission clutch.

The Model WD Tractor

It looked a lot like a WC, but it really didn't act like one. The new WD was loaded with so many new features and improvements that the sales force had to relearn everything except the name Allis-Chalmers. Terms like Two-Clutch Power Control, Single Hitch-Point implements, Traction-Booster, and Power-Shift wheels gave them plenty to talk about and demonstrate. The WD also had about 16 percent more power than the WC, thanks to boosting the compression ratio from 5.0:1 to 5.75:1 and the rated rpm from 1,300 to 1,400.

One of the new features, the Two-Clutch Power Control, provided a continuous power take-off. Disengaging the transmission clutch stopped the forward travel of the tractor while the power take-off continued to run. It was exactly what was needed for the Roto-Baler which required stopping each time a bale was wrapped. The new transmission clutch had been designed into the Models D and F (1944) and this important feature finally found its way into production with the Model WD. The hand-controlled clutch ran in oil and could tolerate a considerable amount of abuse with minimum service. It became a speed control device as users allowed it to slip almost continuously when a harvesting machine needed a slower speed in heavy crops. Rugged, reliable, and convenient, this clutch was important to the success of the WD and WD45. Much to everyone's satisfaction, the notorious hand-operated brakes were replaced by foot pedals.

The first Model WD to come off the assembly line in April 1948. (Left to right) Walter Strehlow, chief engineer; Aage Jorgenson and Sam Ouweneel, manufacturing; and Fred Worley, works manager.

Single Hitch-Point implements were unique in that they followed the natural draft lines around contour curves and over uneven

Single hitch-point implements followed natural draft lines for better depth control.

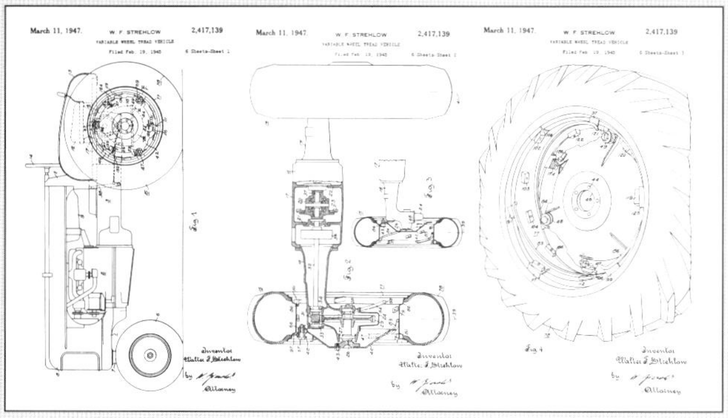

Tractor power spun wheels in or out on spiral rails to change treads without jacking the tractor. Power-shift wheels were invented by Walter Strehlow.

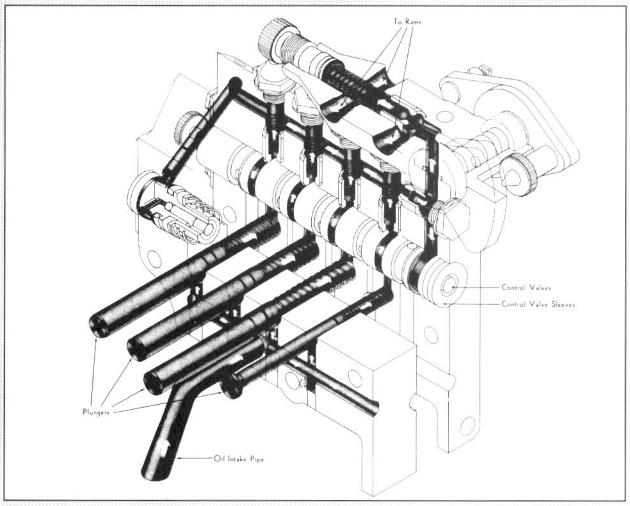

The WD hydraulic system had a four-piston high-pressure pump operating at 3,500 psi. At the end of the piston stroke, the pump unloaded to a standby pressure of 1,200 psi for Traction-Booster draft control which functioned with mounted, semi-mounted, and A-C transport valve-equipped pull-type implements.

ground. In tough soil conditions the Traction-Booster system applied just enough lift to transfer implement weight to the tractor to reduce slippage. Traction-Booster and the increased power moved the WD into the three-plow tractor class in many conditions. The heart of the Traction-Booster system was a four-piston hydraulic pump, driven by four cam lobes on the main driveshaft between the engine and transmission clutches. The system was live when the engine clutch was engaged. Hydraulic pressure was 3,500 psi, so cylinders were small compared to more typical 2,000-psi systems.

During fairs and shows in 1948, the A-C sales force delighted in demonstrating the Power-Shift rear wheels. This revolutionary new feature was an immediate sensation wherever it was shown. It was not unusual to see a WD in ruts up to the final drives, just from continuous demonstrations. It certainly was a crowd stopper. The ability to change rear wheel tread almost any where, any time, without jacks, blocks, or great physical effort, was nothing short of phenomenal. Walter Strehlow, chief engineer, was issued U.S. Patent No. 2,417,139 for Power-Shift wheels on March 11, 1947. It was without doubt one of the most important inventions in the tractor industry and was still in demand after 40 years. A-C licensed the manufacture of Power-Shift wheels to Motor Wheel, a subsidiary of Goodyear, and Sankey, a manufacturer in Great Britain.

In Nebraska Test No. 399 (1948) the tractor-fuel-powered WD produced 26.14-belt and 24.31-drawbar horsepower. The gasoline-powered version

(Test No. 440, 1950) developed 34.63-belt and 30.23-drawbar horsepower. This 32 percent difference in power quickly tipped the scales in favor of the more expensive fuel.

Until late 1952, the WD had a four-speed sliding gear transmission. Effective with Serial No. 127007, it acquired a new constant-mesh transmission. Speeds were slightly reduced, but the big benefits were found in the reduced gear noise and easier gear shifting. Tractors with this transmission had a curved gearshift lever, while the older type had a straight one.

During 1953 (Serial No. 136318) the WD traded its magneto ignition for battery and distributor ignition. The WD was offered in dual-front, adjustable front axle, single-front wheel, and cane models. The WD was also the base for the two-row cotton picker. A Model WD with dual-front wheels had a list price of $1,685 in 1947. This rose to $2,060 in 1951 while the adjustable front axle model sold for $2,145. The West Allis plant built 146,125 WD tractors from 1948 through 1953.

The 100,000th Model WD tractor. (Left to right) W. L. Voegeli, Walter Streblow, A. W. Van Hercke, and John Ernst.

This Model WD had an experimental five-speed transmission that was never produced. That extra gear would certainly have been appreciated by WD owners (March 1951).

The B. C. Thomson Co. of Thibodaux, La., modified WD tractors for operation in sugar cane. This one is equipped with a Thomson cane plow.

The Model WD45 Tractor

Most everyone agreed that there was little wrong with the WD that additional power couldn't cure. Some major competitors were already marketing row-crop tractors in the 40- to 50-horsepower range and A-C was beginning to feel the pressure. There was a logical and very welcome solution to this power problem. At Serial No. 146606 the WD production stopped and the WD45 resumed with Serial No. 146607. Still wearing 1939 sheet metal and looking like just another WD, the 45 was a real sleeper and caught everyone by surprise. The WD45 could easily be the most significant tractor ever offered by Allis-Chalmers.

The WD45 was one of A-C's best-known tractors. Nothing in its weight or power class could match its performance, and it was priced right, too.

A WD45 LPG tractor with an Oxnard Snap-Coupler toolbar carrier equipped with chisels. The engine retained the gasoline fuel system so it could run on either LPG or gasoline.

The Snap-Coupler hitch enabled the operator to hitch and unhitch implements from the tractor seat. It was an A-C exclusive.

The Power-Crater gasoline engine installed on the WD45 boosted the power to 43.21-belt and 37.84-drawbar horsepower (Test No. 499), 25 percent more than the WD, and enough to put the WD45 solidly in the four-plow class. The added power was achieved by the newly designed Power-Crater combustion chamber, increasing displacement from 201 to 226 cubic inches, and raising the compression ratio to 6.45:1. The added cubes came by increasing the stroke from 4.00 to 4.50 inches, while the bore remained 4.00 inches and the rated speed held at 1,400 rpm. The LPG model had a 7.25:1 compression ratio and was about one horse more powerful than the gas model.

Dealers and company people took the new WD45 to the field and came back all smiles. Then the demonstrations started. The extra power, combined with the proven Traction-Booster system and mounted implements, had produced a veritable 'barn-burner.' Those field demonstrations evoked only two questions from the spectators, "Where does that little tractor get all its power?" and "How soon can I get one?"

The single hitch-point implements introduced with the WD worked very well once they were hitched. The problem occurred while trying to line up the hitch pin, since it was hidden under the tractor. Bill Tanke, chief engineer at LaCrosse, changed all that when he invented the Snap-Coupler and its lift-arm latches. Utter simplicity, it had a funnel which guided the implement tongue into the spring-loaded latch. The operator backed the tractor over the implement tongue, listened for the "snap" of the coupler, attached the lift links, and was off to the field. It was true in-the-seat hitching, something the three-point hitch could not do for several years. In 1954, at Serial No. 151381, the Snap-Coupler was made standard equipment on the WD45. Older WD and CA tractors could be updated with low-cost conversions and they became a source of extra business for dealers. (The writer speaks from experience — he could hardly wait to get his hands on one.)

By the early 1950s, farmers had discovered the lower cost of operating diesel-powered farm tractors, something that A-C lacked. A-C had turned to General Motors for diesel engines for the Springfield crawler line, but even the smallest GM engine was much too large for a 45-horsepower row-crop tractor. A-C needed its own diesel engines for its wide variety of mobile equipment. The solution came in 1953 with

This July 1953, prototype WD had the Buda 6BD230 diesel engine. Note that the hood does not have the characteristic bead and the lack of the Snap-Coupler.

With many changes the diesel prototype became the WD45 diesel in October 1954, 15 months after the original installation. The diesel turned 1,625 rpm, 225 faster than the gasoline and LPG tractors.

Weight transfer with pull-type implements was an A-C innovation. A chain yoke connected the lift arms to the implement hitch. The Traction-Booster system could apply pressure on the yoke to transfer implement weight to the tractor when more traction was needed. When the system was placed in lift, full pressure bypassed the transport valve to extend the cylinder and lift the implement on its transport wheels.

The WD45 with 15-28 R2O rice and cane tires for rice field operation. The tractor had plenty of clearance for straddling rice levees.

the acquisition of the Buda Company of Harvey, Illinois, a well-known manufacturer of diesel and gas engines, lift trucks, and railroad equipment.

West Allis engineers quickly installed the six-cylinder 6BD-230 in the WD and by mid-July 1953, a tractor was operational. In October 1954, the WD45 diesel was offered to an impatiently waiting organization and 350 diesel tractors were built during the rest of the year. The first WD45 diesel was Serial No. 181341.

The 6BD-230 engine had a 3.4375-inch bore and 4.125-inch stroke, displaced 230 cubic inches, and was rated at 1,625 rpm — 225 faster than the gasoline and LPG models. It had the precombustion chamber design (Lanova System) and a 15.5:1 compression ratio. Test No. 563, in 1955, rated the diesel at 43.29-belt and 39.50-drawbar horsepower.

With its increased power, the WD45 had outgrown its WD-sized mounted implements. This brought about the development of the "Big Capacity" implements, which included four-bottom semi-mounted plows and larger wheel-type disk harrows that could effectively utilize the Traction-Booster system for improved traction and outstanding field performance. The transport valve was invented by W. L. Voegeli, assistant general manager of the division. It allowed the Traction-Booster system to transfer weight to the tractor while the implement was working and withheld oil from the implement cylinder until full pressure bypassed the valve during lift.

In 1956, the WD45 became the first A-C farm tractor to have power steering, a feature that was well received by dealers and customers alike. The WD45 was offered in dual-front, adjustable axle, single front-wheel, and cane models. It, too, became the base for two-row cotton pickers. The adjustable front-axle production outnumbered the dual-front style for the first time in 1954. The last WD45 was built during the summer of 1957, after a run of 90,382 tractors. In 1956 a gas dual-front WD45 had a list price of $2,380. It was an outstanding value so when it was replaced by the D17 with a higher price, the remaining inventory sold like hotcakes. Everyone was reluctant to see it go, but it had to make room for a more modern tractor.

"Old WD45's Never Die, They Just Lose Compression."

The Model CA tractor with adjustable front axle drilling grain with an A-C 9x7 mounted fertilizer-grain drill. This highly maneuverable outfit was ideal for small fields and irregular plots.

The Model CA

No sooner had the WD been launched than the sales department and dealers set up a clamor for a Model C-sized tractor equipped with all the features of the new WD. Certainly the Power-Shift wheels were no problem, just a matter of downsizing the 28-inch wheels to 24 inches. The high pressure Traction-Booster pump and control valve went back on the power take-off housing with a drive from the power take-off shaft, like a Model C.

Two-clutch power control was another matter. There was no way to install a transmission clutch ahead of the transmission in the torque tube, since it would also interrupt power flow to the power take-off. A unique solution was found. The right rear-axle housing was redesigned to accept a clutch between the differential and final-drive pinion. Disengaging this clutch effectively 'broke' the right axle, stopping the right wheel. This allowed the differential to operate, stopping power to the left wheel. Perhaps it was a little unorthodox but it worked. A pilot run of 10 Model CA tractors was built in 1949, followed by full production in late 1950.

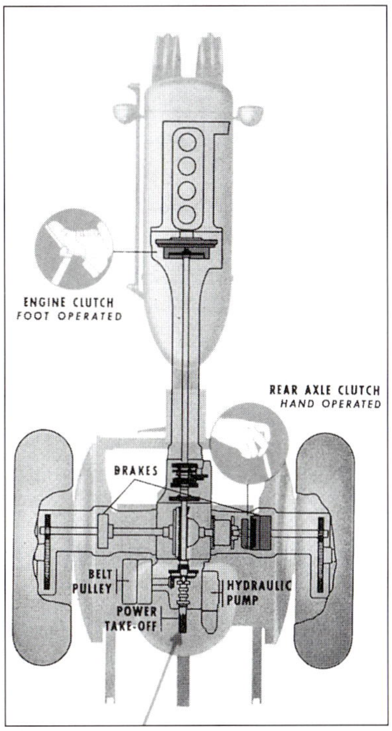

To provide a continuous power take-off the CA had a clutch in the right rear axle which, when disengaged, allowed the differential to rotate, thus stopping the tractor's forward travel.

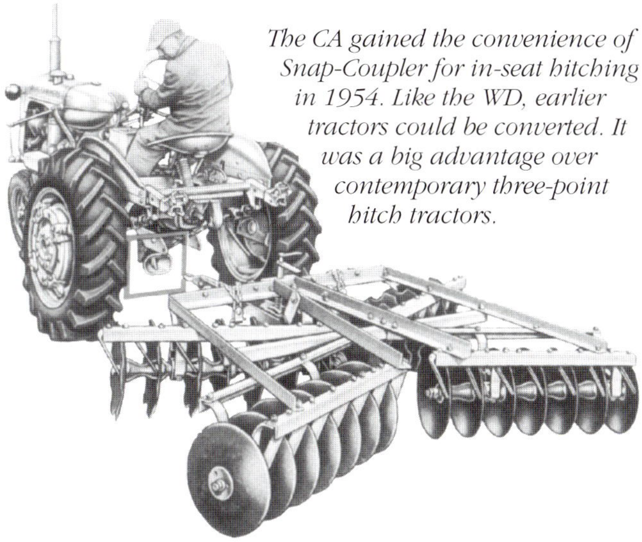

The CA gained the convenience of Snap-Coupler for in-seat hitching in 1954. Like the WD, earlier tractors could be converted. It was a big advantage over contemporary three-point hitch tractors.

The CA was powered by the proven and economical A-C 125-cubic inch engine found in the C, but with improvements. The compression ratio was increased from 5.75:1 to 6.25:1 and the rpm were boosted from 1,500 to 1,650. That was enough to bring belt power up to 25.96 and drawbar power to 22.97 (Test No. 453), slightly over 11 percent more than the gasoline-powered C. With these improvements, the CA was well positioned in the light two-plow class.

One extremely important feature on the CA was the four-speed constant-mesh transmission. It provided easier shifting and was quieter, while it had speeds of 2.0, 3.5, 4.5, and 11.25 mph. With another field working speed and a faster road gear, the CA was a big improvement over the C.

The CA had a full line of front-mounted cultivators, planters, and bedders available. Good operator visibility was emphasized in advertising and promotion.

The experimental Model CA30 had about 30 hp. Never produced, it was apparently abandoned in favor of the 101 and 108 projects which led to the D14.

Everything was up front with a Model G. The 12-in. moldboard plow came in one-way and two-way models. The single 24-in. disk plow also came in one- and two-disk versions.

Several different variations of the Model CA were available. It was offered in dual-front, adjustable front axle, and single-front wheel versions. A cane model appeared in the production records, but no photographs or additional information has been found. Cane usually implied higher crop clearance. In 1954, engineering developed a Model CA30, which had about 30 horsepower. It was never produced, probably due to priority being given to the 101 and 108 projects (discussed later). The Snap-Coupler became available on the CA at about the same time as it did on the WD45. Distributor ignition became standard at engine Serial No. CE149840.

Last published prices:

CA tractor, 10-24 rear, dual front	$1,900
CA, adjustable front axle	$1,950
CA, single front wheel	$1,950

Throughout its production life, approximately 39,500 Model CA tractors were built. Production ended in the summer of 1958, and it was replaced by the new D10 and D12 tractors in early 1959. Like its big brother, the WD45, the CA was a good performer and an outstanding value — a tough act to follow.

The Model G . . .
The Cart Before The Horse

"The tractor with foresight" — So began a 36-page sales manual for the Model G and its matched equipment for sales people.

"The new Model G and implements are designed for all jobs on some farms and some jobs on all farms. A completely new principle of tractor and implement design is introduced, embodying a tractor with a rear engine and an open frame in front. All front-mounted implements are 'mated' to the tractor to provide greater visibility and ease of control."

The Model G and its full line of matched equipment was a radical departure from the conventional designs being offered to the machinery-starved farmers of the post-war 1940s. It arrived on the scene just about the same time as the revolutionary WD. In fact, some midwestern dealers were more than a little disgruntled because they had to order a G to get what they really wanted, a WD. However it wasn't long until the Model G found its niche in the market, which wasn't in the big cornfields of the Midwest. That rightful home was the truck farms, nurseries, and specialty-crop applications where the G was unequalled for precision planting, seeding, fertilizing, and cultivation of vegetables, seedlings, and berries.

A February 1945, mock-up of the Model G had a straight I-beam main frame and full steering wheel.

The complete system that hit the market in 1948 wasn't much like the concept that was mocked-up in February 1945. That version had a straight I-beam for a frame. Another design a few months later had a tubular, but still straight, main frame. The final concept of the arched, twin-tube main frame which allowed more flexible implement design and adaptability, as well as unequalled visibility for the operator, was no doubt the touch that made everything work.

Even by small tractor standards, the Model G was small. With only 10.33-belt and 9.04-drawbar horsepower, it was much better suited to planting, cultivating, or mowing than tillage. Throughout its production life there was a constant flow of requests from dealers and customers to produce a larger Model G of up to 20 horsepower. There was enough interest that, even into the 1970s, several small companies designed and manufactured rear-engine tractors. Most met with limited success, although some were quite well designed.

The Model G was developed by West Allis tractor engineering, but built only in the Gadsden, Alabama plant which also built cotton pickers, mowers, and electrical apparatus.

Since A-C did not produce an engine small enough, the G used a four-cylinder Continental engine. With a 2.375-inch bore and 3.50-inch stroke, displacement was 62 cubic inches. It was rated at 1,800 rpm with a 5.75:1 compression ratio. The four-speed transmission had a special slow speed of 1.6 mph which, at part throttle, permitted running as slow as 0.75 mph for close cultivation of delicate crops. With one of its many special cultivators, it was literally a hoe-on-wheels. With 4.00-12 front and 6-30 rear tires, it could tiptoe between narrow rows. Both front and rear treads were adjustable from 36 to 64 inches for good row-crop adaptability. The G weighed 1,285 pounds and with the fuel and the operator on board, nearly 80 percent of the weight was on the rear wheels for good traction.

A later Model G mock-up had a tubular main frame, cutout steering wheel, real tires, and several changes in the steering gear and front axle components.

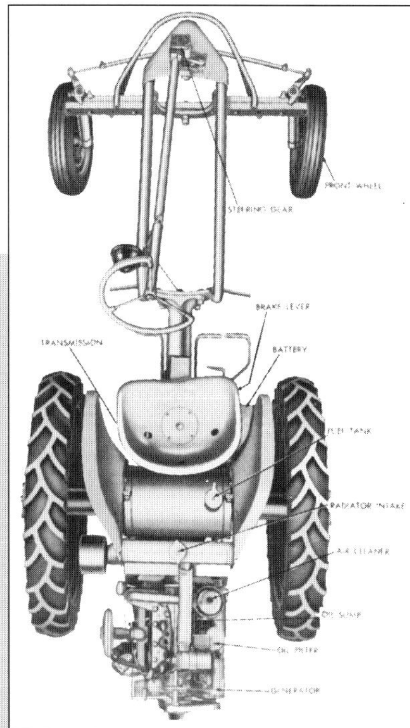

With only two thin tubular frame members ahead, the G operator could see the entire working area and all tools. With a full line of implements, optional hydraulic lift, and belt pulley, the G was ready to tackle just about any farm job.

An operational prototype of the Model G. The arched twin-tube main frame went into production, but hood and rear engine sheet metal was changed substantially.

"The man with the hoe now rides down the row" was the tag line for the Model G and its cultivator. Excellent visibility allowed extremely close work, almost as close as hand hoeing — and a lot more fun.

Hydraulic lift and belt pulley were optional. Even 15 years after the G was out of production, the hydraulic systems were still being requested. In January 1949, the G list price was $760, but by February 1951, it had risen to $900. Just under 30,000 Model G tractors were produced from 1948 through 1955. Sought after by collectors, still actively used by their owners, and always appreciated for its distinctive design, the Model G lives on. It's a farm equipment industry classic.

Evolution of a Tractor

The development of the D-Series tractors was a long complicated process — typical of most major projects. The CA30, built in March 1954, seemed to have elements of the D-Series. The experimental Model 101, November 1954, appeared only eight months after the CA30. Since it followed so soon it is obvious that the two designs were being developed concurrently and that the CA30 had been abandoned in favor of the 101.

Many basic differences existed between the Model CA30 and the 101. The draft-sensing spring had moved ahead of the pivot assembly in the 101. There was a hand clutch, but no indication whether it was a transmission clutch like the WD or a two-speed high-low device like the still-to-come Power-Director. The bulge on the hood back of the muffler probably had gauges in it. Front axle support and oscillation stops were very simple and the adjustable axle was similar to a CA.

The experimental Model 108 of March 1955, was an entirely different tractor than the 101 in only five months. The 108 was longer, indicating a different power train. It appeared to be equipped with the hand-operated, two-speed, high-low on-the-go shifting device which became known as the Power-Director. Slots in the front axle identify a Roll-Shift design, although the front support was still exposed. There was also new sheet metal and fenders. This tractor was so close to the D14 that we wonder why it took another two years to arrive in the dealerships. The real D14 finally arrived in

LaPorte engineering designed this mounted swather, a side-delivery rake, and other push-type harvesting machines for the Model G. None were produced (circa 1950).

The Model GR was built in the Dieppe, France, factory.

The experimental Model 101 of November 1954, appears to be the start of the D-Series tractors, but it is a long way from production. It followed the experimental CA30 by only eight months.

The experimental Model 108 of March 1955, looked much more like a D14. Roll-shift axle was installed, but production was still two years away.

The D14 was the first of an all-new line of A-C tractors that brought many innovations, including Roll-Shift front axle, Power-Director on-the-go shifting, improved operator comfort, and integral power steering.

early 1957. The front support casting had a finished look and the Power-Director lever had changed, but few other differences were apparent.

The several stages of this development program help to illustrate the time span of a major tractor design project. Elements found in the 1944 experimental Model D finally made their way to the customer 13 years later in the Model D14.

1957 — A Vintage Year for A-C Tractors The D Series

A-C tractors experienced a very important year in 1957. Two new models with an all new appearance and loaded with new features were introduced; the D14 in the spring and the D17 in the fall. These tractors provided plenty of new features to talk about and demonstrate: more power, a larger diesel engine, new styling, and better ride for the operator — just to name a few.

The new styling was not only aesthetically pleasing, but functional as well. The hood panels were easily removed for service. The seat, now located ahead of the rear axle with an improved suspension, eliminated the 'catapulting' ride for which the W-line was noted. The Roll-Shift front axle allowed the operator to adjust the front tread without using a jack. Like the Power-Shift rear wheels in 1948, it made a great

There were 50 different models and configurations of the D10, D12, D14, and D17 tractors (only 43 here) built to show the extreme versatility of the D Series in 1959.

demonstration— that is, if you followed the correct procedure. Otherwise, it resulted in the two front wheels pointing toward each other (very excessive toe-in) and a thoroughly confused and embarrassed tractor salesman. Once one learned how, the Roll-Shift axle could be adjusted while others were still looking for a jack.

No doubt the most important new feature of the D14 and D17 was the Power-Director hand clutch with high and low ranges. The hand clutch on the WD45 merely interrupted the flow of power to the transmission, stopping the forward motion of the tractor, and permitting the power take-off and hydraulic pump to continue running. The Power-Director incorporated another clutch for low range, driven at 70 percent of engine speed through two pairs of reduction gears. A neutral position between the forward (high) and rearward (low) lever positions stopped the tractor's ground travel, allowing the power take-off to run continuously. Moving the lever quickly through the neutral position permitted smooth on-the-go shifting with essentially no interruption of power to the drive wheels.

Like earlier models the clutches ran in oil to cool the metal clutch plates, therefore allowing them to tolerate considerable abuse. And plenty of abuse they got, too, with operators slipping them hour after hour to ease harvesting machines through heavy down crops, or to operate the A-C Roto-Baler which required stopping the tractor's forward travel as each bale was wrapped. Combined with a four-speed constant-mesh transmission, the Power-Director provided eight forward speeds plus high-low on-the-go shifting. The improvement in field performance was impressive and demonstrating tractors was fun again. D-Series hydraulics were similar to the WD45, with low-

The 1957-1959 D14 and D17 were painted Persian orange with orange wheels and silver decals. D14 Serial No. 1001-18230, D17 Serial No. 1001-23363.

The Roll-Shift adjustable front axle could be adjusted without using a jack. If the tractor had power steering, engine power moved the wheels in or out.

Grove equipment for the D14 and D17 was dealer-installed and came separately as fenders, fender extensions, low-clearance muffler and grill guard, and operator shield. This D14 has complete grove equipment. List price was $3,770.

volume, high-pressure oil delivered by a four-plunger pump, driven by cams on the driveshaft between the engine clutch and the Power-Director.

A-C designed a new four-cylinder engine, based on a 3.875-inch stroke. The D14 used the first version of this engine with a 3.5-inch bore and 149-cubic inch displacement, but in later tractors it would be offered with 138 and 160 cubic inches. It had the highly efficient Power-Crater combustion chamber like the 226-cubic inch engine. The D17 gas tractor was powered by the proven Power-Crater 226-cubic inch engine which had been upgraded considerably. Compression ratio was upped to 7.25:1, compared to the WD45 at 6.5:1, while the rated rpm was increased to 1,650 (the WD45 was 1,400 rpm). The biggest engine change occurred in the diesel, which now had 262 cubic inches rated at 1,650 rpm, compared to the WD45D with 230 cubic inches at 1,625 rpm.

Both the D14 and D17 were offered in dual-front, single-front wheel, Roll-Shift front axle, high-clearance, and orchard models. A D17 wheatland tractor with full crown fenders was available but it was not quite as large and powerful as most grain farmers wanted, so few were sold.

The D17 tractor was the base for the Model 500 two-row cotton pickers. The D17 cane tractor had 8.5 inches greater clearance under both front and rear axles for working in sugar cane and other tall crops. To obtain the additional clearance, the front axle had taller spindles and 7.50 × 20 tires. The rear outboard final-drive housings were taller and had idler gears between the pinion and bull gears, along with 15.5 × 38 rice and cane rear tires to raise the back of the tractor.

In 1959, the D14 (Serial No. 19001) and the D17 (Serial No. 24001) received a revised radiator grill with a black border and bars. This styling change coincided with the introduction of the D10 and D12 tractors, which shared the new look. In 1960, Allis-Chalmers changed the original Persian orange to a more brilliant shade, and called it Persian orange No. 2. The tractor line was spruced up with cream-colored wheels, grill, steering wheel, and name plates. It was a welcome change and the entire farm line benefited from it.

The D17 with the new colors had some other minor improvements and the serial number sequence began at 32001. Later this

Full crown fenders, larger tires on fixed rims, and heavy-duty swinging drawbar were included in D17 wheatland equipment. Few were sold as most grain farmers wanted more power. By 1960 the wheatland tractor was no longer in the price list.

The D17 cane tractor had over 8 more inches of clearance than a standard tractor. Three-point hitch eliminated the Snap-Coupler for greater clearance under the power train.

The 1960-1962 D17 was unofficially the Series II after the Series III came out and differentiation was needed. It was painted Persian orange No. 2 with cream wheels and grill, short cream metal name plate with black letters, and round aluminum muffler. Serial No. 32001-41540.

Wisconsin's favorite daughter, Alice in Dairyland, not only knew all about cows and dairy products, but she also had good taste in tractors. This is a 1959-1960 D17 black bar grill model in the Serial No. range of 24001-31625. D14 equivalent was Serial No. 19001-24050.

The D17 Series III had Series III on the long cream hood decal, lights on the fenders, and a black oval muffler. Serial No. range was 42001-43358 (1962) and Serial No. 65001-72768 (1963-1964). No tractors were between Serial Nos. 43358 and 65001.

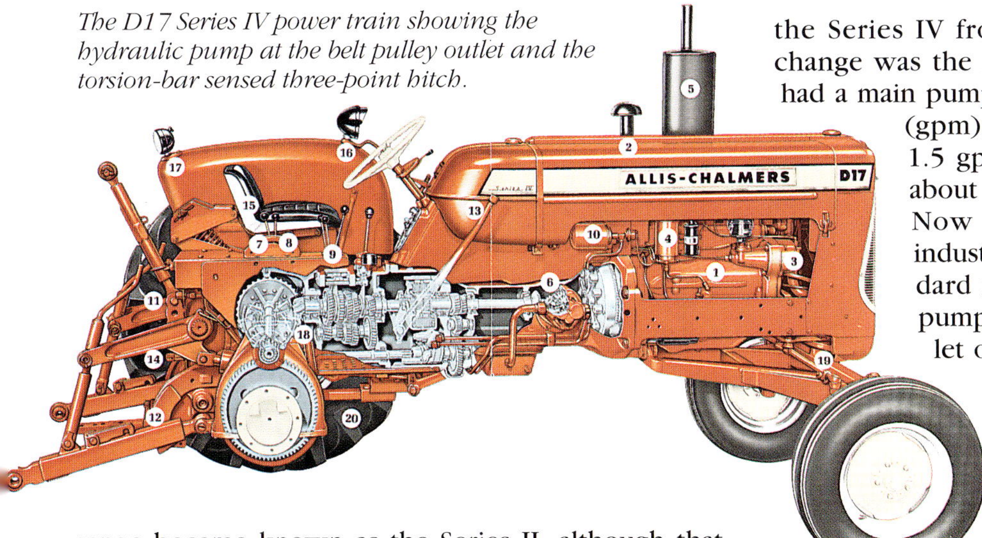

The D17 Series IV power train showing the hydraulic pump at the belt pulley outlet and the torsion-bar sensed three-point hitch.

range became known as the Series II, although that designation never appeared on the tractor or in the sales material. The original orange grill and black bar grill tractors became the unofficial Series I.

The D17 Series III started in late 1962 at Serial No. 42001. The Series III designation appeared on a wide cream-colored hood decal which also included the A-C name and model number. Headlights were relocated to the rear fenders, and there was a new oval-shaped muffler that took some of the bark out of the 226-cubic inch engine. The oil-bath air cleaner was replaced by a new dry air cleaner that was not only a lot more effective at cleaning intake air, but also made servicing a great deal easier and cleaner.

The D17 Series IV, beginning at Serial No. 75001, went into production in mid-summer 1964. Although the styling was similar to the Series III, there was a far-reaching mechanical change that would separate the Series IV from all the preceding D17's. That change was the new hydraulic system, which now had a main pump output of 10.5 gallons per minute (gpm), with a Traction-Booster pump of 1.5 gpm, both operating at 2,000 psi, or about 1,500 psi less than its forerunners. Now the system was compatible with industry standards, and could utilize standard remote cylinders. The two-section pump was located at the belt pulley outlet on the right side of the torque tube, and was driven by bevel gears directly from a hollow shaft attached to the clutch. It was a true live hydraulic system for the first time on a D17.

The D17 Series IV was available with a three-point hitch in lieu of the Allis-Chalmers exclusive Snap-Coupler single-point hitch. The tractor industry had standardized the three-point hitch with its various categories related to tractor power. The smaller D10, D12, and D15 tractors already had three-point hitch available and it was needed on larger models if A-C wanted to sell to competitive tractor owners with a barnyard full of three-point implements. Although it was more convenient and performed better, the Snap-Coupler was on its way to obsolescence. The new category-2 hitch gave the D17 IV sales a real boost. The same three-point hitch was used on the One-Seventy, One-Eighty, 170, 175, and 185, and remained in production until 1981.

The D17 was a very important tractor in the Allis-Chalmers line, proving that the company could keep pace with the rapidly growing American family farm. The last D17 IV, Serial No. 89213, was built on June 16, 1967, just 10 years and 62,540 tractors after Serial No. 1001. It didn't fade away though, since the Model

The D17 Series IV had Series IV on the hood decal and a live hydraulic pump at the belt pulley location. Three-point hitch was offered for the first time. Serial No. range was 75001-89213 (1964-1967).

The D17 Series III and prior models had the high-pressure hydraulic pump on the main drive line and Snap-Coupler hitch. The D14 was similar except that final-drive housings were offset slightly to the rear of the axle housings.

Tractors built in the No. 2 shop were carried on overhead conveyors. These conveyors were abandoned when all final assembly was moved to No. 4 shop with the D21.

The D15 was the first new model to be painted Persian orange No. 2 with cream wheels and grill. It had the same 149 in.^{3} engine as the D14 but with rpm increased to 2,000 and a higher compression ratio. It was called the Series I after the Series II came out. Serial No. range was 1001-8169 (1960-1962).

A standard-clearance D15 with single-front wheel with a vegetable cultivator working in vegetable seedlings. High-clearance models with 38-in. rear wheels and either single-front wheel or adjustable-front axle were available.

One-Seventy tractor, except for the diesel engine, was mechanically similar to the D17IV.

The Model D14 grew up and took the name D15 in 1960, appearing first at French Lick with the new color scheme. Approximately 22,280 D14 tractors were built during its production years.

Some 1960 prices for the D17 and D14:

D17 gas dual-front wheels	$3,705
D17 gas Roll-Shift axle	$3,845
D17 diesel dual-front wheels	$4,600
D17 diesel Roll-Shift axle	$4,740
D14 dual-front wheels	$2,875
D14 Roll-Shift axle	$2,990

The D15

The Model D14 tractor set the stage for a long line of Allis-Chalmers tractors. Not only was it the first of the D Series, but also the physical basis of the D15, the I60 and I600 industrial tractors, and several rough-terrain forklifts. With 34-belt horsepower, the D14 had about the same power as a WD tractor, which is to say, not quite a three-plow tractor. Dealers and customers began asking for more power in the D14, but what they really wanted was a WD45-size tractor with D14 features and price, along with a diesel engine.

The new D15, which was introduced in the fall of 1960, addressed some of the customer's needs. The D15 gas produced about 18 percent more power. The gasoline and LPG tractors still had 149-cubic inch engines, but the compression ratio was increased to 7.75:1 in the gas and 8.9:1 in the LPG. Rated engine speed was increased to 2,000 rpm compared to the D14's 1,650. The D15 diesel was powered by a four-cylinder version of the D17's six. With a 3.5625-inch bore and 4.375-inch stroke, it had 175 cubic inches with a 15.5:1 compression ratio, and also rated at 2,000 rpm.

The Snap-Coupler or lower-link sensitive three-point hitch was optional and the availability of the latter was a sales booster for the D15. Now Ford and Massey-Ferguson owners became potential prospects, since they could use their existing implements. This also meant that Allis-Chalmers had to develop a three-point hitch counterpart to each Snap-Coupler implement, which resulted in duplication of inventories and all the associated costs. The D15, produced from 1960 through 1962, was the first new tractor model to sport the new Persian orange No. 2 color. The D15 was replaced by the D15 Series II in early 1963.

The D15 Series II gas tractor had over 46 PTO hp and outstanding performance. It was painted Persian orange No. 2 with cream wheels, grill and hood decals, and black oval muffler. Serial No. 13001-25419 (1963-1968). No tractors were between Serial Nos. 8169 and 13001.

The D15 Series II

The need for WD45-sized horsepower in the D15 just wouldn't go away. As a result, engineering developed the D15 Series II, which entered the market very early in 1963. It had the power that customers wanted, but only in the gas and LPG models. The Power-Crater engine was given a 0.125-inch larger bore, bringing the displacement up to 160 cubic inches. In Test No. 837 (1963) PTO horsepower increased to 46.18 and drawbar power went to 38.33. In Test No. 838 the LPG model delivered 43.55-PTO and 36.02-drawbar horsepower. Although the diesel received the Series II designation because of various other changes, it remained at 36.51 horsepower. Allis-Chalmers didn't have a four-cylinder diesel engine for a 45-horsepower tractor.

Appearance changes on the Series II included the wide cream-colored hood decals, fender-mounted headlights, oval muffler, and taller fenders to accommodate larger tires. Under the hood was a dry air cleaner that was more efficient and easier to clean. There were improved brakes but the hydraulic system was still basically the high-pressure, low-volume system pioneered by the WD.

The D15 Series II wasn't quite in the same league as the WD45, but it served its owners and Allis-Chalmers pretty well while other projects were top priority in the engineering department.

The D15 Series II with a No. 500 planter in action on a Wisconsin farm. The Series II handled four-row equipment with ease, making it an ideal second tractor on the farm.

The Birdcage

The D14 and D17 orchard models with full wheel shielding met with limited success because the operator still sat fairly high in the normal position. A lower operator position was needed for the orchard models to be accepted by users. This led to the development of the D15 diesel grove tractor. The operator's station was relocated in back of the tractor's rear axle and was quite low. Large 18.4-26 R3 tires provided flotation in sandy groves.

The unique shield-fender assembly gently guided tender fruit and leaves over the tractor and its operator. The front part of the assembly over the radiator and engine was

The D15 diesel grove tractor had full shielding and low rear operator's station for work in low-hanging branches and dense groves. The "birdcage" was only 70-in. high at the top of the operator's shield.

The D15 grove tractor shielding was spring-balanced for easy access to the tractor, or it could be removed completely for open-field work. At only 36.5 PTO hp, it lacked the power needed for large PTO-driven sprayers and few were sold.

made of expanded metal, which shortly led to the tractor being nicknamed the "birdcage." The entire assembly was hinged at the rear axle, and with the help of counterbalancing springs was easily tilted up for servicing the tractor or it could be removed if it was not needed.

Functionally sound, the D15D grove tractor really needed about 15 to 20 more horsepower (it had 36.51) at the power take-off to effectively operate the larger sprayers being used in the groves. With D17 power, it could have been a real winner, but instead only a small number were built in late 1964.

The D10 and D12 Tractors

D Series design finally reached the low end of the tractor line when the Models D10 and D12 went into production in May 1959. Replacing the Model B and Model CA, these were light two-plow, one- and two-row tractors. The only difference between the two tractors was wheel treads, due to the D12's longer rear-axle sleeves and wider front-axle center section, with appropriate changes in the steering linkages. Dual-front and single-front wheel tractors were no longer in demand, so the D10 and D12 had only adjustable front axles. High-clearance models with 4 inches more crop clearance were offered in both tractors.

The D10 and D12 were powered by the same basic engine previously used in the D14, but with a 0.125-inch smaller bore, resulting in 138.7-cubic inch displacement. In Nebraska Test No. 724 the D10 delivered 28.51-PTO and 25.73-drawbar horsepower. In Test No. 723 the D12 produced 28.56-PTO and 23.56-drawbar horsepower.

Like the Model CA, the pair had four-speed constant-mesh transmissions with forward speeds of 2.0, 3.5, 4.5, and 11.4 mph, and a reverse of 3.5 mph. The new little brothers offered a choice of two hydraulic systems. The well-known Snap-Coupler with Traction-Booster was similar to that of the Model CA, with a piston-type pump driven from the power take-off shaft at the rear of the tractor. It was a high-pressure system and was live when the engine clutch was

The D10 and D12 replaced the Models B and CA with D-Series styling, new features, and more power. The black-bar grill identified 1959-1960 models in the Serial No. range of 1001 and prior to 1950.

This high-clearance D12 was painted Persian orange No. 2 with cream grill and is typical of the 1960-1961 D10 Serial No. 1950-3262 and D12 Serial No. 1950-2919.

The 1961-1962 D10 and D12 tractors had the 149-in.^{3} *engine. Draft-sensitive three-point hitch was offered. They had cream grills without bars but otherwise looked like the previous models. Serial No. ranges were D10, 3501-4511 and D12, 3001-3638.*

engaged. The alternative was a lower-pressure lift-lower system which had an engine-driven gear-type pump, in conjunction with a non-draft-sensitive three-point hitch. This system was used primarily on the high-clearance tractors, since it added clearance under the tractor by eliminating the Snap-Coupler.

The twins were introduced with Persian orange paint with black bar grills. In late 1960, they received the new Persian orange No. 2 with cream wheels along with the rest of the line, starting at Serial No. 1950 for both tractors.

Beginning with the D10 at Serial No. 3501 and the D12 at Serial No. 3001 (1961), the displacement was increased to 149 cubic inches, the same as the D14. In Test No. 812 the D10 output increased to 33.46-PTO and 28.78-drawbar horsepower, while in Test No. 813 the D12 power rose to 33.32 PTO and 29.43 drawbar. Following these tests power remained unchanged throughout the rest of their production lives.

At the same time the three-point hitch was converted to a draft-sensitive hitch. Spring-loading for the hitch was supplied by a torsion bar to which the lower links were attached. This arrangement was similar to the D15.

Significant power-train changes occurred in these tractors. A separate drive for an independent power take-off and live gear-type hydraulic pump was provided by a shaft splined directly into the engine flywheel. This shaft ran inside a hollow (quill) shaft to the transmission. Power to the quill shaft was controlled by the engine clutch. The live hydraulic system operated at a lower 1,500 psi, and was compatible with most other standardized systems.

With the increased power, independent power take-off, and improved hitch and hydraulics, the D10 and D12 were now serious contenders in the market-place. In early 1963, the Series II designation was added at D10-6801 and D12-5501, but no significant mechanical changes occurred at that time. Series II appeared on the decal under the fuel tank.

The Series II tractors had Series II on the decal under the fuel tank but otherwise no other appearance changes. Big mechanical change was the independent power take-off. Built in 1963-1964, the Serial No. ranges were D10, 6801-7850 and D12, 5501-6144.

The D10 and D12 Series III tractors had Series III on the decal under the fuel tank, fender-mounted lights, and black oval muffler. A range transmission and front rockshaft were optional. They were built in 1964-1968, D10 Serial No. 9001-10100 and D12 Serial No. 9001-10172.

During 1964, at Serial No. 9001, both tractors became known as the Series III. Appearance changes included an oval muffler and fender-mounted headlights. A new mid-mounted lift shaft (optional) permitted the independent control of front and rear cultivator gangs. Front-mounted cultivators continued to be an A-C advantage. The most significant change affecting performance was the availability of an optional range transmission, which effectively doubled the speeds to eight forward and two reverse. Now the D10-D12 tractors were extremely competitive and had everything an operator could require, except the one thing that would really make dealers and customers happy and boost sales off the chart — a diesel engine. That one serious shortcoming led to a dwindling market share and the ultimate demise of these dandy little gas tractors.

Tractor engineering made a number of valiant attempts to design and build a small diesel tractor, but the economics always killed the projects. There was just no way to provide the required specifications at the right list price using domestic components, and still leave a profit for the company. Competitors were already bringing in complete tractors or assembling tractors with high European content and diesel engines that could be sold for very reasonable prices. The Japanese weren't even in the picture yet.

The Model D19

During the late 1950s it was becoming increasingly apparent that A-C customers' power requirements would soon outgrow the D17. The Model D18 was developed as an entry in the 60-horsepower class. It had straight-through rear axles with sliding hubs and 34- or 38-inch rear tires. Many felt that A-C had reached the limit of power that could be transmitted efficiently with 28-inch tires, and had long been calling for 38-inch tires to meet the competition who were quick to point out that A-C's rear tires were among the smallest in the business. But the D18 had two big problems — appearance and power. Imagine the front of a D17 diesel with the engine, frame, front axle, sheet metal, and fuel tank joined to the rear of a D19 on 15.5-38 tires. That was the appearance of the D18 and it just was not as rugged as it needed to be. The PTO horsepower was pegged at about 60, and that was neither competitive nor enough more power than the D17 to attract their owners to trade up. The D18 would have been powered by the same D262 diesel as the D17, with the G262 in the gasoline and LPG models.

The D18 was less than one plow bottom larger than the D17, or approximately 60 horsepower. In the meantime, the horsepower race had gotten off to a fast start and other major companies were bringing out larger row-crop tractors. Allis-Chalmers had to

The D18 tractor program was quite far along with many advertising photographs already taken. Many were retouched and used, as was this one. The D18 was recognizable by its lower hood line, about 4 in. below the top of the rear fender. This one had 18.4-34 rear tires, changed from 16.9-34 when power was increased for the D19.

The D19 was introduced in late 1961, and as the largest tractor in the line it was 35 percent more powerful than the D17. It was just what dealers and customers wanted.

compete in the power race, too, and needed 65 to 70 horsepower to do it.

An eleventh-hour program was initiated to increase the D18's horsepower. There was no problem moving the G262 gasoline engine up to the needed output, but the naturally-aspirated diesel could not reach that objective at the required 2,000 rated rpm. Two alternatives were available to get more power. One was to bore the 262 out to 290 inches and the other was to turbocharge it. The turbocharger won and it proved to be the right choice.

It is important to differentiate between a turbocharger and a supercharger. Both serve the same function, that of forcing more air into the cylinder during the intake stroke. Superchargers are driven mechanically by either a belt or gear set from the engine crankshaft, and therefore consume some engine power. Superchargers are widely used on high-performance automotive and aircraft engines, but few have found their way to the farm tractor. The turbocharger, on the other hand, enjoys the reputation of having a "free ride." It utilizes energy remaining in the exhaust gases which would otherwise be dissipated into the atmosphere to drive a turbine (hence "turbo"). This in turn drives a compressor on the same shaft to charge the intake manifold and the cylinders with an increased quantity of air under pressure, therefore greatly improving the engine's volumetric efficiency.

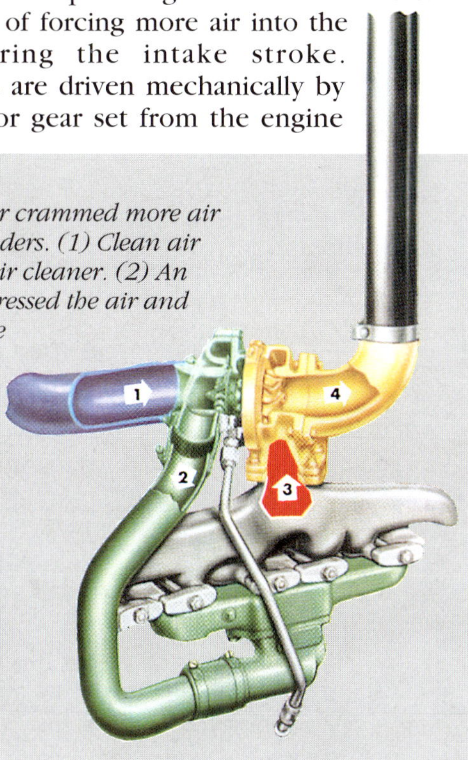

A turbocharger crammed more air into D19 cylinders. (1) Clean air from the dry air cleaner. (2) An impeller compressed the air and pressurized the intake manifold. (3) Energy in hot exhaust gas drove the turbine at about 80,000 rpm. (4) Exhaust was quieted and went out the stack.

This greater supply of air to the combustion chamber promotes more complete burning of a slightly larger amount of fuel, normally producing up to 25 percent more power, but with a substantial improvement in fuel efficiency. This is why turbocharged diesels deliver more horsepower-hours per gallon in Nebraska tests and have long been standard design practice for medium to large horsepower diesels.

Sometime during this time period, the model designation was changed from D18 to D19. To provide the necessary, more powerful appearance, the sheet metal was changed to raise the hood line which also made room for a deeper fuel tank. Even then it was hardly big enough for the gasoline tractor.

The Model D19 went into production in December 1961. It burst on the American farm scene proclaiming another important Allis-Chalmers first — the first farm tractor to have a turbocharger as factory-installed standard equipment. When the D19 was tested, its power ratings were 10 to 15 percent better than the D18 could have delivered. The gas D19 had 71.54-PTO and 63.91-drawbar horsepower and was the powerhouse of the three. The turbocharged diesel produced 66.92-PTO and 62.05-drawbar horse-

The D19 diesel was the first farm tractor to have a factory-installed turbocharger for increased power and fuel economy. This one was pulling a Model K offset disk harrow at the sales training center in Racine county, Wisc.

The D19 gas had nearly 5 more PTO hp than the diesel. It and the 70 Series three-bottom spinner plow were a well-matched team. The two-way plow left no dead furrows.

power, and the LPG had 66.19-PTO and 58.29-drawbar horsepower. Displacement was 262 cubic inches with 2,000 rated speed on all models. Everyone was happy to get the extra power.

The Power-Director three-position hand clutch and four-speed constant-mesh transmission were similar to the D17, providing high-low on-the-go shifting, continuous power take-off, and eight forward and two reverse speeds. The straight-through axles with sliding hubs and standard 34- or 38-inch tires were welcomed as much as the increased horsepower. Power-shift rear wheels, an A-C feature since 1948, were offered in both sizes. The Roll-Shift adjustable front axle was standard equipment, but dual-front wheels and a single-front wheel were also offered. Power steering was standard equipment for the first time. The hydraulic system was the familiar high-pressure low-volume type with Traction-Booster. Both Snap-Coupler or category 2 three-point hitch were offered.

The power-train layout of the D19 was similar to the D17 back to the differential. Straight-through axles and 34- and 38-in. tires were welcome improvements.

Traction-Booster Drawbar

The field performance of the WD45 and D17 tractors, with the close-coupled Traction-Booster hitch and pull-type implements, was so successful that the concept was expanded with the Traction-Booster drawbar. This drawbar was connected to the Snap-Coupler and to the lift arms by a non-parallel linkage which could swing or be stabilized on center by inserting bolts. A large rugged

The D19 was often matched with a 400 Series four-row cultivator but had plenty of power for a six-row cultivator as they became more widely used.

pivot kept the tongue rigid vertically, but not horizontally, from the Snap-Coupler to the implement. It worked the same as the Traction-Booster hitch, but allowed very short turns with large implements. The T-bar hitch, as it soon became known, was used on D17 and D19 tractors with large pull-type implements such as disk harrows and chisel plows and thus enabled them to equal or better the field performance of larger and heavier tractors.

Even as late as 1960, Allis-Chalmers continued to supply Model UC tractor components to Thomson Machinery, in Thibodaux, Louisiana, who built them

The Traction-Booster drawbar permitted sharp turns. Weight transfer was inactive when the implement was lifted because there was no draft load.

Special 23.1-26 R2O rear tires and larger 7.50-16 front tires equipped the D19 forward in the rice fields. Fender extensions kept mud off the platform.

The D19 high-clearance tractor had 15 in. more daylight than the standard tractor. Dropbox final drives, tall front spindles, and 20-in. front tires took the D19 to new heights. There was 37 in. of clearance.

into their own line of cane tractors and equipment. This was fine for Thomson, but other dealers and company people could not benefit from this approach and wanted an A-C high-clearance tractor to sell. Furthermore, A-C wanted to discontinue building the old UC components since they were not compatible with the farm tractor line then in production.

Engineering developed a high-clearance model based on the D19. The new tractor had 37 inches of clearance under the front axle, or 15 inches more than the standard D19. The extra height was accomplished with drop boxes on the rear axles, and much

A prototype D19 high-clearance tractor with a standard tractor for comparison. Production high-clearance tractors had adjustable front axles. Presumably the B-1 is present to display minimum clearance.

taller spindle supports on the front axle. During 1962 and 1963, 70 diesel and 9 gasoline tractors were built.

The D19 tractors were built from December 1961, until mid-1964, with 10,591 units coming off the line. The serial number range was 1001-16266 with no tractors being built between Serial Nos. 7331 and 12001.

The 100-Horsepower Barrier Was Broken — The D21

The last of the D Series tractors, the Model D21, broke a lot of new ground for A-C with many big changes for the tractor line:

• New styling. It was the first tractor to be designed by a newly created Industrial Design Department. The modern, rugged styling signaled recognition that appearance was very important in creating the right product image.

• The D21 was the first Allis-Chalmers farm tractor with over 100 horses — at 103.06-PTO and 93.09-drawbar horsepower. A-C, ever committed to the family farm, had accepted the fact that family farms also needed big power and higher productivity.

• First use of the new A-C direct-injection open-combustion-chamber diesel engine in a farm tractor. Besides the advantages of direct injection, these engines featured cross-flow head design for improved volumetric efficiency, alternating valves for more even distribution of heat throughout the cylinder head, and many other improvements. Called the Model 3400, the new engine had a 4.25-inch bore and 5.0-inch stroke, displaced 426 cubic inches, and had a 16.0:1 compression ratio.

• An all-new power train that was much larger and heavier than the D19. The constant-mesh transmission had four forward and one reverse gears, doubled by a two-speed range, giving eight forward and two reverse speeds.

The 103-hp D21 moved A-C into the big tractor market and helped erase the company's image as just a small tractor manufacturer. Besides the new styling, the D21 brought many new features that would eventually spread across the entire tractor line.

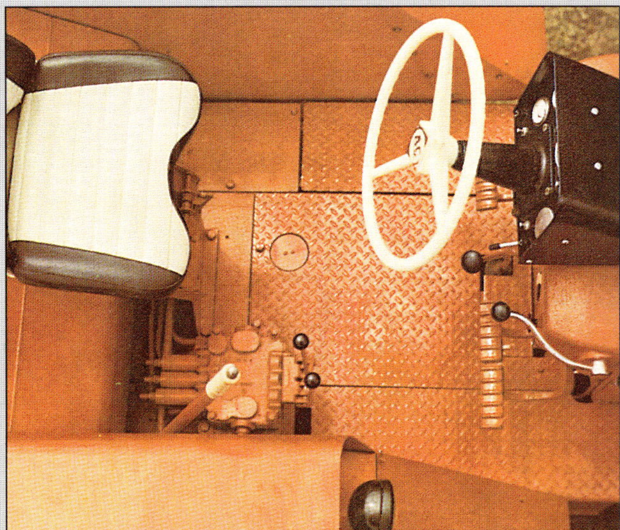

Big-as-a-ballroom platform, full fenders, and hydrostatic steering with tilt steering wheel and instrument panel vastly improved the operator's comfort and convenience. The 52-gal rear fuel tank allowed about 10 hours operation at 75-percent load.

With standard front tires and 18.4-38 tires, the D21 was all set for row-crop work. Dual wheels were added for heavy draft operations. Six- and eight-row rear-mounted planters and cultivators were available but no front-mounted equipment was offered.

• First of the larger A-C tractors with an independent power take-off designed into the power train. Earlier D10 and D12 were a redesign into an existing power train.

• First A-C tractor with an engine-driven hydraulic pump in an open-center system operating at standard pressure.

• First A-C tractor with hydrostatic steering with a tilt steering wheel that also tilted the instrument panel for better viewing while standing.

• Very large operator's platform, with full fenders that extended forward over the rear wheels to keep mud off the platform.

• Large 52-gallon rear fuel tank. The D21 could run nearly 10 hours at 75-percent load on a tank of fuel.

The D21, with its sizable horsepower, required a line of larger implements and the LaCrosse Implement Plant responded to that need. Moldboard plows were still big business in the 1960s, so the 9000 series semi-mounted and pull-type plows with up to seven bottoms were developed. The larger semi-mounted plow was designed so that both rear wheels of the tractor could run up on the land and still have acceptable draft lines. On-land hitching permitted the use of wide single tires, provided equal traction for both rear wheels, and gave the operator a level ride. It was certainly important in the industry, so every company felt obliged to do it.

Later, A-C marketing set about to promote the D21 as a row-crop tractor with 18.4-38 rear tires instead of the popular large 24.5-32 singles. One of the demonstrations was to operate a D21 on 24.5-32 tires pulling a seven-bottom semi-mounted plow with an on-land hitch, stop and unhitch the tractor, convert the hitch to in-furrow operation, and replace the tractor with another D21 equipped with 18.4-38 tires set to run in the furrow. To everyone's amazement and consternation, the in-furrow tractor could easily and consistently pull the plow one gear faster, although both tractors were developing the same horsepower. There seemed to be no logical explanation until one of the company's plow experts noted that the line of draft, previously straight forward, was now offset toward the furrow. This reduced the landside pressure at each plow bottom and substantially decreased overall power requirements. Needless to say, enthusiasm for on-land hitching quickly waned.

The new 9000 Series seven-bottom semi-mounted plow was produced for the D21. Hitch could be adjusted for on-land or in-furrow plowing. It pulled easier with in-furrow hitching due to reduced landside pressure.

The D21 Series II

With the introduction of the One-Ninety XT in early 1965, the D21 found itself with a musclebound little brother that could easily outperform it. Although rated at 93.64 horsepower, the XT diesels usually delivered somewhat more power than that, particularly after a "laying of hands" by a reasonably proficient diesel mechanic, which was not an

The D21 Series II had a turbocharger to boost power to nearly 128 hp. It had 28 percent more power than the D21 but burned only 7 percent more fuel.

uncommon occurrence. It quickly became obvious that the D21 must have more power to retain its flagship position in the A-C tractor line. Solving the problem wasn't really difficult, however, just use the same recipe that produced the Model D19 — add a turbocharger.

The turbocharged version of the 3400 engine, the 3500, was waiting to go to work and it soon got the opportunity in the D21 Series II. The Series II went into production on May 19, 1965, at Serial No. 2201. Power take-off horsepower jumped to 127.75 while drawbar horsepower grew to 116.41. According to Test No. 904, PTO horsepower was 24 percent more, but fuel consumption increased only 7 percent.

The D21 Series II could easily handle this 12-yd elevating scraper. The tractor was promoted for industrial applications but needed a power-shift transmission to be really successful.

As tractor power increased throughout the industry, engineers kept searching for ways to better utilize it. The tire companies had produced larger single tires, already 24.5 inches wide, and and they soon grew to 30.5 inches. Duals were an obvious solution, and soon displaced the wide singles in most applications because of lower cost and greater versatility.

The D21 Hydrostatic Front-Wheel Assist

Another approach to utilizing engine power was four-wheel drive. During the mid- and late-1960s the larger four-wheel drive tractors gained acceptance among big acreage operators, especially in the wheat growing areas of the western United States and Canada. Many large two-wheel drive wheatland models could also be equipped with mechanically-driven front-wheel assists. Some after-market axles, both mechanically- and hydrostatically-driven, became available to convert existing two-wheel drive tractors, and found customer acceptance. This created a demand from A-C dealers and customers for a driven front axle on the D21 II.

The rear main housing of the D21 had not been designed for a mechanical drive and modification would have been prohibitively expensive. As a result, a hydrostatically-driven front axle was developed. The pump was mounted in the front support housing, with a hydraulic motor, mounted vertically where the steering king pin would normally be located, driving each front wheel through bevel and reduction gears.

Ten D21 Series II tractors with hydrostatic front-wheel assist were built in 1967. Not successful, all but one was converted to two-wheel drive. The 10th is still out there somewhere.

The tractor operator could vary the system flow from the seat to provide more or less power to the front wheels. Later it was discovered that too much flow caused wheel spin and excessive tire wear, while too little meant that the tractor had to push the front wheels, negating the advantages of the front-wheel assist. For this reason, the hydrostatic front-wheel drive met with only limited success on tractors and was eventually replaced by mechanical designs.

The sales department was happy to get the new feature and was poised to aggressively market it, but fortunately senior management was far less enthusiastic. Eventually a program to build and lease 10 units was implemented. The tractors were built and shipments began in late January 1967. As a part of the lease, Allis-Chalmers agreed to convert each tractor back to two-wheel drive if the company determined that the program was not completely successful and that no more units would be produced (which proved to be the case). Nine tractors were eventually converted, but one user steadfastly refused to part with his hydro, although he knew that A-C would be unable to supply parts in the future. So, somewhere out there in someone's collection, fence corner, or wrecking yard, is a one-of-a-kind D21 II with hydro-FWA — happy hunting.

Unfortunately the owner's name of this "winningest" D21 has been lost. He brought home a nice collection of trophies for Allis-Chalmers. Tractor pullers liked the D21 Series II because the 426-in.^{3} engine could tolerate high-boost pressures for the needed power.

In any event, A-C gained enough experience in this project to discourage any further use of hydrostatic drive on a farm tractor, although the concept proved to be an excellent solution for improving traction for a self-propelled combine. The next attempt at front-

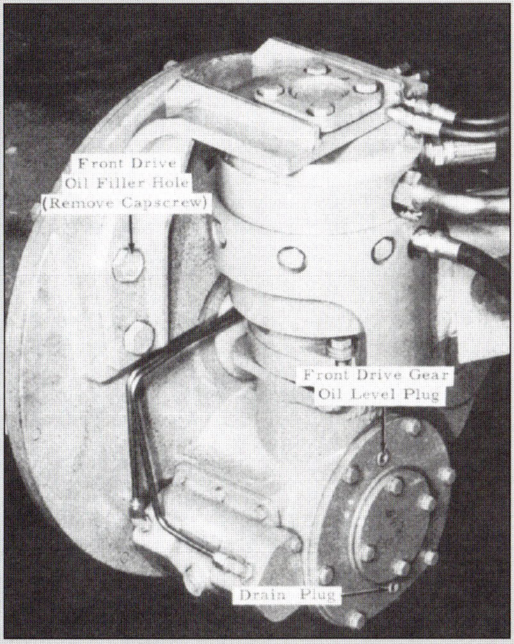

A vertically-mounted hydraulic motor drove each front wheel through bevel and reduction gears. The hydro-front-wheel-assist was a good idea that didn't work very well.

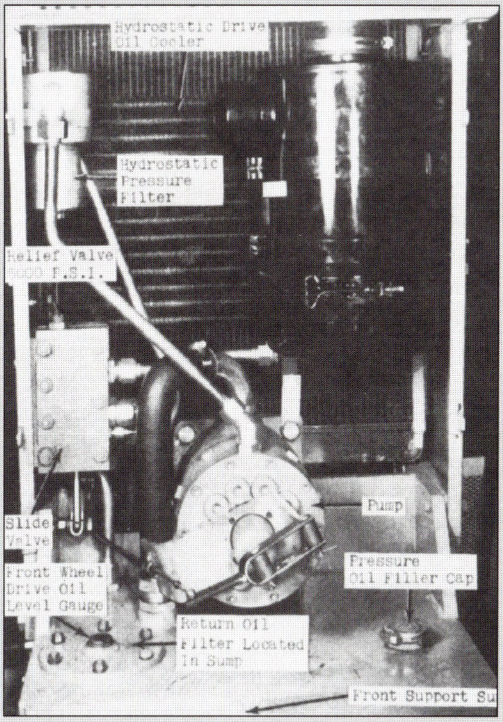

The D21 hydrostatic drive pump was driven from the engine crankshaft. Control valves were to the left of the pump and the operator controlled wheel speed from the seat.

wheel assist would be a mechanically-driven axle on the Model Two-Twenty in 1970.

The D21 turned out to be the sweetheart of tractor pullers. The power train was light enough that a stripped D21 could pull in lightweight classes, while the big 426-cubic inch engine could tolerate multiple turbochargers, intercoolers, and the resulting high boost pressures. It was a real powerhouse, winning more than its share of trophies. In 1969 the D21 Series II was replaced by the somewhat more powerful Model Two-Twenty.

Serial numbers and production data:

Tractor	Serial Nos.	Years Built
D21	1001-2129	1963-1965
D21 II	2201-4609	1965-1969

This prototype model had a lower and wider grill and hood that was changed to the nearly square production design for better row-crop visibility.

New Look, New Power, New Performance — The Hundred Series

The modern styling of the D21, introduced in 1963, was well accepted and soon prompted a groundswell of requests for new styling on the rest of the tractor line. The original D Series styling was in its seventh season and most everyone seemed ready for a change. In the meantime, the industrial design department, which had styled the D21, had conceived an entirely new concept which they proposed for the replacement of the D19. By existing tractor design, it was radical — a real attention-getter. The hood was sloped even more than the final One-Ninety styling, with a grill that was only about two-thirds as tall as its width, which was easily 50 percent wider than the grill on a D21. However, the extreme width tended to obscure visibility of row crops being cultivated, and it was modified to the appearance which gave the One-Ninety its classic good looks.

Like the D21, the One-Ninety featured a large flat operator's platform, tilt steering wheel, full fenders, seat mounted on an inclined track, and a large rear fuel tank that held 48 gallons. At one time, some thought was given to naming the tractor the D19 Series II, but this soon gave way to a strong desire for an all-new line designation to support the modern styling and features. There was a perceived need to retain the '19' in the model number, and after much deliberation, One-Ninety (always spelled out) was selected.

The One-Ninety brought all-new styling that would identify the entire A-C line by 1969. Many features initiated in the D21 were carried into the One-Ninety.

The engineering department adapted many of the features already in production on the D21 to the new tractor. These included the low-pressure high-volume hydraulic

The turbocharged Model 2900 engine gave the One-Ninety XT diesel the performance needed to be a class winner. Improved versions powered A-C tractors and Gleaner combines until the end of the A-C line.

system with an engine-driven pump and a new direct-injection diesel engine from the engine division. This was the first application of the 2500-2800 series engines in a farm equipment product. This family of engines was developed with two bores and two strokes to provide engines with three displacements:

Bore (in.)	Stroke (in.)	Displacement (in.^{3}) 4 cyl.	6 cyl.
3.75	4.00	177	265
3.875	4.00	190	283
3.875	4.25	200	301

The original plan called for most sizes to be built in gasoline, LPG, and diesel versions, but all those proliferations proved to be unnecessary to serve the market. For example, instead of building three sizes of six-cylinder diesels, it was more practical to build only the 301 and provide lower horsepower engines by restricting the fuel flow.

The One-Ninety gas and diesel versions were introduced first. The general design and features were readily accepted but their 75.37- and 77.20-PTO horsepower ratings were met with great disappointment. Major competitors already had models with at least 90 horsepower, and nothing less would satisfy the field organization.

The engine division had already developed larger 301-cubic inch turbocharged diesel, gasoline and LPG engines, so there was no problem increasing horsepower. Thus the One-Ninety XT was born. For a time, it was almost named the XP (extra power), but XT finally received the most votes. (XT doesn't stand for anything. We just liked it.) The PTO horsepower of the turbocharged diesel jumped to 93.64, exactly what the sales crew wanted. Introduced in late 1964, production began the following January on the turbocharged diesel. At the same time the One-Ninety XT gas (89.53-PTO hp) and the One-Ninety XT LPG (85.25-PTO hp) rounded out the line. The tractor was an immediate success, and for many months, the factory couldn't build them fast enough.

The 2900 turbocharged diesel was a ruggedly built engine that was only getting warmed up at 93 horsepower and would tolerate vast amounts of increased boost and fuel input. Extra horsepower came cheap and easy. Even in tractor application, which was always conservative, the same basic engine as the 649I was later rated at 123-PTO horsepower in the 7020.

The spring of 1965 found dealers and company people demonstrating the hottest tractor they ever had to sell. Field performance was phenomenal, and the XT would easily outperform the D21, which presented another marketing problem.

The field organizations quickly discovered the potential horsepower and it was not unusual to hear of 120-horsepower XT diesels. If the user chose to utilize his new-found power by pulling the same implements a gear faster, things worked out pretty well. If he chose to add a lot of weight in order to pull a larger implement, reduced power-train life could be expected, and in too many cases, this is exactly what happened. As a result, engineering programs were initiated to improve power train reliability.

The XT diesel became a tractor puller's favorite. Weighing less than 8,000 pounds, it could be stripped to pull in the 7,000-pound class and with all that potential power it brought home trophies.

What had been a mechanical Power-Director clutch was now hydraulically actuated. It was controlled by a short lever on the Console Control, which also held the throttle and all hydraulic con-

The unique reversible PTO shaft made speed changing easy and it couldn't get lost. There was no oil loss during the change as in some designs.

Dual-front wheels were available but few were built as row crops went to narrow rows. Dual-front tractor had the interchangeable front end with internal steering and could be converted to adjustable front axle.

trols. The One-Ninety had a four-speed constant-mesh transmission. Combined with the Power-Director, there were eight forward and two reverse speeds.

The Power-Director stopped the tractor and allowed the power take-off to run continuously. An optional hydraulic power take-off clutch permitted independent power take-off operation. The original One-Ninety tractors were offered with either a 540 or 1,000 rpm power take-off. In early 1970, an optional two-speed 540-1,000 rpm power take-off went into production and was A-C's first use of the reversible power take-off shaft. It was a badly needed feature that helped attract new customers who owned power take-off-driven equipment with both speeds.

The One-Ninety had an open-center hydraulic system with an engine-driven pump that powered the power steering and Traction-Booster system, and supplied 12 gpm at 2,000 psi to the remote valves.

Both the exclusive Snap-Coupler hitch and category 2 three-point hitch were offered, but the Snap-Coupler was discontinued in the late 1960s as the company standardized on the three-point hitch. A wide-swinging drawbar tractor met the needs of the wheat country.

Two types of Roll-Shift adjustable front axle were available on the One-Ninety. The standard Roll-Shift axle had an external steering cylinder, while the interchangeable Roll-Shift had internal steering cylinders in the front support housing and could be equipped with adjustable front axle, dual-front wheels, or single-front wheel. By the time the Series III was introduced in December 1968, customer demand for the interchangeable type was so low that it was dropped. The dual-front tractor was now extinct.

Rear tires on either Power-Shift or non Power-Shift rims included 15.5-38, 16.9-34, and 18.4-34 sizes. The One-Ninety was adapted to the adverse conditions of the rice country with 23.1-30 R20 tires and to wheatland operation with 23.1-30 R1 tires. Both high-clearance and cane versions were offered on the One-Ninety. The only differences were the braces on the front axle spindles and R20 rear tires on the cane model.

The One-Ninety's large 48-gallon fuel tank behind the operator was very popular, especially among D19 users who had been limited to only 23 gallons of fuel. (There was a standing joke that it was difficult, if not impossible, to drive a D19 gas past a fuel pump.)

The interchangeable front end could be equipped with single-front wheel, but large tractors were hard to handle. High-clearance adjustable front axles became widely accepted and replaced single-front wheel models.

The One-Ninety XT with 23.1-30 R20 tires could ride the levees of the rice fields. These conditions required rugged front axles.

The One-Ninety and One-Ninety XT high-clearance tractors had 37 in. of clearance under the centerline for very tall vegetable and nursery crops.

The One-Ninety and XT cane tractors were similar to the high-clearance tractors but with front axle braces and R20 rear tires. Dropbox final drives raised the rear of the tractor an additional 13 in.

Huge platform, fully-adjustable seat, tilt steering wheel, and Console Control made the One-Ninety an operator's tractor.

The rear fuel tank location was ideal for the XT LPG 52-gal fuel tank. Turbo-charged diesels won the battle and LPG tractors were discontinued by 1969.

The three-point hitch Traction-Booster drawbar could reduce wheel slippage up to 40 percent with a large disk harrow. It came in category 2 and 3 sizes.

The operator's platform at nearly 10 square feet was the largest in the industry at that time and proved to be an important selling feature. With the huge platform, adjustable steering wheel, and fully adjustable seat, the One-Ninety was an operator's tractor in those days before cabs were the rule.

Traction-Booster Drawbar for Three-Point Hitch

A new type of Traction-Booster drawbar was developed for the One-Ninety and other category 2 and 3 three-point hitch tractors with lower-link sensing. This consisted of a heavy steel fabrication that connected to the lower hitch links and the swinging drawbar, which was unpinned at its hitch point and allowed to slide fore-and-aft as draft load changed. The implement hitch was near the front of the hitch plate, and as draft increased, the plate would rise under the implement tongue to transfer weight to the tractor. Used with the One-Ninety, One-Ninety XT, and D21 tractors, performance could be improved dramatically.

Tests were performed on the One-Ninety XT diesel with 18.4-34 tires and duals to determine the advantages of the Traction-Booster drawbar. With liquid

A One-Ninety XT with Traction-Booster drawbar and an A-C disk harrow could out-perform much heavier tractors and also burn less fuel.

A factory-installed cab became available in 1965. The air-conditioner compressor was enclosed and driven by the unused belt-pulley drive.

ballast in the main tires, dry duals, and front-end weights of 430 pounds, the tractor weighed 11,500 pounds. The implement, an Allis 20-foot 2000 Series disk harrow, was hitched with the T-bar hitch. A revolution counter was attached to a rear wheel to record wheel revolutions. The results were astonishing. Over a measured distance with the disk harrow lifted, the tractor wheels turned 47 revolutions. Disking with the Traction-Booster engaged, the wheels turned 52 revolutions or an acceptable 10-percent slip. Disking without Traction-Booster produced 66 revolutions and slippage of 40 percent — an unacceptable loss of efficiency. The Traction-Booster drawbar enabled the sales organization to hold extremely effective demonstrations and, more importantly, make sales.

Shortly after the introduction of the One-Ninety, the factory-installed cab became popular in the industry and it was necessary to offer one to remain competitive. Allis-Chalmers, working with Stolper Industries, Menomonee Falls, Wisconsin, designed a cab which complemented the tractor's lines. Factory-installed air conditioning was available with this cab. Since belt pulleys were obsolete, the belt pulley drive in the torque housing was available for the compressor drive. The first tractor with factory-installed air conditioning, Serial No. 10962, was built in May 1965.

The first styling change on the One-Ninety tractors occurred in late 1967. The Models One-Seventy and One-Eighty had been introduced during the summer of 1967 and were the first tractors to have headlights on each fender. Their grills lost the horizontal bars and now sported the new A-C logo. In order to retain the family appearance, the One-Ninety and One-Ninety XT tractors also received this treatment. Although the designation did not appear on the trac-

The Series II tractors had two lights in a new flat-top fender and no grill bars.There was no Series II decal. Serial numbers started 19001 in late 1967.

The One-Ninety XT Series III had major power train improvements. It began at Serial No. 23001 in December 1968, and was produced until it was replaced by the Model 200 in 1972.

The One-Ninety and One-Ninety XT power train had a two-speed power clutch with a four forward, one reverse constant-mesh transmission. A hydraulic clutch permitted independent PTO operation. Disengaging the engine clutch stopped both tractor travel and power take-off.

tor, it was unofficially called the Series II, beginning at serial number 19001 and continuing to Serial No. 22153. Production of the Series II ended on December 1, 1968.

The Series III began production on December 6, 1968, at Serial No. 23001. The Series III decal appeared only on the XT diesel, and along with other improvements, was known primarily for its four-pinion differential. The XT diesel remained in production until late 1971, ending prior to Serial No. 31056. None were built in early 1972 when it was replaced by the Model 200.

The One-Ninety diesel remained in production through January 1973, with 23 units being built that month, starting with Serial No. 31118 to about Serial No. 31140. The 185, introduced in 1970, had found its niche in the marketplace and crowded out the One-Ninety.

Models One-Seventy and One-Eighty

After the One-Ninety and One-Ninety XT tractors were introduced in late 1964, the styling of the remaining D-Series tractors was badly dated. The field organization was calling for modern styling on the D17 Series IV and smaller models.

Engineering was working a project that was called, for lack of a better name, the D17 V. In sales planning it was called the One-Seventy, then later the One-Eighty. This was a tractor of around 60-PTO horsepower in both gas and diesel models with a choice of front-end styles, three-point hitch, and sheet metal to match the One-Ninety. The rear main housing and final drives were similar to the D17 with 28-inch rear tires and the torque tube was common with the One-Ninety.

The diesel engine chosen for the D17 V was the 283-cubic inch version of the Model 2800, with a bore of 3.875 inches and a stroke of 4.00 inches. In the meantime, however, the engine division had decided to build only the largest 301-cubic inch version and restrict the fuel flow to provide the required 60-plus PTO horsepower. This is why the One-Eighty diesel had the same displacement as the One-Ninety diesel.

For the One-Eighty gas, the horsepower of the West Allis 226-cubic inch gasoline engine was increased by raising the rated speed from 1,650 to 2,000 rpm. The engine had no difficulty in delivering the required 60-PTO horsepower, but being a four-cylinder engine with no counterbalancing, it had considerable vibration. After much experimentation, engineering was able to produce a prototype gas tractor which met vibration specifications. The first One-Eighty diesel (Serial No. 1007) was built on May 24, 1967. The gas model was delayed but it eventually went into production during the fall of that year.

The One-Seventy replaced the D17 Series IV and the One-Eighty was a new horsepower size in the line. They were introduced in the summer of 1967.

Nearly 100 One-Eighty four-cylinder gas tractors had been built when manufacturing audits discovered that platform vibration at high idle would put one's feet to sleep. Production was halted and an investigation was quickly mounted to determine how this could possibly happen. The prototype was retested, and it was just as smooth as it had been the day it was approved — but it was the only one.

The several dozen gas tractors were converted to diesels at considerable cost and the search for a suitable gasoline engine began. For cost reasons, the G262 gasoline engine, as in the D19, was favored but the engine division was discontinuing that engine and replacing it with the G2500. The result was the use of the G2500 — a 265-cubic inch engine with a smaller carburetor than the One-Ninety gas. All of this testing took several months before the One-Eighty gas, with the six-cylinder engine, went into production on February 5, 1969, at Serial No. 6644.

The original One-Eighty gas tractor had the West Allis four-cylinder 226-in.³ engine turning 2,000 rpm. Production was halted after excessive vibration was discovered in first several dozen tractors.

The power train of the One-Eighty was part D17 IV and part One-Ninety. It shared the Power-Director and optional power take-off clutch with the One-Ninety. While the four-speed constant-mesh transmission, power take-off, differential, brakes, and final drives were closely related to the D17 IV. The hydraulic system was similar to the One-Ninety with an engine-driven pump delivering hydraulic power to the hydrostatic power steering, Traction-Booster system, and the three-point hitch and remote valves.

Front-end styles included Roll-Shift front axle in both interchangeable and non-interchangeable types, high-clearance Roll-Shift axle, dual-front wheels, and single-front wheel. The most popular tire equipment was 18.4-28 Power-Shift rear and 7.50-16 front tires.

This December 1965, prototype D17-V was to have been the Model One-Seventy until the D17-IV 1/2 project was added. After that the D17-IV 1/2 became the One-Seventy and the D17-V became the One-Eighty. Confused?

Ignore the decal. The D17-V high-clearance prototype of August 1966, became the One-Eighty. The pleated and perforated grill did not go into production until the Model 200.

The prototype One-Eighty diesel went into production with only minor changes and soon became one of A-C's most popular tractors. It had the 301-in.³ six-cylinder diesel, the same as the One-Ninety diesel, but rated at 2,000 rpm and with restricted fuel flow.

The One-Eighty and 185 power train had the One-Ninety hydraulic clutch with D17 type four-speed constant-mesh transmission and offset final drives.

The One-Eighty gas with the 265-in.³ six-cylinder engine finally went into production in February 1969, 21 months after the diesel. It was worth the wait.

For high-clearance applications, the 15.5-38 rear and 7.50-20 front were most common, and 23.1-26 R20 tires were used in rice farming. Later 18.4-30 tires on Power-Shift rims were offered and they quickly became the high-volume tire.

When the D17 V was planned, it was expected to replace the D17 Series IV, which would have left a 14 horsepower gap between the D15 Series II gas and the next larger size, and an even larger gap between the diesels. The D 15 Series II gas had 46.18 horsepower and the diesel had only 36.51 horsepower and it was a big step to the new 60-horsepower tractor.

The D17 Series IV had been doing very well for A-C. Many dealers and customers shared the opinion that it was one of the best tractors Allis-Chalmers ever put its name on. The marketing department was reluctant to give it up and the plant could build it efficiently. So, many asked, why don't we restyle it like the One-Eighty and make everybody happy? Enter the D17IV 1/2 project. Yes, that's what it was called during its development. Everything was feasible except for one small detail — no diesel engine. The D262 six-cylinder diesel was no longer in production, and a new 2500 — 265-cubic inch diesel was far too costly for a 50-horsepower tractor. This meant turning to an outside supplier. The engine selected was the Perkins 4.236 diesel, with a bore of 3.875 inches and a stroke

A-C offered the Ansel cab for the One-Eighty as a dealer-installed option. Later the Full-Vision cabs were supplied for both the One-Seventy and One-Eighty.

The One-Seventy gas tractor was the "same old girl in a new dress," mechanically the same as the D17 Series IV. That was a compliment because the D17 IV was one of A-C's best tractors. Rated rpm was increased 150 rpm and compression ratio was 8.0:1.

The D17 IV 1/2 project became the Model One-Seventy in production. Note the pleated grill which was not used in production.

of 5.00 inches, displacing 235.9 cubic inches. It proved to be a good choice.

The first One-Seventy tractor, Serial No. 1005, was built on August 4, 1967. It was a good thing the One-Seventy gas was available, considering the long delay involved with the One-Eighty gas. Adding the One-Seventy made it necessary to realign the horsepower ratings. Since both engines in the One-Eighty were easily capable of 77 horsepower in the One-Ninety, it was no trick to set them wherever needed. The One-Eighty was set at 64 and the One-Seventy settled at 54 horsepower.

Mechanically, the One-Seventy gas was similar to the D17 Series IV gas. The compression ratio was raised from 7.25:1 to 8.0:1 and the rpm increased from 1,650 to 1,800. Gear ratios remained unchanged, so field speeds at rated rpm were slightly faster in the new model.

Using a four-cylinder diesel engine permitted building both tractors on the same 95.5-inch wheelbase and using the same sheet metal, etc. This also simplified the adaptation of front mounted equipment such as loaders.

The Perkins diesel had a Lanchester balancer which consisted of opposed counterweights geared to the crankshaft and running at twice crankshaft speed, providing reduced vibration and smoother operation.

Dual-front wheels and single-front wheel were offered for both the One-Seventy and One-Eighty tractors, but few were sold as adjustable front axles took over in narrow rows.

The One-Seventy diesel had a Perkins four-cylinder 236-in.^{3} engine so it had the same wheelbase as the gas tractor and was shorter than the D17 diesel.

Another important change was the power steering. The D17 IV had mechanical steering with optional power steering. The new tractor had hydrostatic steering as standard equipment. The power steering hydraulic pump was incorporated into the main hydraulic pump which was mounted in the belt pulley drive outlet like the D17 Series IV.

During the production life of the One-Seventy there were major changes at Serial No. 7500 in late 1971. One-Seventy became 170 on a black hood decal with an orange grill like the 175. The front support casting was changed to a hollow casting and the hydraulic pump moved from the belt pulley drive to this location, driven directly from the engine crankshaft. The hydraulic pump ran cooler and the drive was much less complex, therefore lower cost. The front axle was changed so that the wishbone attached to the front of the axle pivot pin instead of back near the flywheel housing. With the wishbone removed, the steering cylinder and linkage could be simplified. The change in the front support also allowed the use of the more convenient suitcase-type front weights.

The One-Eighty gas was discontinued during early 1970, that market having been taken over by diesels. At the end of 1973, the One-Seventy-170 diesel and One-Eighty diesel tractors were discontinued, replaced by their more powerful siblings, the 175 and 185. Beginning in early 1974, the 170 gas was revised and became the 175 gas.

The One-Seventy and 175 power trains were similar to the D17 Series IV. The hydraulic pump was located in the unused belt pulley outlet and the Power-Director was mechanical.

At Serial No. 7500 the One-Seventy became the Model 170, got a new front axle, and the hydraulic pump was relocated to the hollow front support casting. The dual-front and single-front wheel options were dropped at that time.

The Crop Hustlers

The year was 1969 and the Model Two-Twenty Landhandler was introduced at 136 horsepower to replace the four-year-old D21 Series II. The A-C tractor business was going well, but still could use a shot in the arm. Everyone was looking for opportunities to expand the tractor market share, and the engineering department developed the Crop Hustler plan.

The 62-hp Model 175 and the 75-hp Model 185 were powered-up One-Seventy and One-Eighty diesels. A lot more power for only a little more money. Everybody won.

It was based on the concept of high-speed farming, that is, utilizing tractor horsepower more efficiently by pulling the same implement faster rather than replacing it with a larger implement. The larger implement would require a heavier tractor that would only be less efficient on light loads, which were always a large part of any farmer's jobs. Other companies were promoting this concept, and Allis-Chalmers was already selling this approach in the One-Ninety XT to a great extent, but not saying it in just so many words.

The result was the Crop Hustler Models 175 and 185. The name said it so well that it was immediately accepted. To provide the products, the One-Seventy's diesel horsepower was increased to approximately 62 (just under the One-Eighty) by raising engine rpm to 2,000. The One-Eighty diesel horsepower was raised to 75 (just under the One-Ninety diesel) with 2,200 rpm. Only diesels were offered, since the market for gasoline-powered tractors had deteriorated. To preserve power-train integrity, engineering insisted on maximum weight limits, which were pegged as follows:

	Tractor Rear Tires		
Tractor	**28-in.**	**30-in.**	**38-in. (high cl)**
175 Diesel	8,400 lb	6,750 lb	–
185 Diesel	9,400lb	9,000 lb	8,000 lb

At first, this weight limit was perceived to present a marketing problem, but there really wasn't any place to hang that much weight on either tractor and it wasn't desirable anyway, so the restriction did not seem to affect either performance or sales.

To give these tractors some identity of their own, the cream-colored hood stripe was changed to black, the grill to orange, and black front suitcase weights (like the Model One-Sixty) were offered. Racing stripes were very much in vogue at that time, so racing stripes were duly placed on the front hood panels. They added a little flair and carried out the high-speed farming theme.

Not the least of the impetus to bring these models into the line was to raise the price. Tractors have always been priced primarily on the basis of their horsepower while

The 175 moved well into the four-plow class, giving A-C a more competitively price entry. Starting with the Perkins 236-in.^{3} diesel, it received the larger 248-in.^{3} engine at Serial No. 1459. About 6,500 were built.

The 185 quickly shouldered its way into the line and eventually replaced both the One-Eighty and One-Ninety diesels. Nearly 15,000 units were built.

casting long sidewise looks at competitive tractors, especially the market leaders. The pricing philosophy for the Crop Hustlers was to provide a lot more horsepower for only a little more money, making the tractors more attractive to the customer. Obviously the concept was accepted by the marketplace because the 175 and 185 replaced the One-Seventy and One-Eighty in about three years.

The 175 tractor was essentially a One-Seventy diesel with the 236-cubic inch Perkins engine running at 1,800 rpm. It was tested (Test No. 1043, 1970) with this engine and delivered 62.47-PTO horsepower. It was decided that more reserve was needed and that the Perkins 4.248 would be a much better engine for the tractor. The 4.248 had a 3.975-inch bore and 248-cubic inch displacement. It replaced the 4.236 in December 1970 at Serial No. 1459. This tractor was not retested. The Model 175 received the new front axle and the hydraulic pump in the front support housing, like the Model 170, at Serial No. 2035 in 1973.

Even though the diesel engine had taken over the market in larger tractors, there was still demand for gasoline-powered tractors in the lower horsepower sizes. The Model 170 was upgraded to the 175 gas at 60-PTO horsepower. To get the additional power, the compression ratio was increased from 8.0 to 8.2:1, and the carburetor diameter enlarged from 1.125 to 1.25 inches. Rated rpm remained at 1,800. The 175 gas started at Serial No. 2151 in December 1973, and was produced until March 1976. It was the last Allis-Chalmers gasoline-powered farm tractor.

The 175 and 185 diesels continued to carry the banner for Allis-Chalmers in the 60- to 80-horsepower

The 170 gas was upgraded to the 175 gas in December 1973, at Serial No. 2151. The 175 diesel had already received the new front axle and crankshaft-driven hydraulic pump at Serial No. 2035.

The Model Two-Twenty Landhandler was A-C's largest wheel tractor to date with nearly 136 hp. It was almost 1,500 lb heavier than the D21 Series II, much of it due to a larger rear main housing and axles.

classes for several years. The 175 was discontinued in September 1980, when it was replaced by the new Model 6060. The 185 carried on until September 1981, being built concurrently with its replacement, the Model 6080, for a year.

About 42,500 Model One-Seventy, One-Eighty, 170, 175, and 185 tractors were produced from 1967 through 1981. They served Allis-Chalmers, its dealers, and its customers well — very well indeed.

The Model Two-Twenty Tractor

An entry in a log of product change proposals dated September 20, 1967, reads: "It is proposed that the D21 tractor be restyled and modified as indicated on the attached marketing specification. This is needed to reduce cost and extend the life of the D21 until the new line of tractors becomes available in 1971." Thus began the life of the tractor which would eventually be designated the Model Two-Twenty. Production began on March 17, 1969, at Serial No. 1004. Some creative person at the advertising agency suggested the name, "Landhandler." It was immediately accepted, and the Two-Twenty became the first of the Landhandler tractors. After serving well in this capacity, the name Landhandler became the title of the Allis-Chalmers direct mail magazine.

Another "ignore the decal." The prototype Model Two-Twenty was shown with One-Ninety decals because the right ones were not yet available, a common practice. It sure didn't act like a One-Ninety.

The Two-Twenty completed the transition to the Hundred Series styling and once again the West Allis-built tractors shared a family appearance. The grill was surrounded by a huge casting instead of sheet metal. This gave the front end an unusually rugged appearance. Fenders, platform, seat, and controls were very much like the smaller One-Ninety XT tractors, and the 51-gallon rear fuel tank permitted about seven hours of operation at 75-percent load.

The familiar Allis-Chalmers 3500 turbocharged diesel now delivered 135.95 horsepower at the power take-off, compared to 127.75 in the D21 Series II. The Two-Twenty had 117.21-drawbar horsepower and was about 1,500 pounds heavier than the D21 Series II. Much of the added 1,500 pounds in the Two-Twenty was found in the rear main housing, with substantially larger final-drive gears, now straddle-mounted to provide constant alignment and improved gear-tooth contact under extremely heavy loads. Rear axle diameter was increased to accommodate the greater stress encountered with on-axle dual wheels. Larger pinion and final-drive gears meant greater shaft-center distances, therefore a longer rear main housing which resulted in increasing the wheelbase about 4 inches. Larger dry disk brakes and an extra heavy-duty six-button cerametallic clutch also contributed to improved power-train life and performance.

Like the D21 Series II, the transmission in the Two-Twenty was a four-forward, one-reverse speed constant-mesh unit combined with a two-speed range box. In late 1970 and early 1971, some optional gear ratios were offered to better adapt the tractor to customers' special requirements. The "fast low range" provided a good plowing speed of 5.8 mph in fourth and permitted a one lever downshift, and a speed of

The Two-Twenty with mechanical front-wheel assist was an effort to crack the lower end of the four-wheel drive market. Only 100 were built in 1970. Here it's pulling a No. 3000 seven,16-in. bottom semi-mounted plow.

The United States Air Force "Bare Base" program was a plan to establish fully operational air bases overnight at any of 1,400 abandoned airstrips throughout the world. This military Model 220M with a compactor was one of a family of air-transportable airport construction machines all of which were powered by the A-C 3500 engine. The family included a 645M3 articulated loader, M100M2 grader, modified HD11 crawler, and an Oshkosh dump truck.

The Two-Twenty and Two-ten tractors were available with wide-swinging drawbar for wheatland operation. The 1,000-rpm power take-off could be deleted. Both 24.5-32 and 30.5-32 large single tires were offered.

6.3 mph for tillage. The "slow first and fifth" gave the user a slow first gear of 1.6 mph for heavy power take-off work.

During this period some major manufacturers were successfully marketing large tractors with mechanical front-wheel assist (FWA), as well as large four-wheel drive tractors. Quite a bit of exploratory work had been done toward acquiring a four-wheel drive tractor from an outside source but nothing was imminent in late 1969 and early 1970.

To provide an entry into the four-wheel drive market, the Model Two-Twenty FWA was developed and in 1970 a total of 100 tractors were built. The first Two-Twenty FWA was Serial No. 2223, built on June 3, 1970. The tractor performed well, but the market just did not materialize and the last of the 100 four-wheelers were difficult to sell. It would take the energy shortages of 1973 and 1979 with their rapidly escalating fuel prices to spark a significant demand for the increased efficiency that front-wheel assist tractors could deliver.

The Two-Ten gave A-C a much needed entry in the strong 120-hp tractor market for two years until the new family tractor came out in early 1973.

The QCS cab, introduced in 1972 on the Model 200, was offered for the Model Two-Twenty as well, but Two-Twenty production was winding down so very few cab tractors were built. The Two-Twenty was produced from June 1969, until December 1972, with Serial Nos. 1004-2866.

The Model Two-Ten

The 42-horsepower gap between the One-Ninety XTD and the Two-Twenty was wide enough to let through a lot of competitive tractors. Allis-Chalmers badly needed a tractor at about 120 horsepower and it was going to be a long dry spell before the New Family tractors would hit the market. The XT diesel couldn't reach that high, so that left only the Model Two-Twenty as a basis for the needed new model. Engineering went to work to see if Two-Twenty tractor cost could be reduced to provide a 120-horsepower tractor at a competitive price and still leave a small profit for the company.

The result was the Model Two-Ten. The most obvious change in this new tractor was the use of the One-Ninety grill and some other minor sheet metal changes to accommodate it. Standard tires were reduced to 10.00-16 front and 18.4-34 rear to help reduce the base price, but few were ever sold on the smaller tires.

The 122.4-hp Model Two-Ten was all Two-Twenty under the sheet metal except the fuel system. It was a real value and Two-Twenty orders took a nosedive.

The 3500 turbocharged diesel turned the wheels, but had a lower-output injection pump set to limit horsepower to the target level. In Test No. 1065 the Model Two-Ten delivered 122.40-PTO and 104.95-drawbar horsepower. Otherwise there were very few mechanical differences between the Two-Ten and the Two-Twenty. That made the Two-Ten a real value, so when it was introduced Two-Twenty orders took a big drop. The power train was the same as the Model Two-Twenty so speeds were similar to the "slow first and fifth" transmission of that tractor. The operator's station and controls, hydraulic system, and other primary components were common with the Model Two-Twenty. Model Two-Ten production began in November 1970, at Serial No. 1004 and ended in December 1972, at Serial No. 2469.

A dualled Model 200 pulling a 20-ft No. 2200 disk harrow near San Antonio, Tex. Both had just been introduced to the dealers in 1972. Field performance was outstanding.

The Model 200

Although they shared the same horsepower ratings, engine, sheet metal, and many other components, the Model 200 was a vastly different tractor from the One-Ninety XT diesel. The XT had carried the A-C banner in the 90-horsepower class since early 1965 and it was time for a new look, along with a number of important mechanical improvements and a new cab. The Model 200 had the black hood decal of the 175 and 185 Crop Hustlers, combined with a new perforated and pleated cream-colored grill. The pleated design allowed the use of much smaller holes whose total open area was greater than the flat grill it replaced. Anything that could pass through the small holes would easily pass through the radiator.

Besides a number of internal improvements, the 2900 engine had a more reliable turbocharger, a resonator for quieter operation, and a slower-running seven-blade fan for better cooling with less noise. A crankshaft-driven oil pump was added at Serial No. 4001 in early 1973. The 200 was not tested but was rated on the One-Ninety XT diesel Test No. 887 with 93.64-PTO and 79.12-drawbar horsepower.

The Model 200 replaced the One-Ninety XT Series III in 1972. The new QCS cab with air-conditioning on an isolated platform greatly improved the operator's environment. Horsepower rating was unchanged at 93.64 PTO hp.

With larger lift cylinders and improved geometry the three-point hitch could lift 4,600 pounds, a 50 percent improvement. Larger ring and pinion gears and wider final-drive gears provided greater power-train reliability, and long rear axles were available for on-axle duals. A necessary option was 18.4-38 tires.

The all-new QCS (Quiet, Comfortable, and Safe) cab was designed by West Allis and LaPorte engineers and built at LaPorte. This cab was the forerunner of several tractor cabs that would set new sound-level targets for an entire industry. Mounted on an isolated platform, the QCS cab was factory-installed on the 200, 210, and 220. It was Allis-Chalmers' first ROPS (Roll-Over Protective Structure) cab designed to protect the operator in tractor rollover. The Model 200 was produced from February 1972, until October 1975. During that time approximately 10,126 tractors were produced.

A Clean Sheet of Paper The New Family of Tractors

The 7030 and 7050 tractors were all new except the tires with features A-C had never offered before. It was the beginning of "The Rising Power in Farming."

During the 1960s the Farm Equipment Division found itself at a crossroads, at least as far as farm tractors were concerned. The line had good performance and reliability, and even competitive pricing, but lacked several important mechanical features that identified a modern farm tractor. An all-new power train with plenty of room for anticipated growth in horsepower was needed. Differential lock, planetary final drives, power disk brakes, and various axle lengths were also required. More forward speeds, synchronized shifting, and power-shift transmissions were now high priorities. Operator comfort called for a quiet, air-conditioned cab to reduce the fatigue of long days. Hydraulic systems needed versatility. Features already found in A-C tractors, such as independent two-speed power take-off, Power-Shift wheels, and Traction-Booster system were mandatory.

Luckily for A-C, engineering had plenty of ideas ready to lay on a clean sheet of paper. Specifications incorporating those features along with horsepower ratings were agreed upon by engineering, manufacturing, and marketing departments.

The following is excerpted from an engineering report dated April 1969:

New Family of Tractors with 100-, 120-, and 140-PTO horsepower

1961-1963 Feasibility studies
1964 Built 100-hp tractor
1965 Allison quoted 135-hp power-shift transmission
1966 Built 110-hp tractor
Designed tractors for Allison transmission
1967 Allison withdraws
Designed 90- 110-, and 130-hp manual and power-shift transmissions
1968 Built 120- and 140-hp manual tractors
1969 Built 120-hp power-shift tractor

The road to the new tractor line was anything but smooth. High cost estimates required redesigning components or rethinking entire concepts. All was not negative, however, as good things also happened. Power-train engineers found a way to install a Power-Director into the manual transmission, using two clutches driving different input ratios into the counter shaft, doubling the number of speeds. This proved to be a major breakthrough, considering the loss of the Allison power-shift which caused a two-year delay.

A 1971 styling proposal was considered too much like the Hundred Series and many felt that a more substantial change was needed to better separate the new line from the old. The all-new tractors needed an all-new appearance to help people appreciate the radical departure from the former line. An outside firm,

The New 7030 and 7050 tractors were introduced to the A-C dealer organization in the Los Vegas Convention Center on February 22, 1973. The long-awaited new models were warmly welcomed.

Fulton and Associates, designed the distinctive sheet metal and cab that gave the new 7000 Series their contemporary appearance.

The 7030 and 7050 Tractors

Throughout the development period many A-C personnel wondered if the New Family of tractors would ever be manufactured — if the company could muster the resources to bring this huge project into reality. Proof that qualified and dedicated people can accomplish great objectives, the Models 7030 and 7050 were introduced to the dealers in Las Vegas, Nevada, on February 22, 1973. That meeting was a happy time for a lot of people. The long-awaited New Family of tractors was launched, and in turn, firmly established Allis-Chalmers as a builder of big, modern farm tractors.

Production began in January 1973, at Serial No. 1001 for each tractor. To build the line at a high quality with cost-cutting efficiency, Allis-Chalmers invested in a computer-controlled flexible machining center, the most sophisticated of its kind at the time.

Although the new tractors had been pegged at 100-, 120-, and 140-PTO horsepower, modifications to the development program between 1969 and 1972 changed the horsepower ratings of the new models to an estimated 130- and 155-PTO horsepower. A Model 7010 prototype tractor at 105 to 110 horsepower was built and tested in 1972, but was not placed in production with the 7030 and 7050, mostly

This 1971 prototype was an early styling proposal. It was one of several proposals, all of which were considered too much like the Hundred Series to represent the new line.

The 7050 tractor's 156 hp was nearly 20 percent greater than the Two-Twenty, a new high for A-C two-wheel drive tractors.

This 1972 Model 7010 was ready for production along with the 7030 and 7050, but high cost put it on the shelf. The Model 7000 filled that slot from 1975 until 1979.

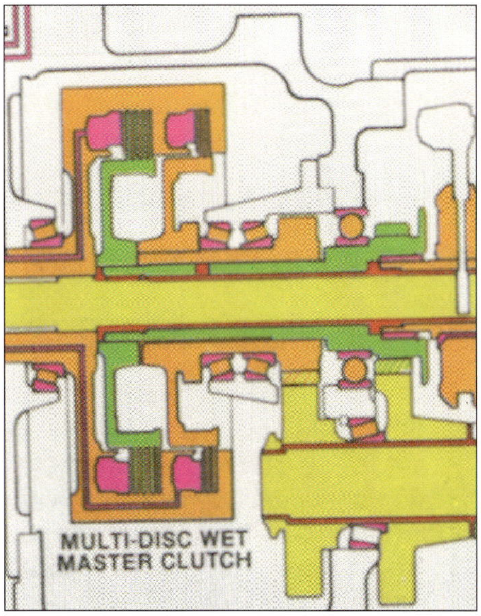

The hydraulically-actuated Power-Director had two clutches geared to the transmission countershaft with high and low ratios. The low ratio was 80 percent of the high ratio. With a 5-forward, 1-reverse transmission and two-speed range, there were 20-forward and 4-reverse speeds.

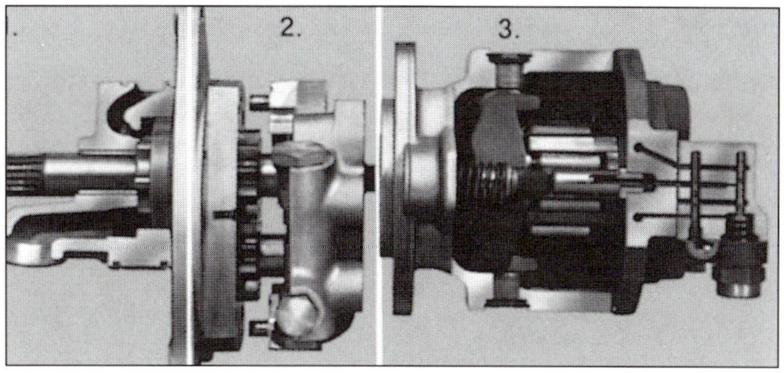

The hydraulic system had three pumps: (1) A 13.5-gpm gerotor pump pressurized the the Power-Director and lubricated the range box. (2) A 9.5-gpm gear pump ran the the power steering, power brakes, and differential lock and pressure-lubricated the the rear main housing. (3) The 22-gpm axial-piston pump was the heart of the load sensitive system and delivered power to the three-point hitch and remote valves.

due to cost. The horsepower rating changes reflected a market preference for larger tractors, as well as engine and power train capabilities. For the first time, when the market called for more power, Allis-Chalmers had the hardware to respond.

The 7030 was powered by the A-C 3500 Mark II, while the 7050 had the A-C 3700 engine, both based on the proven 3500 engine. These were modern direct-injection turbocharged diesels and the 3700 was also equipped with an intercooler, bypass oil filter, and piston cooling. Both engines had 426-cubic inch displacement, 16.0:1 compression ratio, and were rated at 2,300 rpm. The 7030 developed 130.98-PTO and 110.29-drawbar horsepower and the 7050's results were 156.49 and 131.54, respectively.

The Power-Director XX transmission on both the 7030 and 7050 provided 20-forward and 4-reverse speeds by utilizing the two-speed Power-Director clutch, the five-speed constant-mesh transmission, and a two-speed range transmission (2 × 5 × 2). Forward speeds on these models ranged from 1.6 to 20 mph.

The 7030 and 7050 tractors were offered in platform, Roll-Over Protective Structure, and cab models. This 7030 with ROPS was most likely to be sold in mild climates or fleet operations.

Load-Sensitive Hydraulics

The completely new load-sensitive hydraulic system was an important feature of the 7030 and 7050. The heart of the system was a variable-displacement axial-piston-type pump controlled by a pressure compensator which regulated both pressure and flow to match the load, thereby increasing efficiency. When no hydraulic power was required, the system stood by at a very low pressure; consuming very little horsepower. Flow to the remote valves could be varied from about 2 up to 17 gpm. The ability to control the rate of flow was especially important in running hydraulic motors for chemical application pumps. The 7030 and 7050 tractors were the first farm tractors to have this capability and it wasn't until 1978 that other major tractor builders began to adopt load-sensitive systems.

Record-Setting Acousta Cabs

Operator environment was a hot topic in the tractor industry during the early 1970s. Temperature control and ventilation were important but even more

The Acousta-Cab tested a record setting 79.5 dbA and was welcomed by operators' ears everywhere. The tilt and telescope steering wheel and the fully-adjustable seat let drivers of all sizes find real comfort.

crucial, medical science had discovered that exposure to high sound levels for extended periods of time could cause hearing impairment. Typical big tractor cabs could exceed the 90 dbA per eight-hour criteria for exposure.

A well-designed cab with low sound levels was a primary requirement and that is exactly what the A-C engineers delivered. When they were tested, the new Acousta-Cabs generated an ear-pleasing 79.5 dbA, the lowest in the industry at that time. Fuel tanks were now located under the cab floor and could easily be filled from the ground. Batteries were placed under the cab and could be swung out for service — an A-C exclusive. The 7030 and 7050 tractors launched Allis-Chalmers into advanced tractor technology, but they were just the start of a complete line of two-wheel drive and four-wheel drive powerhouses.

The Models 7040, 7060, and 7080 Tractors

In late 1974 the 7030 became the 7040 and the 7050 became the 7060, each with an additional 5 horsepower. The 7040 had 136.49 and the 7060 had 161.51-PTO horsepower. The big news came, however, in the new Model 7080, which was at the time the largest two-wheel drive tractor in the business at 181.51-PTO horsepower.

The 7080 was powered by an A-C 3750 Mark II engine. This was an upgraded version of the familiar 426-cubic inch engine, which was turbocharged, intercooled, and equipped with a counterbalanced crankshaft. Rated speed was 2,550 and the brake mean effective pressure was increased only slightly from 148 psi in the 7050 to 153 psi in the 7080. Many changes provided improved lubrication, piston ring and valve life, and reliability at the higher rated speed. A larger air cleaner, more radiator capacity, and a larger fan were matched to the big engine.

The 7080 power train was based on the proven 7050 design with modifications to accommodate the added power. The Power-Director clutch was upgraded with thicker low-range plates and six clutch-piston return springs for higher torque loads. The power take-off shaft was the larger 20 spline, 1 3/4-inch size required for tractors of more than 160 horsepower.

The 181-hp Model 7080 was the most powerful two-wheel drive tractor in the business when it came out in 1974, a record it held for several years.

The final-drive ring gears were nearly 4-inches wide, twice that of the smaller tractors. Inside, the planetary reduction gears were wider and the ratio increased from 5.41:1 to 6.57:1 to provide speeds similar to the 7060. The rear axles were increased to 4 inches in diameter. To put its power to work, the brawny beast was equipped with the largest farm tractor tires available. Although not the biggest seller, the 7080 was the flagship of the line until the four-wheel drive 7580 came out in 1976.

The 7040 had five more horses than the 7030 and also had the larger axles. The long axles permitted on-the-axle duals, a popular option.

The Power-Shift Transmission

The original product plan for the 7030 and 7050 tractors included a power-shift transmission (PST). Problems in the development program moved the power-shift transmission to the 7040 and 7060, and then was delayed several months past the introduction of those two models in late 1974.

At the inception of the New Family tractor program in the mid-1960s, Allis-Chalmers entered into an agreement with the Allison Division of General Motors for a nine-speed power-shift transmission. Allison quoted specifications and prices for a 135-horsepower unit in 1965 and A-C designed tractors to accept it during 1966. During this time, Allison encountered design problems and cost overruns that would result in a price to Allis-Chalmers that was

The 7050 grew 5 hp to become the 7060. New 3.75-in. axles were standard with 4-in. diameter long axles optional.

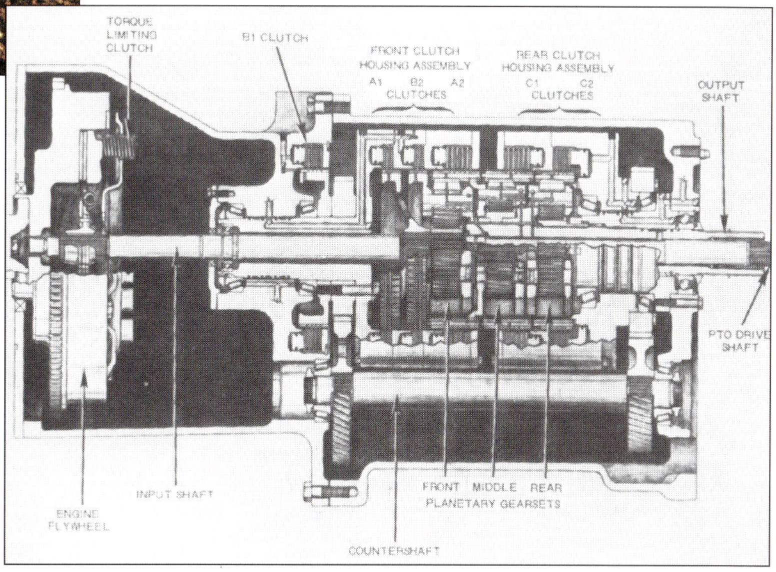

The power-shift transmission had six clutches and three planetary gear sets. Single swaps provided smooth shifts.

Gear	*Clutches*
Reverse	*A2-C2*
1	*A2-B1*
2	*C1-B1*
3	*C1-A2*
4	*C1-B2*
5	*C1-A1*
6	*A2-A1*

nearly double the original quote. The increased cost rendered the project infeasible and Allison withdrew. Since the availability of a power-shift transmission was considered necessary to the success of the new tractor line, it was a great blow to the whole project. Although at the time it seemed disastrous, the withdrawal of Allison proved to be a blessing in disguise. It forced Allis-Chalmers to launch a program to develop its own power-shift transmission. The company was fortunate to have the engineering expertise available to meet the challenge.

To meet target costs it was decided to build a seven-speed power-shift transmission, utilize the range transmission from the Power-Director model, and provide a 14-speed tractor. By 1969 the design was complete and a 120-horsepower, power-shift transmission tractor was built. Later this design was designated the phase I. Like many new designs, the phase I power-shift transmission left much to be desired — speeds and shifting characteristics were unacceptable.

Phase II, a simplified six-speed design, eliminated some of the problems, but high inertia loads of heavy internal components caused rough shifting and reverse was too fast for safe hitching. After much further study, phase III was conceived. It was right. Phase III had only six clutches and three planetary gear sets but these could provide six-forward and one-reverse speeds. Smooth shifting was made possible by "single swapping" clutches during each shift. For instance, in any gear, two clutches were engaged, then during the shift one clutch remained engaged, while the other was disengaged and a third clutch engaged. This overlap, plus progressive internal component speed changes, resulted in smooth shifting even under full load. Coupled with a manually-shifted range transmission, the tractor had 12-forward and 2 - reverse speeds. It was a long-awaited, welcome day when the prototype phase III was driven in the field.

After all the delays, the power-shift transmission for the 7040 and 7060 tractors was introduced on an extremely hot and dry day in a field east of Wichita, Kansas, during the summer of 1975. The Model 7000 tractor arrived at the same time. It made the day under that scorching sun worthwhile.

As one might expect, customer acceptance of the power-shift transmission was slow at first. As the field discovered its good performance and reliability, however, the power-shift transmission was ordered at an increasing rate:

Year	PST%	Tractors
1975	10	7040, 7060
1976	43	7040, 7060
1981	57	7010, 7020, 7045, 7060
1985	83	8010, 8030, 8050, 8070

The A-C power-shift transmission was as efficient as the Power-Director. When the 7000 tractors were tested the five power-shift tractors averaged 12.99-drawbar horsepower hours per gallon, slightly better than the 12.88 for their Power-Director counterparts. Along with efficiency and fuel economy, the Power-Shift transmission enjoyed a fine reputation for good performance and high reliability.

The 7000 Series Black Chassis Tractors

A-C held major tractor introductions in 1977. At a national dealer meeting in Kansas City, Allis-Chalmers made big news by introducing five new tractor models, a new cab for the large farm tractors, and a new color scheme for all models. The 5020 and 5030 compact diesels with their implements were also launched, while the new 253-horsepower, four-wheel drive 8550 was the hit of the show.

However, the new tractors which helped the most dealers and put the most money in the cash registers were the 7020 at 123 and the 7045 at 146 horsepower. With the Model 7000 at 106 and the 7040 at 136 horsepower, the power spacing was too wide to meet the market's needs, especially when many manufacturers were selling 120-horsepower tractors in

At 123 hp, the 7020 filled the wide gap between the 7000 and the 7040. The 301-in.^{3} engine was turbocharged and intercooled.

large numbers. Just like the wide spacing between the One-Ninety XT and the Two-Twenty in 1970, Allis-Chalmers needed a 120-horsepower tractor. This time the company had all the pieces on the shelf to build one.

The 301-cubic inch turbocharged diesel had plenty of room to grow and the power train of the former 7030 was more than adequate. With an intercooler, the engine was named the 649I (6 cylinders, 4.9 liters, turbocharged and intercooled). This engine, the 7030 power train, along with other changes became the Model 7020. It tested at 123.79 horsepower and nicely filled the gap. This left only 13 horsepower between the 7020 and the 7040, so the 7040's power was boosted to 146.18 horsepower to become the 7045. The engine in the 7045 was known as the 670T (6 cylinders, 7.0 liters, turbocharged) and easily produced the needed horsepower.

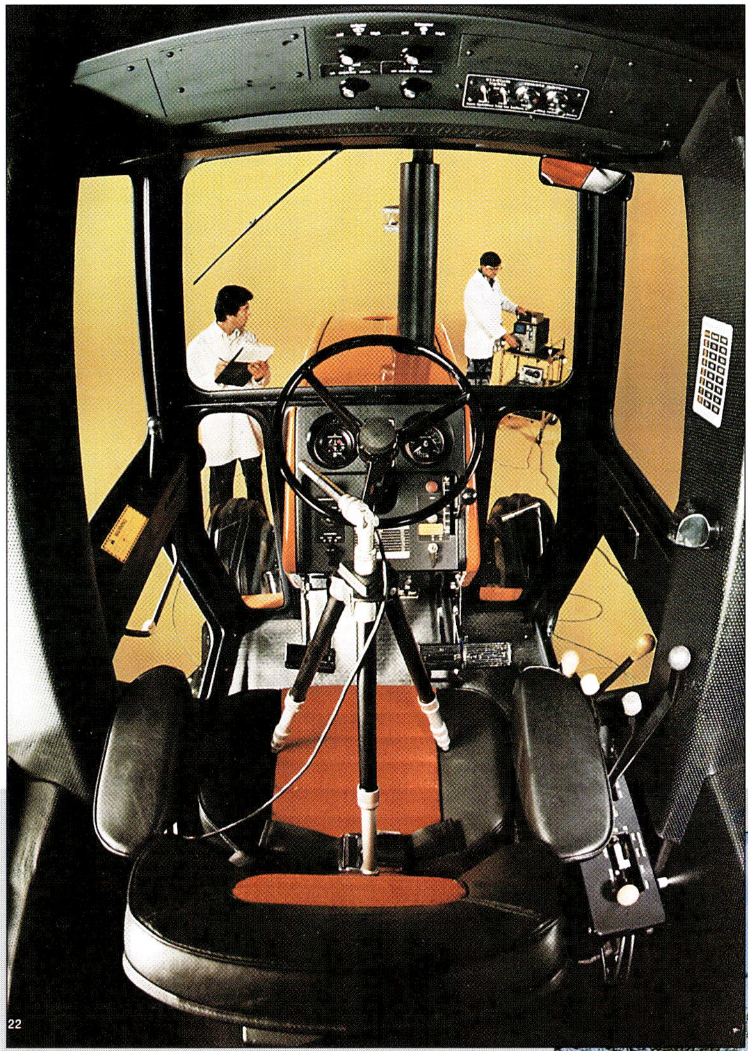

The Acousta-Cab II continued the low sound levels of the original cab. All 7000 Series tractors except the big 7080 were under 80 dbA.

The Acousta Cab II was a major improvement in the 7000 Series. The offset in the front of the cab was replaced by one continuous surface from top to bottom which simplified manufacture. The transmission control levers were removed from the cab floor and relocated to the console. To select a gear, the gear lever was twisted either right or left, followed by the normal fore-and-aft motion. A new luxury seat was offered as well. Stadium lights, which added two flood lights on the front and three work lights on the rear of the cab, became instantly popular.

The new color scheme gave the line a fresh new look. Corporate orange, more brilliant with slightly more red, replaced Persian orange 2. Since black and white were part of the corporate color standards, the maroon chassis became black, and the cream on the wheels and cab top became white. Model decals were updated with bolder numbers. The 7020, 7045, and 8550 were first to use the new paint scheme in 1977.

The 7060, 7080, and 7580 tractors received the new Acousta Cab II and the corporate colors in January 1978. There were serial number breaks

The 146-hp 7045 had the 7060 power train with a four-pinion differential, differential lock, and final-drive ratio.

The 106-hp Model 7010 finally joined the line in 1979 after waiting in the wings for seven years. Now the entire two-wheel line over 100 hp shared the same mechanical features.

The black-chassis 7080 kept its 181-hp specification and gained the updated colors and new cab. Stadium lights with remote-controlled mirrors were a popular option.

with the 7060 starting at Serial No. 6001, the 7080 at Serial No. 3001, and the 7580 at Serial No. 2001.

The next important event in the 7000 Series was the addition of the 7010 in 1979, replacing the Model 7000. Many features, taken for granted in the larger models, were not available on the Model 7000. This size tractor needed load-sensitive hydraulics, differential lock, planetary final drives, wet power disk brakes, and power-shift with more speeds. To design them into the older power train was neither practical nor economically sound. The 7010 was built as a prototype tractor during 1972, along with the impending 7030 and 7050 tractors, but was shelved in favor of the Model 7000 tractor project (1975) because of cost. As the Model 7000's sales volume gradually dropped, component costs increased, finally allowing the use of the larger power train without a cost penalty. The new 7010, with New Family features, joined the line in 1979 at an introduction held at the Loretta Lynn Ranch at Hurricane Mills, Tennessee.

The Tall Ones
7000 Series High-Clearance Tractors

Allis-Chalmers had long maintained a presence in the sugar cane country, especially Louisiana, by building cane tractors or by supplying components to Thomson Machinery Company, at Thibodaux, Louisiana. As Thomson had phased out of using Model UC pieces around 1960, A-C began producing high-clearance tractors for that market. First the D19 high-clearance units, followed by One-Ninety XT, and D17 Series IV tractors. Carryover One-Ninety XT cane models met the need for a time, but soon dealers in the cane country were calling for a cane version of the 7000 Series. Although only relatively small numbers of tractors were involved, they were crucial to those dealers.

To meet this need, the 7000 Series high-clearance tractors were developed. New rear axle drop-box final drives, similar to the One-Ninety XT, and extend-

Beginning in 1978 the 7020, 7045, and 7060 high-clearance tractors were offered. Power-shift transmission and cab were standard equipment.

The crop-toppers had braces on the front-axle spindles to withstand the high stresses of crossing cane rows. There was 37 in. of clearance under the front axle.

Special rear dropboxes put 32 in. of clearance under the drawbar. Hitch lower links attached to the dropboxes for proper hitch geometry.

ed adjustable front-axle spindles gave an additional 13 inches of clearance under the tractor. Clearance under the front axle was about 37 inches with 9.50-20 tires and 32 inches under the drawbar with 20.8-38 rear tires.

High-clearance versions of the Models 7020, 7045, and 7060, all with power-shift transmission and cab, were produced. Production records are non-existent, since cane tractors were combined with conventional units. Probably fewer than a couple hundred were built, starting in 1978 until the 7000's phased out at the end of 1981. The 7000 Series represented the core of the A-C tractor enterprise until the end of 1981. The 7000's were responsible for putting Allis-Chalmers back in the tractor business.

The Model 7000 Tractor

As soon as the 7030 and 7050 made their debut in 1973, the Model 200 showed its age. One hundred or more horsepower, more speeds, and a modern cab were needed in this important market. The 7010 had been developed to fill this slot, but proved to be too costly to build at that time. As a result, a project to modernize the Model 200 was initiated and eventually called the Model 7000. The 2900 engine, with its 301-cubic inch displacement, had plenty of reserve. After all, there were many One-Ninety XT's and 200's in the fields running at 120 horses, all without the A-C's blessing.

The new 7000 Series Acousta Cab and sheet metal required only slight modification for this modernized model. The fuel tank was mounted under the left side of the cab with the swing-out batteries under the right side.

More forward speeds with added power-shifting were needed. Engineering developed a new torque tube section of the power train. The 200's Power-Director had two clutches, one for the low range underdrive, and the other for high-range direct drive. A third clutch for an overdrive was added, providing power shifting through three speeds. Combined with the four-speed forward, one-reverse constant-mesh transmission, there were 12-forward and 3-reverse speeds. Compared to the 200, it looked pretty good. With a slower first gear for power take-off work, a higher road gear, and more speeds in between, the 7000 was a big step forward. Other power-train changes included an independent 540-1,000 power take-off, larger spiral bevel set, wider final-drive gears, park lock, and 3.25-inch rear axles.

Hydraulic output was increased to 25 gpm to power the hydrostatic steering, Traction-Booster, three-point hitch, and remote valves in the Model 7000. Chassis changes included a new adjustable front axle with the forward wishbone similar to the larger 7000 Series tractors. With PTO horsepower increased from 94 to over 106, the Model 7000 fit well in the new Allis-Chalmers line, where it took its place in mid-summer 1975. Now all the two-wheel drive tractors over 100 horsepower shared a common appearance.

The Model 7000 joined the 7040, 7060, and 7080 in 1975. It shared the modern cab and styling but there was a lot of Model 200 under the sheet metal.

Black Chassis Model 7000

In May 1978, the Black Belly 7000 went into production, modified with the corporate colors and Acousta Cab II which were already installed on the larger 7020-7080 tractors in 1977. The cab included the new console which eliminated the gear lever on the floor. Other changes included stadium lights, electrically controlled outside mirrors, luxury seat, and other minor improvements.

The most significant change was in the Power-Shift. The clutching function previously handled by a dry engine clutch was now controlled by the low-range (underdrive) wet hydraulic clutch in the Power-Shift. Inching or load starting always occurred in the underdrive for safe-

The black-chassis 7000 came in 1978. It had the Acousta-Cab II with transmission lever in the console and optional stadium lights and mirrors. The 7000 was replaced by the 7010 in 1979.

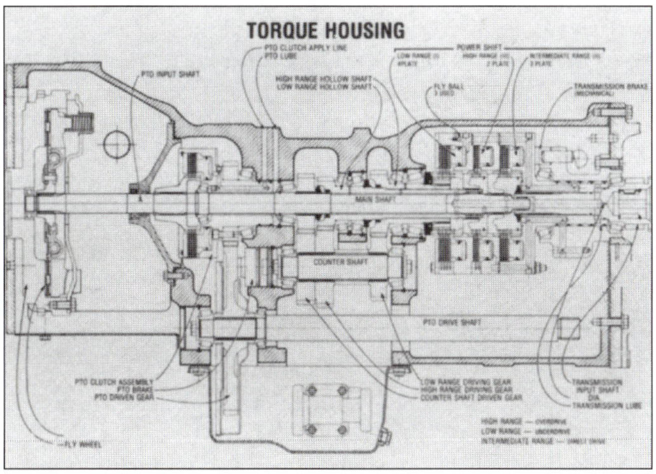

In-house the black-belly 7000 was known as the wet-clutch 7000 since it used the low range clutch in the three-speed power-shift for a master clutch. A torque-limiter replaced the engine clutch.

ty and longer clutch life. The dry engine clutch was replaced by a torque-limiter which reduced power-train shock loads. There were no gear ratio or speed changes. The 301-cubic inch 649T (six cylinders, 4.9 liters, turbocharged) power plant was basically unchanged and PTO horsepower remained at 106.44. This tractor was not retested.

Even with all these changes and improvements, the 7000 still lacked many features found on the larger tractors, which limited its acceptance in the marketplace. The Black Belly 7000 remained in production a short 13 months, from May 1978 through June 1979. It was replaced by the more modern 7010 which was originally planned for this horsepower slot in 1972.

New Mid-Range Tractors

While the New Family of large tractors was being developed, there were many planning sessions on a new line of 50- to 90-horsepower tractors. Though the D17 V and D17 IV tractors were in development and would be introduced in 1967 as the One-Eighty and One-Seventy tractors, the long-range product plan called for developing and manufacturing a line of mid-size tractors with features comparable to the larger models.

A-C engineering considered several projects, but feasibility studies showed that product, development, and tooling costs could not be recovered by projected sales and target list prices. As a result, many good designs and innovations were abandoned. This was during the late 1960s when North American tractor builders had to face the fact that domestic manufacture of mid-range tractors would not be profitable. This occurred because target list prices were competitive with tractors that were imported in their entirety, or had a great deal of European content and could be sold for lower prices.

Since A-C couldn't profitably build the mid-range tractors entirely at West Allis, the only alternative was to buy components or complete tractors from an out-

This mock-up of a 6000 Series tractor is dated July 1976, over four years from actual production. At that time it was still planned to build the power train in West Allis.

This 6000 Series styling proposal was close but fenders and platform were changed considerably.

The front-wheel assist moved the 83-hp 6080 up to Model 7000 field performance levels. In poor traction conditions it could actually out-work the 7000 two-wheel drive tractor.

side source, probably European. This would not be a new experience. The company had already acquired the Models One-Sixty and 6040 from Renault of France beginning in 1969. Although it had been hoped that Renault could also supply larger units, this possibility was lost when the dollar-franc exchange rate turned unfavorable for Renault and they withdrew, terminating the 6040.

During the mid-1970s the company had been negotiating with Fiat of Italy for the 50-horsepower tractor which would become the 5050. Using Fiat power trains for the mid-sized tractors was logical.

The 6000 Series Tractors

Following negotiations, Fiat agreed to supply complete power trains, along with the three-point hitch and driven front axles, in two sizes. These were similar to the Fiat 780 and 980 tractors and had the torque capacity to utilize the Allis-Chalmers 433T and 433I engines.

The availability of complete power trains at an acceptable cost saved the company a huge investment in development and tooling, and enabled engineering to move quickly into the test program, engine installation, and design of the rest of the tractor. As a result, 6060 and 6080 production began in September 1980, and was up to speed by December. Production of the 175 ceased in September 1980, while the 185 was built until September 1981.

The 6000's were immediately accepted since they had features the customers wanted. Turbocharged engines, planetary final drives, wet disk brakes, differential lock, straight-through axles, larger rear tires, and front-wheel assist resulted in modern tractors. Above all, they had performance. The Allis-Chalmers 433T engine, turbocharged in the 6060, easily delivered 63.83-PTO horsepower and had a phenomenal 35-percent torque rise. The turbocharged and intercooled 433I in the 6080 developed 83.66-PTO horsepower and had an outstanding 27-percent torque rise. These numbers meant good field performance.

Front-wheel assist (FWA) was enthusiastically received by the sales people. The 5050 FWA had

The 6060 and 6080 in both two-wheel and front-wheel assist models were introduced in the fall of 1981. The 6060 had 63.83 hp and the 6080 developed 83.66 hp. Both had outstanding torque rise for exceptional field performance.

been around for some time, but it was used mostly for loader operation. The 6000's, especially the 83-horsepower 6080, were large enough to be the primary field power on many farms.

Front-wheel assist tractors soon made up over 35 percent of the total 6000 Series production. Many users said that the 83-horsepower front-wheel assist 6080 could easily outwork a 106-horsepower Model 7000 in the field. This field performance increased the demand for putting front-wheel assist on the 7000 Series tractors.

The 6060 had a partially-synchronized transmission with eight-forward speeds from 1.4 to 13.9 mph. The 6080 had a fully-synchronized transmission with 12 well-spaced forward speeds from 1.0 to 18.0 mph. Both tractors had independent power take-off and the 6080 had a 540-1,000 rpm PTO, easily changed by swapping the output shafts. The live hydraulic pump was engine-driven and delivered a total of 19 gpm with 11.4 gpm available to the remote valves and three-point hitch. This output was plenty for a loader, and a 6000 with front-wheel assist and a 460 loader made a good loading rig.

The original plan called for an air-conditioned cab to be introduced with the tractors, but work on the 8000 cab and other priorities delayed it until early 1982. Styled like the larger 8000 cab, it had good looks, lots of glass, visors for operator shading, and plenty of room. Sound levels were acceptable at 79 db(A) for the 6060 and 81 for the 6080. Cabs soon were being installed on over 35 percent of the tractors.

The 6080 was doing well in the marketplace, but the 6060 needed help. Customers wanted larger tires and the transmission lacked the versatility of the 6080. Furthermore, a price increase was needed, but the competitive situation wouldn't permit it unless value was added to the tractor.

The 6000 cab had all the features of the larger 8000 cab in a slightly smaller version. It was relatively quiet at 79 (6060) and 81 (6080) dbA.

This 6060 with a 460 loader was a good combination in a dairy operation. Easy maneuvering and short cycle times provided high productivity.

The 6070 with front-wheel assist, 16.9-34 rear tires, and cab was a popular version of the 70-hp tractor. Only 972 Model 6070 tractors were built.

The 6070 Tractor

The solution to the sales problem with the 6060 was to provide a 12-speed fully-synchronized transmission, to increase the power to 70 horses, and to offer optional 16.9-34 rear tires — enter the 6070! The increased power and four more speeds could be added for reasonable cost and the list price could be set comparable to other 70-horsepower tractors and that improved the profitability. As a result, everybody won.

The 6070 was exactly what was needed. Along with the 6080, it was produced until December 6, 1985, when a 6070, Serial No. 1972, became the last tractor to come off the West Allis assembly line — a very dubious honor.

What Did the Future Hold?

Where was the 6000 Series headed in future product plans? Actually the line was stable. One proposal would provide a 90- to 95-horsepower tractor by installing a 649T engine in the 6080 chassis. It would have been called the 6095 and offered only in the front-wheel assist configuration. This approach to providing more power and performance in a specific tractor had been used successfully by other major tractor manufacturers. The 6095 could have been a winner. It might have been another One-Ninety XT in performance, but many were concerned that the 649T was just too much engine for the power train, even with front-wheel assist. Only testing would tell.

Time To Make Some Changes

Good products must be constantly improved if they are to survive in the rough and tumble of the marketplace. Therefore, it was time to make some changes in the 7000 Series tractors. Armed with the experience of many years, thousands of hours of customer use, files full of service reports, a long wish list from the marketing department, and dozens of good ideas of their own, the engineering department started a project to make the 7000's, already good products, even better.

Model 6070 Serial No. 1972 was the last tractor off the West Allis assembly line on December 6, 1985. A sad day in West Allis.

There were several important requirements:

- A new cab which would have more room, better visibility, more convenient controls, more air-conditioning capacity, a more comfortable seat, more filtering capacity, and a new, more modern appearance.
- The need for greater air-conditioning and filtering capacity became obvious when a large fleet of 7060 and 7080 tractors was used by Paloma Ranch, Gila Bend, Arizona. Ambient temperatures up to 116° F and heavy

A major preliminary styling proposal for the 7000 Series replacement which was renamed the 8000 Series. Although sheet metal was changed, cab design was already firm. Design was by William Schmidt and Associates of Detroit, Mich.

The 8000 Series incorporated the improvements and refinements needed in the 7000 line, added front-wheel assist, new styling, and an all-new cab.

concentrations of dust really tested their climate control systems.

• Front-wheel assist was needed to make the tractor more productive and to meet the competition. This feature had originally been in the specifications for the New Family of large tractors, but was deleted in order to move the project along more quickly. The Model Two-Twenty FWA of 1970 was only partially successful, so there was little to support the need for front-wheel assist at that time. By the late 1970s the market for a large two-wheel drive tractor with front-wheel assist had matured and any new tractor should have it available. Necessary features included on-the-go engagement, a wide range of tread adjustment, and short turning radius.

• Power-shift transmission in a 7080-size tractor. Users liked the power of the 7080 and wanted the convenience and productivity of power-shift as well.

• Minimum 60-inch rear tread to allow centering 18.4-38 rear tires in 30-inch rows.

• An electronic .digital instrument system that could display true ground speed for accurate calibration of application equipment and wheel-slip to help minimize slippage.

These major requirements, plus a substantial number of other product improvements, set the stage for a new line of large tractors. Once the new cab was designed, a new look for the hood and grill was needed, so they were restyled to be more compatible with the cab. Somewhere in the process, the 7000 Series was no more. The new tractor looked so different that a new name was needed.

The 8000 Series Tractors

A new tractor's name is probably the least important of hundreds of decisions. One of the more significant decisions, however, is the horsepower rating for each model. That major selection is not necessarily a free choice, since the iron itself has built-in limits. The smallest tractor would be about 105 to 107 horsepower like its predecessor, the 7010. The determining factors were the engine, power train, and their costs. Less horsepower would dictate a lower list price, and more would make a larger gap in the range above the 83-horsepower 6080. The largest tractor was determined by the torque capacity of the power-shift transmission (PST). The dealers wanted a 7080-size tractor, 181 horsepower, with a power-shift

The 8000 Series with their ultra-roomy cab, efficient front-wheel assist, and classy modern styling were a great leap ahead of the already established 7000 Series.

transmission. The A-C power-shift transmission in the 7060 would still have the required reliability at 170-PTO horsepower, so that set the top of the horsepower range.

With a 65-horsepower range it was evident that five models, as in the 7000 Series, were not needed. Spacings of 20 to 25 horsepower were about right in the 100- to 200-horsepower range, so the need for two models around 130- and 150-horsepower was obvious. All this quickly defined the model designations — thus the 8010, 8030, 8050, and 8070.

The 8010 was mechanically similar to 7010. The 8030 received the 670T (426 cubic inches, turbocharged) engine because the 649I of the 7020 had

The inside story of the 8000 Series was displayed in this highly accurate drawing. This unit had the power-shift transmission which was specified on 83 percent of orders and front-wheel assist which was ordered on 35 percent of 8010's to 90 percent on 8070's.

The 8070 FWA had outstanding performance and equal to the 4W220 in most conditions, but for $10,000 less at list price.

reached its practical limits. Power-Director and power-shift transmissions were established. Since the new 8030 was similar to the original 7030, rear main-housing components such as two-pinion differential and differential lock were a logical choice. It would also use the same final drives and axle diameters. The 8050 had basically the same engine and power-train components as the former 7060. The attitude was: "If it ain't broke, don't fix it."

The 8070 was introduced only as a two-wheel drive tractor. The 670I engine speed was reduced from 2,550 to 2,400 rpm for better economy as well as noise reduction. Rear main housing and axles were essentially the same as a 7080. Since it had the same planetary final-drive ratio, speeds were similar to the 7080 except for the reduction in engine speed. The 8070 FWA was delayed until early 1983 while power-train testing continued. This testing confirmed that the narrow planetary final drives of the 8050 were strong enough for the 8070 FWA, so this explains why 8050 and 8070 FWA field speeds were the same. The use of the narrow finals also permitted 60-inch rear-wheel treads. The 8070 FWA turned out to be the hot tractor in the series and it could perform with a 4W220 in the field for $10,000 less — good-bye, 4W220.

Even further testing of an 8070 two-wheel drive confirmed that the narrow final drives were adequate in this application. Beginning at Serial No. 2663, they went into production. Since orders for the two-wheel drive tractor were only about 10 percent of total 8070 orders, it didn't affect many tractors.

Of course, developing a tractor required a lot more than just picking parts from the bin. Designing, testing, and tooling the drive out of the rear main housing, selecting and testing drivelines and front axles, and designing the control system for the front-wheel drive was a major project and was one of the most significant in developing the 8000's.

A Room With A View
The 8000 Series Cab

The marketing specification for the new cab read like a salesman's wish list — more of everything. More room, more glass, more cooling, more ventilation, more seat adjustment, more convenient controls, more filtering capacity — all of which would add up to more operator comfort, convenience, and satisfaction. The new cab for the 8000 Series met those requirements.

To provide more cab room, the width at shoulder height was increased from 49 (the narrowest cab) to 60 inches. The wheel well width was reduced to permit a 60-inch rear tread with 18.4-38 tires. A right door was added, and the door width was increased by 9 inches to 40 inches to better accommodate those college halfbacks who had come home to the farm.

Glass area was increased 24 percent from 42 to 52 square feet. To limit heat radiation, the roof was made larger with visors all around to shade the operator. The wide visors on each side provided space for two filters, double the capacity, and both were easily reached from the cab steps.

An improved seat with a new suspension improved the ride, which was later enhanced with an attenuator. A spring-loaded frame on a track, it allowed the seat to move fore-and-aft slightly as the tractor's front wheels encountered bumps in the field. This reduced the jerky seat movement in rough conditions and provided a big improvement in operator comfort.

Control levers were located in the right console, but the throttle remained in the cowl, a long reach for the operator. This evoked many complaints and eventually the throttle was moved to the console. Conversions were made available for existing tractors.

Climatic environment was vastly improved in the new cabs. Blower output was increased 55 percent to 700 feet per minute for better pressurization and the heater output went to 25,000 British thermal units. The air-conditioner capacity was increased 45 percent from 22,000 to 32,000 British thermal units to assure the specified 35° F temperature differential (ambient to cab temperature) required in hot climates.

A half-and-half tractor with a new cab let everyone compare a production tractor with various sheet metal and decal proposals.

The 8000 cab had the most shoulder room and the largest glass area in the industry. Air-conditioning capacity was increased 30 percent.

Ground-level refueling eliminated climbing for greater convenience and safety. Cab filters could be serviced from the cab steps.

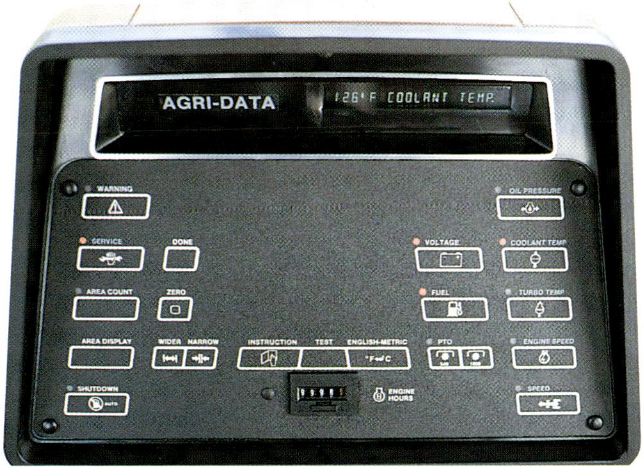

The Agri-Data electronic monitoring system could display dozens of operating, warning, and service functions. Only 30 were built and most were converted to conventional instruments when sensors in the tractor proved unreliable.

The Agri-Data Electronic Monitoring System

The Agri-Data system could display eight operating parameters: engine speed, oil pressure, coolant temperature, turbo temperature, battery voltage, fuel level, theoretical ground speed, and power take-off speed. It could display one or two Operation functions alternately or scan up to eight continuously. There were 14 Warning functions which would alert the operator to any impending problems, such as low engine oil pressure or high coolant temperature. The Service function monitored 18 tractor conditions, such as hours since last oil change, to remind the operator of upcoming service. The Acre Estimator function could, when the implement width was entered, provide an estimate of acres covered based on theoretical ground speed. The Unattended Shutdown function monitored critical factors such as oil pressure and coolant temperature during stationary power take-off operation and stopped the engine if a problem arose.

Agri-Data did not have radar so it could not display true ground speed or wheel slippage, both considered mandatory in an electronic instrument system. This deficiency was not the downfall of Agri-Data, however. Function sensors throughout the tractor were unreliable, causing the panel to display erroneous or no information.

Initial plans called for 50 tractors with Agri-Data, but only about 30 were actually built. General unreliability caused most of those to be converted back to conventional instruments and the $1,000 list price was refunded to the customer. When tractor engineering was dissolved in 1985, alternative electronic systems were being evaluated.

8000 Series Cane-High Clearance Tractors

As soon as the 8000's hit the market, dealers in cane country called for a cane model. Since production of these tractors had given the tractor plant little volume and less profit, management was reluctant to build them. However, when plant personnel found a substantial number of raw castings for cane axle sleeves and drop housings about to be scrapped, that decision was reversed. Thus the 8010, 8030, and 8050 cane-high clearance tractors joined the line. Specifications were similar to the 7000 Series cane models.

Two-Wheel Drive Versus Front-Wheel Assist

The sales organization greeted the 8000 Series front-wheel assist tractors with open arms. However, many farmers just couldn't see spending the extra

Cane/high-clearance models models were offered in 8010, 8030, and 8050 power sizes. This one had the Roll-Over Protective Structure which was a stripped cab.

The 8050 FWA and 8050 two-wheel drive tractors were compared for fuel economy and acres worked using an A-C 11-shank Min-Til coulter-chisel conservation tillage tool. The front-wheel assist covered 25 to 34 percent more acres on the same amount of fuel.

$7,500. Demonstrations were needed and dealers held them often. They set a big chisel plow or coulter-chisel deep enough to load the tractor about right, then switched off the front-wheel assist on the run. The tractor would nearly bury itself. When the front axle was re-engaged the tractor would pull right out of its ruts. Very impressive, but was that worth $7,500?

Engineering and marketing conducted tests to demonstrate the advantages of front-wheel assist. The tractors used in these demonstrations were an 8050 two-wheel drive with duals, and an 8050 FWA with both singles and duals, all on 20.8-38 tires. The load was an Allis-Chalmers 11 shank (13.75 ft) Min-Til conservation tool with 23-20-inch coulters and 11-4 inch twisted shovels, set to run about 9-inches deep. The field was oats stubble and the soil was fairly moist from recent rains, although traction conditions were good. A wheel revolution counter was used, and a no-load count of 117 revolutions in 2,000 feet was established. The differential lock was engaged on all runs. The 8050 FWA covered 25 to 34% more acres on the same amount of fuel. See complete test results on page 176.

Bright Future – No Future

The 7000 Series tractors had established Allis-Chalmers' credentials as a builder of large modern farm tractors and the new 8000 Series not only upheld that reputation but enhanced it in many ways. They had performance, reliability, and features equal to any in the industry, and were gaining acceptance as customers learned more about them. There was one crucial factor, however, that the 8000 Series didn't have — that of good timing.

The 8000's were introduced in early 1982, shortly after the start of the most severe recession to hit American agriculture since the '30s. Many companies, both large and small, were not able to weather the economic disaster. Had A-C survived, there were plenty of projects in the mill to create some real excitement. Among them were:

• The 8095 (planning designation): A 200-PTO horsepower front-wheel assist tractor. Imagine an 8070 FWA with an extended frame and wheelbase of 120 inches with dualled 20.8-42 rear and 18.4-28 front tires. Four prototypes were built, all powered by 673-cubic inch Komatsu diesel engines, and equipped with Power-Director 20-speed transmissions. Scheduled for 1986 production, the 8095 would have been the industry's first 200-horsepower front-wheel assist tractor. With a tremendous torque rise, it could start a load at part throttle in about any gear. It was a locomotive without rails. The tractors were dismantled, however, when tractor engineering was closed in the summer of 1985.

• Power-shift transmissions: One project would have revised the power-shift transmission to narrow the steps between gears to about 25 percent average, improving the first to second shift smoothness and speed spacings. Another project would have built a power-shift transmission big enough for the 8095 and 4W305.

• Three-speed range: Engineering found a way to squeeze in a third speed into the range transmission. A three-speed range plus the revised power-shift transmission would have provided 18-forward and 3-reverse speeds, all well-spaced and more useable. When it went into production, the Power-Directors would have been discontinued.

• Electronic instrumentation: The revised Agri-Data system would have had all the functions needed for complete instrumentation, true ground speed, and wheel-slip sensing.

Ignore the decal again. Four 8095 prototypes were built in 1984. They had 673-in.³ six-cylinder diesels and could develop 200 PTO hp. All were dismantled when the proving ground was closed.

• Engines: During the latter years, the company had determined that the sales volume generated by the engine division was insufficient to justify the substantial investment required to modernize the factory and to develop new products. More powerful, lightweight high-speed diesels were needed to power the larger tractors and combines in the planning stage. Therefore the company was considering outside sources for an entire line of engines, among them both domestic and foreign suppliers. At that time (1984) two domestic diesel lines were being evaluated.

Of course, there were all the other product improvements, blue-sky projects, and just plain good ideas, but the major concepts outlined here pretty well indicate that tractor engineering was alive and well — right up to the end.

All Wheels Pulling Four-Wheel Drive Tractors

After World War II several innovators began to build big drawbar power into tractors with four equal-sized rubber-tired wheels. These tractors were quickly accepted in the wheat country, especially where the land was reasonably flat or with only moderate slopes. Later, after some years of development, the four-wheel drive (4WD) tractors even took over the steep slopes of the Pacific Northwest and heavy soils of California's central valleys.

The Springfield crawlers filled the company's needs for big power in both construction machinery and farm markets. The track layers performed well in the wheat country of the northern plains, the hills of the Palouse, and in the broad valleys of California and the southwest. With good crawler sales, there was little interest within the company in the four-wheel drive tractor.

By the early 1960s, many A-C dealers were clamoring for an Allis-Chalmers four-wheel drive tractor, while some of the more impatient ones began selling Steiger and Versatile tractors. During this period A-C farm crawler sales declined to a very low level and the outlook was bleak. It was time to get into the four-wheel drive business and keep dealers loyal to the company.

The T16 Sugar Babe

With four-wheel drive loaders in production at the Deerfield, Illinois, plant near Chicago, it seemed that the company had a good basis for a four-wheel drive farm tractor. Using the TL16 loader the Deerfield engineers developed the T16 Sugar Babe, a 150-engine horsepower tractor. The T16 Sugar Babe and the D21 were both introduced in 1963.

The T16 was powered by an A-C 7000 (no relation to the tractor) turbocharged diesel engine of 344 cubic inches rated at 2,200 rpm. With 150-maximum, but only 117-rated engine horsepower, it had little if any more power than the D21 with 103-PTO horsepower. The 7000 engine was based on an older model that was designed to run about 1,800 rpm and it really didn't like 2,200 rpm. Field work and long hauls meant continual high engine speeds and engine reliability problems arose. A good solution, the 3500 engine, was waiting to take over the job in the T16. At Serial No. 1910, the Sugar Babe got a new engine and a new lease on life.

This prototype T16 tractor started out as a TL16 loader built in the Deerfield, Ill., plant. It had 150-engine hp and an Allison six-speed power-shift transmission.

The T16 was not tested at Nebraska. A few were sold in the northern plains wheat country, but were repurchased by the company because the tractor proved to be unreliable in that application. The tractor was sold mostly in the Florida cane country, with perhaps a few in Louisiana. The applications were soil preparation and pulling cane wagons to the sugar mill. With four-wheel drive and high-traction differentials, it could pull cane wagons out of soft fields, then shift up to road speed for the haul to the mill. It had an Allison Torq-matic power-shift transmission which would shift through six forward speeds from 3.1 to 16.5 mph. A torque converter permitted starting big loads easily in the muddy fields and it had a converter lock which was actuated automatically. When the tractor pulled implements in the field, lockup was seldom attained, so fuel efficiency was reduced and excess heat was generated in the transmission. It was a great hauling tractor, although somewhat less successful as a field tractor.

The T16 was painted either Persian orange No. 2 or industrial yellow. Sugar Babes were yellow, but the T16 was painted orange for farm sales. The Sugar Babe was a Construction Machinery Division product, so production, serial numbers, and sales by that division are not known. The Farm Equipment Division sold 12 in 1963, 3 units in 1964, and none after that. The T16 was not a very impressive start for A-C in the four-wheel drive business.

The production T16 had been restyled and was nicknamed the Sugar Babe since it was targeted to the cane country in Florida and Louisiana.

Beginning with Serial No. 1910, the T16 Sugar Babe was powered by the modern 3500 six-cylinder 426-in.^{3} turbocharged engine which was better suited to the non-stop operation of field work.

The T16 with a Model L offset disk harrow ready to work deep in big fields. The T16 was painted orange for its entry into the farm market.

Several attempts were made to get a four-wheel drive tractor project off the ground. An in-house project was proposed in 1966 but work on the New Family and other programs were of much higher priority. Next, a delegation was sent to visit the folks at Steiger. They had little extra assembly capacity and were not interested in selling to an Original Equipment Manufacturer (OEM) account. Talks were then held with Mississippi Road Machinery (M-R-S), Jackson, Mississippi. They were interested but when the quotes came in, there was no profit at competitive list prices.

In July 1970, there was a proposal to purchase the White Model A4T-1600 four-wheel drive tractor for resale. It was to be evaluated with both the White and the A-C 3500 engine and restyled to match the Model Two-Twenty. Target list price was $15,000 and estimated first year sales were 250 units totaling $3.7 million. That deal didn't work out either — probably just as well.

The Model 440

Just when it looked as if A-C could not acquire a four-wheel drive tractor, the climate improved at Steiger. Ownership had changed, they had moved to Fargo, North Dakota, and they wanted outside business. Steiger agreed to sell their Bearcat to A-C at a price that both companies could make a little money. Engineering tested both the Caterpillar and Cummins

The Model 440 was A-C's first competitive entry in the four-wheel drive segment. It was powered by a Cummins 555-in.^{3} V8 and had 165-drawbar hp.

The Model 440 and a Model 3000 semi-mounted eight-furrow plow could plow up to 60 acres per day.

After Serial No. 1460, the 440 had a black grill and new decals to match the West Allis line. Over 1,000 Model 440's were sold between 1972 and 1976.

engines, and selected the Cummins. It became the Model 440 and was introduced in 1972. A-C was finally in the four-wheel drive business.

The 440 was well received, although a few people wanted more power, but that was nothing new. The tractor could handle a seven- or eight-bottom moldboard plow or a 30-foot chisel plow, which was enough for most customers. The well-known Cummins 555C engine (the triple-nickel), a naturally aspirated V8, was smooth and responsive. It was rated at 2,800 rpm. With a 17.0:1 compression ratio and two intake and two exhaust valves per cylinder for good volumetric efficiency, the 555-cubic inch engine developed 208 horsepower at the flywheel.

A Dana Spicer 14-inch two-plate cerametallic clutch delivered power to the Fuller TA4510 10-forward, 2-reverse speed transmission. Speeds ranged from 2.5 to 20.0 mph. The Steiger drop box could swing inside the tractor center line in turns, reducing the angularity of the drive-line universal joints and improving its life. The axles were Dana (later Franklin) outboard planetary-reduction units with treads of 73.5 or 79 inches.

The hydraulic system of the 440 had a Cessna dual pump. The steering section delivered 16 gpm at 1,500 psi, while the main pump for the three remote valves and three-point hitch supplied 21 gpm at 2,000 psi. A rugged category 3 three-point hitch and quick hitch were optional. The lack of a power take-off was a constant complaint, but probably didn't limit sales much since most other builders didn't have it available either.

The cab was ROPS certified. A fully-adjustable, pivoting seat with an easy-ride suspension and air-conditioning made long days easier to endure.

In early 1974, at Serial No. 1460, the paint scheme was revised. The grill was changed to black and the cab to orange with a white roof. Decals matched the 7000 Series tractors and mechanical changes were minor.

The company purchased over 1,000 Model 440's from April 1972, through June 1976. It filled an important slot until the Model 7580 was ready. Many 440's were exported, mostly to Australia, where nearly 10 years later a used one in good condition would bring its' original list price.

Home-Grown is Always Better West Allis Four-Wheel Drive Tractors

Just like Mom always said, "Home-grown is always better." When we were kids on the farm, that applied to fried chicken, sweet corn (roastin' ears in Indiana), sliced tomatoes, and strawberry shortcake — especially when she gave it her special touch. The same theory applied to big four-wheelers, too, after A-C engineers applied their special touch. Of course, like Mom, they started with some pretty good ingredients. Some 73 percent of the components in the new Model 7580 were common with the two-wheel drive 7000 Series. The engine was an upgraded version of the 3750 Mark II from the 7080. The rear main housing, Power-Director transmission, and most of the hydraulic system came from the two-wheel drive tractors. It was necessary to design a new two-speed dropbox and front differential housing for the power train. New fabrications for the hinge, front frame, and a pair of 85-gallon fuel tanks were required. Twelve percent of the new parts were made on existing tooling. All of these factors resulted in a 7580 that was very closely related to the rest of the family.

About the time of the great Bicentennial celebration in the summer of 1976, A-C announced the 7580. A-C people had a double celebration that summer with the introduction of their own four-wheeler. The 222-engine horsepower 3750 Mark II (later called the 670HI) was turbocharged, intercooled, and had a counterbalanced crankshaft. It had 426 cubic inches, a 16.0:1 compression ratio, and was rated at 2,550 rpm. At the Nebraska Testing Labs, the 7580 cranked out 186.35-PTO and 153.70-drawbar horsepower.

The 7580 got the Acousta Cab II and black chassis at Serial No. 2001 in early 1978. Power and specifications remained unchanged.

The proven 7080 Power-Director 10-speed transmission, coupled to the 2-speed dropbox, provided 20-forward and 4-reverse speeds. The dropbox also housed the power take-off drive and the drive for a dual hydraulic pump. The pumps handled steering, clutch pressure, lubrication pressure, and cooling functions. Most other hydraulic components were shared with the two-wheelers. The 7580 had the power wet-disk brakes from the 7080, but operated by one pedal.

The Model 7580 developed 186-PTO and 153-drawbar hp and had many components in common with the 7000 Series two-wheel drive tractors. The 1976 and 1977 models had the maroon power train.

This 1976 prototype 8550 had a maroon power train. The unusual air intake and filter were not destined for production, although this arrangement became standard practice by the late 1980s.

The 8550 Tractor

The 7580 was a needed and welcome addition to the tractor line, but this was the 1970s when "only bigger is better," so it wasn't long before the call for more power was heard. This time A-C was ready. The summer of 1977 saw the Model 8550 rolled out, along with the 7020, 7045, 5020, and 5030. All had the new colors: corporate orange, white wheels, and black chassis. The 8550 Super Beast was the star of the show. That name was tacked on the tractor in an advertising campaign, but fortunately it didn't stick. One of the marketing guys named it the "Queen Mary" for its two exhaust stacks. By any name, the 8550 was the biggest news from Allis-Chalmers in a long time.

The 8550 was powered by the A-C 6120T (six cylinders, 12 liters, turbocharged) engine. It was based on the big 844-cubic inch A-C engine used in the HD21 crawler, but with a shorter 5.625-inch stroke which reduced the displacement to 731 cubic inches. To match the displacement, the single large turbocharger was replaced with two smaller ones, thus the twin stacks.

The Power-Director transmission was similar to the 7580 design except for heavier gears and other components to handle the 300-horsepower input. The dropbox had originally been designed for the higher input, and the rest of the power train was essentially 7080 componentry. Hydraulics were similar to the 7580 except for the three-point hitch. The 7580 used 7080 iron and could lift 6,680 pounds at 24 inches in back of the hitch points. A new much heavier category 3-4 hitch with external lift cylinders allowed the 8550 to lift 9,740 pounds.

Cab, controls, fuel tanks, etc., were common with the smaller unit. The 8550 was a huge machine. According to a University of Nebraska press release, it had to be shoe-horned into the test lab by removing an exhaust hood. Once there, it spun the needle to 253.88-PTO horsepower, and pulled 217.92-drawbar horsepower on the track. That kind of performance put A-C in the big leagues of tractor builders.

The 253-PTO hp 8550 put A-C in the big leagues of tractor power. The 731-in.^{3} six-cylinder engine had four valves per cylinder and twin turbochargers for high volumetric efficiency and fuel economy.

The 4W220 and 4W305 Tractors

The 7580 and 8550 four-wheelers were ready for their share of the innovations and improvements that engineering had conceived for the 8000 Series. A project to develop new four-wheel drive (4WD) models ran parallel to the 8000 Series program.

To make the January 1982, deadline, high priority was given to building prototypes and the 4W220 and 4W305 were shown to the dealers in Reno, Nevada. It was a great meeting! Talk about drawing a full house — four new 8000 Series models, two new four-wheel drive tractors, new 6000 Series cabs, and the new 40-horsepower 6140. Some of us thought we had suddenly gained admission to a tractor salesman's paradise. Since the 6000's were less than 18 months old, A-C really had an all-new line from 40- to 250-PTO horsepower for the first time in Allis-Chalmers history.

The most obvious changes in the powerhouses were the cab and sheet metal. Like the 8000 cab, the 4W Series cab was designed like a truck cab, in that it had a solid fire wall and cowl that were part of the cab, not part of the tractor. The instrument cowl, all controls, and the seat were preassembled in the cab, then it was mounted on the tractor in a body-drop operation. Controls were flexible to reduce noise transmission into the cab. As the widest cab on a four-wheel drive tractor, it also had the most glass for unexcelled visibility. Cab improvements made in the 8000 Series were carried over into the 4W220 and 4W305. New seat and suspension, greatly improved climate control, and well-appointed interior all contributed to a first-class operator environment.

Mechanical changes, although less obvious, were significant. Rated speed of both engines was reduced from 2,550 to 2,400 rpm, while torque was increased to maintain the same horsepower ratings. The reduced engine speed helped control noise and vibration. The 670HI engine got a new injection system which improved performance and reduced smoke.

The hinge of an articulated tractor receives tremendous stress loads in many applications, particularly those involving earth-moving equipment. To improve component reliability in this area, the hinge was completely redesigned with a rugged casting replacing the former weldment. Drivelines were simplified and the power take-off driveline speed was reduced about 50 percent for greater reliability. These are just a few of many changes made to maintain A-C's quality and reliability standards.

Changes often came quickly in the big tractor market. The 8070 FWA came out in 1983, filling an important slot in the line. As soon as a few of them were in dealers' hands and demonstrations began, 4W220 orders dropped to zero. The field discovered that the 8070 FWA could do anything the 4W220

A styling-decal proposal for the 8550 before the 4W305 model designation was selected. The pronounced hood side panel flares were abandoned for a less expensive design.

The new 4W220 and 4W305 tractors had many major improvements compared to the 7580 and 8550. Although power ratings were unchanged, engine rated speed was reduced 150 rpm.

The 4W220 cab had more room, more glass, and more air-conditioning capacity. The 4W220 was displaced by the 8070 FWA and it was phased out in 1984.

The 4W305 was the top of the line with 305 engine, 250-PTO hp. Like the 8550, the engine displaced 731 in.^{3} and had two turbochargers.

could on a little less fuel and for $10,000 less. Production of the 4W220 ceased in mid-1984 and no one seemed to notice. That's what it's all about — finding a way at to do the job at a lower cost.

The 4W305 proved to be a good, reliable tractor and was produced through 1985. Had the 4W305 been continued, significant changes included a new engine from an outside source and a new high-torque capacity six-speed power-shift transmission. When Allis-Chalmers tractor engineering and manufacturing disappeared, some fine tractors went with them.

If You Can't Lick'em Join'em Allis-Chalmers Imported Tractors

By the early 1960s it was certain that the D10 and D12 could not get their share of the small tractor market. To compete, dealers in the southeastern states badly needed a small diesel of about 40 horsepower. A-C's D175 diesel engine was too big and expensive. The A-C International Division was building the ED40 in their Essendine, Lincolnshire, England plant. Some were imported into eastern Canada in 1963 and they seemed to be working well. Management decided to release the ED40 tractor to nine branches in the eastern United States and 450 tractors were imported in 1964. The field sales crew was glad to get the tractor and most thought the small diesel problem was finally solved.

The ED40 shared the styling of its colonial cousins, the D Series. The Standard (brand name) diesel had 137.9-cubic inch displacement, turned 2,250 rpm, and developed 33.7-PTO horsepower. There were eight-forward and two-reverse speeds. The power take-off was similar to the Model CA with a clutch in the rear axle that stopped forward travel and allowed the power take-off to run continuously. The ED40 also had a differential lock, a feature that was gaining popularity.

The crankshaft-driven hydraulic pump delivered 5.35 gpm at 2,650 psi. The category 1 three-point hitch had top-link sensing and could lift 2,200 pounds at the hitch points. Only adjustable front axle with manual steering was offered. Front 6.00-16 and 12.4-28 rear tires were standard while 13.6-28 tires on power-shift rims were optional.

The ED40 appeared to be fully competitive and everyone anticipated getting a nice share of the 40-horsepower diesel market which was huge. That might have happened if a serious problem had not

The first shipment of ED40 tractors arrived in an east coast port-of-entry from the Allis-Chalmers factory at Essendine, Lincolnshire, England. A number had been imported to eastern Canada in 1963, then 450 came to the United States in 1964.

The 33-hp ED40 handled a No. 73 three 14-in. plow very well in normal conditions. The Standard diesel engine, eight-speed transmission, and Depth-o-matic hydraulics were the specifications for good performance.

arisen. A hydraulic malfunction caused the lift cylinder bellcrank to break through the lift arm housing. It was very expensive and a temporary fix was hastily devised. In January 1966, a new hydraulic valve was released but by that time enthusiasm for the ED40 was nearly non-existent. After the 450 tractors in 1964, no more were imported.

The French Connection The One-Sixty Tractor

The search for a 40-horsepower tractor continued abroad. In 1967 a team of engineering, product planning, and marketing people went to Europe to contact possible suppliers and evaluate tractors. Visits to David Brown, soon to be purchased by Case, and British Motors, who built the Nuffield (Leyland) tractors, were made. The huge German manufacturer, Zanradfabrick Friedrickshofen (ZF), was contacted for a power train. ZF had just settled a wildcat strike. Management had been intent upon removing the beer-vending machines from the production area and the workers were not about to tolerate that move — the vending machines stayed.

The most successful visit was with Regie Nationale des Usines Renault, the largest manufacturer of automobiles, trucks, and tractors in France, and probably second in Europe at that time. They built a full line of tractors from the smallest vineyard up to two- and four-wheel drive models of about 100 horsepower. The tractor plant in LeMans was modern and efficient, and also produced parts for cars and trucks.

Later that year Renault agreed to supply a 40-horsepower tractor built to A-C specifications. To be expected, communication was sometimes a bit of a problem. Some of the Renault folks had a decent command of the English language, but none of the A-C people knew enough French to even order food. At dinner one evening, the Allis-Chalmers representatives were discussing the transmission, particularly the "low range." Immediately the Renault engineers assured everyone that the tractor would be painted "orange." "Low range" came off as "l'orange" to the listeners. These little misunderstandings were quickly overcome and the project made good progress.

Several tractors were tested and many changes were made to adapt to U.S. applications. The hitch-sensing system was modified to provide the precise control needed for light cultivators in the sandy soils of the southeast. A D15-style Roll-Shift front axle with hydrostatic steering replaced the original manual setup. A redesigned hood and grill was too costly, so some resourceful A-C engineers borrowed a One-Seventy hood and cobbled it on a test tractor. To everyone's happy surprise it fit with only minor modi-

The writer had the pleasure of introducing the One-Sixty to the field sales crew on a farm near Perry, Ga., in 1969. They sure were glad to see it after waiting so long for the small diesel tractor which A-C so badly needed for the southeastern market.

fication. The rear fenders were also adapted from the One-Seventy.

From the start, the Perkins 3.152 three-cylinder diesel was specified because of its good acceptance in the United States. Renault equipped a high proportion of their tractors with the German-built MWM air-cooled diesels. In the early 1970s one was shipped to the United States for evaluation and it was placed somewhere in southeast where it was well-liked. Although against company policy to sell non-production machines, it was sold to the user. Somewhere there is a one-of-a-kind One-Sixty with an M.W.M. air-cooled engine.

The One-Sixty had a 10-forward, 2-reverse speed transmission, differential lock, and continuous power take-off controlled by a two-stage clutch. The engine-driven two-section hydraulic pump delivered 6.7 gpm at 2,300 psi to the hitch and 7.5 gpm at 1,100 psi to the power steering. The category 2 three-point hitch had lower-link sensing and, after changes, provided good implement control. Power-shift rear wheels were standard, with either 13.6 or 14.9-28 tires. Later 11.2-38 rear tires for high-clearance work were offered. A narrow 48-inch rear tread model was available for tobacco work. For the highway mowing market, an LCG (low center of gravity) model with 7.50-10 front and 18.4-16A rear tires became available in 1970.

This Renault gasoline test tractor had a One-Seventy hood cobbled on and was the prototype for A-C styling on the Model One-Sixty.

The One-Sixty LCG tractor was an effort to crack the highly-competitive highway mowing market. The tires were 7.50-10 front and 18.4-16A rear to lower the tractor for safer operation on highway banks.

A shipload of One-Sixty's in the port of Milwaukee awaited shipment with West Allis tractors to dealers in the central and western states.

Before roll-on roll-off automotive vessels were widely used, the One-Sixty's were unloaded by sling. It was slow and extensive sheet metal damage was common.

The high-clearance One-Sixty with taller front spindles and 11.2-38 rear tires was adapted to vegetable cultivation.

The Model 160 with black decals that matched the home-grown tractors came out in 1972. Mechanically there were few changes.

A prototype on the One-Sixty industrial tractor. It proved to be infeasible because imported industrial tractors carried high import duty while farm tractors incurred none.

The 1971 basic and intermediate versions of the Model 160 had manual steering, Renault front axle and wheels, and French electrical system for less expensive entries into a very competitive market.

The 6040 had new styling and maroon chassis to match the 7030-7050 paint scheme. Only 499 units were imported in 1974.

Two revised models, the Basic and the Intermediate, were brought out in 1971 to provide lower-priced entries. They had the original Renault front axle, front wheels and manual steering, non-power-shift rear wheels, and French electrical system. The Basic also deleted the differential lock and creeper gears. That reduced the price, but neither were very popular and were soon discontinued. The Model One-Sixty with cream hood decals became the Model 160 with black decals in early 1972. In early 1974, the 6040 replaced the 160. It had a maroon chassis and new hood and grill to match the 7030 and 7050 tractors, but underneath the sheet metal the specifications were standard 160.

As the dollar/franc ratio worsened, Renault could not continue to sell tractors at the negotiated price and A-C couldn't pay more and still remain competitively priced. Another good product was lost to the harsh realities of economics.

The 5040, 5045, and 5050 Tractors

With the 6040 gone, it was necessary to find another source of 40-horsepower tractors. Several eastern European countries were already exporting tractors to the United States, among them Zetor of Czechoslovakia and Uzina Tractorul Brasov (Universal Tractor Brasov or UTB) of Romania. UTB agreed to supply a 40-horsepower tractor and a prototype 5040 was shown to the dealers in the fall of 1974. Shipments began in late 1975.

UTB built tractors under license from Fiat of Italy, but there was little Italian content. Except for sheet-metal changes for A-C styling, there were few changes in the original tractor. The 5040 was offered only as a utility tractor with an adjustable front axle. A 48-inch rear tread conversion adapted it to tobacco cultivation.

The 5040 had 40.05–PTO horsepower, just meeting the specification. The three-cylinder diesel had 142.8 cubic inch displacement, 17.0:1 compression ratio, and turned 2,400 rpm. The original 6- and 9-(creeper) speed transmissions were replaced by partially synchronized, 8- and 12-speed gearboxes in 1979. A two-stage clutch provided a continuous power take-off.

A 5.75 gallons per minute (later 6.23 gpm) pump supplied the power steering, category 2-1 three-point hitch, and remote valve. The hitch had upper-link sensing and could lift 2,645 pounds at the hitch points. The 5040 was sold from 1975 through 1980, when it was replaced by the 5045. About 6,000 total units were imported.

The Model 5040 was a Fiat design built by UTB of Romania. It was A-C's 40-hp entry from 1975 through 1980. The 5040's had a maroon power train.

The 5045 replaced the 5040 in 1981. It was Fiat-built and had a black chassis. A larger engine produced 44-PTO hp.

The 5045 Tractor

The 5045 was built in Italy by Fiat and only 499 were imported in 1981. Not tested, it had an estimated 44-PTO horsepower. The engine was similar to the 5040 engine, but displaced 158 cubic inches. This was the engine that powered the first few 5050 tractors.

Only the 8-speed transmission was offered, since the 12-speed creeper did not sell well. The rest of the power train was similar to the 5040. It looked like the 5040 except for a black chassis and the flat-top rear fenders which matched the 5050. After 1981, it was replaced by the 6140.

The Model 5050

The last 170 diesel tractor was built in December 1973, and the last 170 gas a month later, leaving a gap of over 22 horsepower between the 5040 and the 175. A 50-horsepower tractor was needed. Again the company went outside to get a competitively-priced tractor and turned to Fiat as a supplier. The 5050's began to arrive in July 1976, two years after the 170 was discontinued. The 5050 was styled like the 5040 except for flat-top rear fenders, and came in two-wheel and front-wheel assist (FWA) models.

The direct-injection three-cylinder diesel engine had 158.6 cubic inches rated at 2,600 rpm. After the first shipment or two of tractors, the bore was increased, bringing the displacement up to 167.8 cubic inches. Rated speed increased to 2,700 rpm. It tested 51.46 horsepower on the dynamometer and pulled 43.29 horsepower on the track. An engine-driven hydraulic pump supplied 6.47 gpm to the steering, hitch, and remote valve. The category 2 hitch had a lift capacity of 3,200 pounds.

The 5050 FWA was a good performer when equipped with a loader. A relatively short 130-inch turning radius provided good maneuverability in the feed lot. The 5050 Low-Profile model was offered for grove and orchard work. With 24-inch rear and 9.00-

The 5050 came in two-wheel and front-wheel assist versions to give A-C entries in the 50-hp class after the 170's were phased out.

The early low-profile 5050 had a maroon chassis, smaller diameter tires, horizontal muffler, and flat-top fenders. It was 6 in. lower than the standard tractor.

In 1981 the 5050 tractors got the black chassis to match the West Allis tractors. The 5050 remained in the line until mid-1983.

The 5050 low-profile chassis was painted black and shell fenders replaced the flat-top style in 1981. Otherwise it was the same as the earlier maroon chassis.

10 front tires, the tractor was six inches lower than a standard tractor. It was fine for mowing and tillage operations but more power was needed for power take-off-driven sprayers, and sales were limited. The 5050 tractor filled the 50-horsepower slot from 1976 to 1983 when it was discontinued without a replacement. About 6,500 units were imported.

The Nippon Connection 5015, 5020, 5030, and 6140 Tractors

In the late 1960s and early '70s there was an invasion by a swarm of small tractors from the Land of the Rising Sun, and soon the hum of small diesel engines was heard throughout the land. The U.S. tractor industry had abandoned small tractors, except lawn and garden tractors, and the Japanese were quick to fill the void. The success of Kubota and others precipitated a rush by domestic companies to acquire small diesels from Japan. A-C joined the rush, negotiating with Toyosha who manufactured and sold tractors under the Hinimoto brand name. An agreement was reached and the 5020 and 5030 tractors were announced in 1977. The sales organization was elated.

The 5020 came in two-wheel and front-wheel assist models and the 5030 was only two-wheel drive. Implements, including mowers, tillers, snowblowers, and loaders, were sourced in the United States. Toyosha two-cylinder diesels powered the pair. The 5020 had 77.0 cubic inches and the 5030 had 90.2 cubic inches. They tested 21.79- and 26.42-PTO horsepower, respectively. Early models had 8-forward and 2-reverse speeds, which were later replaced by 12-speed creeper gearboxes. The 580-rpm power take-off had an overrunning clutch to eliminate the flywheel effect of a rotary mower.

The live hydraulic pump supplied 5.44 gpm at 1,780 psi to the category 1 three-point hitch and remote valve. The hitch had position control, but

The 15-PTO 5015 hp was the smallest of the compact diesels and had a smooth-running three-cylinder engine. With front-wheel assist it was adapted to a broad range of jobs in both farm and commercial markets.

The 5030, 5020 FWA, and 5020 were highly-reliable entries in the compact diesel tractor class. They were built by Toyosha in Japan.

The full line of compact tractors included the 5030, 5020, 5020 FWA, 5015, and 5015 FWA. They were revised in 1985 and continued as the Deutz-Allis 5200 Series.

was not draft-sensitive. The 5020 could lift 1,100 pounds, while the 5030 lifted 1,130 pounds. Turf and agricultural tires were available, but only agricultural were possible on the 5020 FWA.

The neat little 5015 was added in 1982. This 15-horsepower mighty midget had a new three-cylinder diesel of 61.3 cubic inches rated at 2,500 rpm. The nine-forward, three-reverse speed transmission was synchronized. A front-wheel assist model was offered, with other available features that were similar to its big brothers.

List prices in 1985 for the 5020 ranged from $9,060 to $9,980, while the 5030 list was $10,100. The 5015 prices ranged from $7,259 to $8,059. All had ROPS and a canopy. The Deutz-Allis 5200 Series replaced the 5000 line in 1985.

The 6140 Tractor

Experience with the Toyosha-built 5000 Series tractors had been very good, with high customer acceptance and extremely low warranty cost. It only seemed logical to work with them on larger tractors. The 40-horsepower 6140 joined the team in early 1982. It was all-new to Toyosha and built to A-C specifications. Styling matched the existing 6000 and 7000 Series tractors.

The 6140's three-cylinder engine was new in the Toyosha lineup and the swirl-chamber combustion system was similar to the smaller tractors. With 142.0 cubic inches and 2,500 rpm, it produced 41.08-PTO and 33.31-drawbar horsepower. It was an extremely responsive little powerhouse.

The five-forward, one-reverse speed transmission was fully synchronized, even in reverse. With a two-speed range there was a good choice of field speeds. On-the-roll shifting was easy and loader operation was enhanced. The 6140 was introduced in both two-wheel and front-wheel assist versions. The front wheel-assist was delayed for redesign after testing with a loader revealed a need to strengthen the front axle. The front wheel-assist proved to be a good loader tractor.

A 10-gpm main pump handled the three-point hitch and remote valves, and could operate a loader without an auxiliary pump. A separate 3.2-gpm pump powered the integral power steering. The 6140 had a

The 6140 with a 450 loader was a handy rig on an Indiana dairy farm. Hydraulic output was plenty for good loader operation at 10 gpm.

The 6140 FWA handled a three-bottom plow, 10-ft disk harrow, or a 10-ft field cultivator to power the small farm or help out in the big operation.

The 6140 with 9.00-10 front and 14.9-24 R3 tires could tread lightly for golf course mowing and other turf work.

category 1 hitch, but the field needed category 2 capacity, so at Serial No. 1201 a convertible category 1-2 hitch became standard. Lift capacity was 3,190 pounds at 24 inches in back of the hitch points.

After a few months, dealers requested a low-profile model with turf tires. A turf model with 9.00-10 front and 14.9-24 R3 rear tires and horizontal muffler was offered. A 6140 two-wheel drive tractor had a list price of $16,300, a front wheel-assist for $20,434, and a turf tractor for $17,150. The 6140 was discontinued when the company changed hands. About 1,900 were imported between 1982 and early 1985.

Future Models

Had the A-C tractor line continued, some more tractors were needed in the under-100 horsepower line. When dealer inventory of 5050 tractor was depleted, A-C again had a wide model spacing between the 40-horsepower 6140 and the 70-horsepower 6070. There was a proposal to turbocharge the 6140 for a 50-horsepower tractor but that was not considered a viable option.

The working relationship with Fiat was successful and it was logical to turn once again to that source for tractors. Engineering began testing two Fiat utility tractors at 50- and 60-PTO horsepower in 1984. With two-wheel drive and front-wheel assist they would have nicely filled out the line.

New 50- and 60-hp utility tractors from Fiat would have filled the horsepower gap between the 6140 and the 6070. A-C styling to match the 6000 Series was planned.

Models WD, WD45, and CA Serial Numbers and Production Data

Serial Numbers:

	Model WD	Model WD	Model WD45	Model CA
1948	7-9249	---	---	---
1949	9250-35444	---	---	---
1950	35445-72327	---	---	14-321
1951	72328-105181	---	---	322-10538
1952	105182-126931	127007-131242	---	10539-22180
1953	---	131243-146606	146607-160385	22181-31423
1954	---	---	160386-190992	31424-32906
1955	---	---	190993-217991	32907-37202
1956	---	---	217992-230294	37203-38617
1957	---	---	230295-236958	38618-38976
1958	---	---	---	38977-39513*

Production:

	Model WD						Model CA				
	Dual	AFA	SFW	Cane	Copk†	Total	Dual	AFA	SFW	Cane	Total
1948	9,144	0	28	50	14	9,236	---	---	---	---	---
1949	22,348	3,518	209	100	25	26,200	10	---	---	---	10
1950	30,154	6,250	266	100	100	36,870	305	---	---	---	305
1951	26,026	5,512	230	138	643	32,549	8,289	1,882	47	---	10,218
1952	15,724	8,751	196	30	1,330	26,031	6,678	4,809	157	---	11,644
1953	8,962	5,815	145	10	307	15,239	3,902	5,158	120	63	9,243
1954	---	---	---	---	---	---	428	999	29	28	1,484
1955	---	---	---	---	---	---	1,321	2,834	92	49	4,296
1956	---	---	---	---	---	---	474	915	6	20	1,415
1957	---	---	---	---	---	---	3	330	4	10	347
1958	---	---	---	---	---	---	0	512	5	20	537
Total						146,125					39,499

	Model WD45 Gas					Model WD45 Diesel				
	Dual	AFA	SFW	Cane	Total	Dual	AFA	SFW	Cane	Total
1953	7,626	5,960	170	21	13,777	---	---	---	---	---
1954	14,437	15,401	363	31	30,232	63	332	0	1	396
1955	8,830	14,351	291	34	23,506	599	2672	21	27	3,319
1956	3,671	6,125	0	0	9,796	532	1804	26	22	2,384
1957	2,332	3,893	0	0	6,255	46	357	0	7	410
Total					83,536					6,509

Note: Some special purpose units and OEM products are not included above.

* Approximate last serial number.

† Cotton picker.

Nebraska Test Results and Specifications for Models WD, WD45, and CA

	WD Tr. fuel	WD Gas	WD45 Gas	WD45 Dist.	WD45 LPG	WD45 Dsl.	CA
Test No.	399	440	499	511	512	563	453
Year	1948	1950	1953	1953	1953	1955	1950
Serial No.	WD21	WD48595	WD147753	WD149901	WD150743	WD199022D	CA12
Fuel	Tr. fuel	Gas	Gas	Dist.	LPG	Diesel	Gas
Max. belt hp	26.14	34.63	43.21	33.01	44.13	43.29	25.96
Max. dbr. hp	24.31	30.23	37.84	29.49	38.53	39.50	22.97
Weight (lb)							
Unballasted	3,388	3,721	3,955	3,955	4,149	4,285	2,763
Ballasted	5,042	6,313	8,005	7,185	8,065	8,305	5,045
Shipping wt.*	4,000	4,000	4,465	4,465	4,640	4,730	2,850
Engine	A-C	A-C	A-C	A-C	A-C	A-C	A-C
Cylinders	4	4	4	4	4	6	4
Rated rpm	1,400	1,400	1,400	1,400	1,400	1,625	1,650
Bore (in.)	4.00	4.00	4.00	4.00	4.00	3.4375	3.375
Stroke (in.)	4.00	4.00	4.50	4.50	4.50	4.125	3.50
Displ'mt (in.3)	201	201	226	226	226	230	125
Comp. ratio	4.5:1	5.75:1	6.45:1	4.75:1	7.2:1	15.5:1	6.25:1
Carburetor (in.)	1	1	1	1	1	DNA	.875
Tires:							
Front	5.50-16	5.50-16	5.50-16	5.50-16	5.50-16	5.50-16	4.00-15
Rear	11-28	11-28	13-28	13-28	13-28	13-28	10-24

* Shipping wt. from A-C specifications.

Model G Serial Numbers

Year	Serial Nos,	Year	Serial Nos.
1948	6-10960	1952	25269-26496
1949	10961-23179	1953	26497-28035
1950	23180-24005	1954	28036-29035
1951	24006-25268	1955	29036-29976

D Series Models and Production Dates

D14	1957-1960
D17	1957-1960
D17 Series II	1960-1962
D17 Series III	1963-1964
D17 Series IV	1964-1967
D10-D12	1959-1963
D10-D12 Series II	1963-1964
D10-D12 Series III	1964-1968
D15	1960-1962
D15 Series II	1963-1968
D19	1961-1964
D21	1963-1965
D21 Series II	1965-1969

The D17 Series III.

Serial Numbers and Production Data for Models D14, D15, and D15II Tractors

Serial Numbers:

	D14 Orange Grill	D14 Black Bar Grill	D15	D15 Series II
1957	1001-9399	---	---	---
1958	9400-13262	---	---	---
1959	14900-18230	19001-21799	---	---
1960	---	21800-24050	1001-1899	---
1961	---	---	1900-6469	---
1962	---	---	6470-8169	---
1963	---	---	---	13001-16927
1964	---	---	---	16928-19680
1965	---	---	---	19681-21374
1966	---	---	---	21375-23733
1967	---	---	---	23734-25126
1968	---	---	---	25127-25419

Production:

	D14				D15 Gas		D15 Diesel			
	Dual*	RSFA†	SFW	HiCl‡	Dual*	RSFA†	Dual*	RSFA†	Grove§	Total
1957	1,761	6,664	---	---	---	---	---	---	---	8,425
1958	1,084	4,402	40	---	---	---	---	---	---	5,526
1959	1,307	4,808	14	---	---	---	---	---	---	6,129
1960	510	1,690	---	6	128	555	15	201	---	3,105
1961	---	---	---	---	594	3,321	67	580	7	4,569
1962	---	---	---	---	175	1,439	2	74	11	1,701
1963	---	---	(Series II)		246	3,463	7	183	7	3,906
1964					INA II	INA II	INA II	INA II	INA II	2,759
1965					INA II	INA II	INA II	INA II	INA II	1,683
1966					INA II	INA II	INA II	INA II	INA II	2,244
1967					INA II	INA II	INA II	INA II	INA II	1,465
1968					INA II	INA II	INA II	INA II	INA II	299
Total										41,811

LPG tractors included in gas tractor totals:

1957	---
1958	204
1959	518
1960	40
1961	51
1962	31
1963-1968	INA II

* Dual front wheels.
† Roll-Shift front axle.
‡ High clearance.
§ Orchard fenders and shields, not the "Birdcage."
II Information not available.

Nebraska Test Results and Specifications for Model G

Test No.	398
Year	1948
Serial No.	G42
Fuel	Gas
Max. belt hp	10.33
Max. dbr. hp	9.04
Weight (lb)	
Unballasted	1,549
Ballasted	1,749
Shipping Wt.*	1,285
Engine	Continental
Cylinders	4
Rated rpm	1,800
Bore (in.)	2.375
Stroke (in.)	3.50
Displ'mt (in.^{3})	62.0
Comp. ratio	INA†
Carburetor (in.)	0.625
Speeds:	
1st	1.6
2nd	2.25
3rd	3.5
4th	7.0
Rev.	2.0
Tires:	
Front	4.00-12
Rear	6-30

* Shipping weight from A-C specifications.
† Information not available.

Nebraska Test Results Models D14 and D17

	D14 Gas	D14 LPG	D17 Gas	D17 LPG	D17 Diesel
Test No.	623	645	635	644	636
Year	1957	1958	1957	1958	1957
Serial No.	1093	9643	1128	5773	1001
Fuel	Gas	LPG	Gas	LPG	Diesel
Max. belt hp	34.08	31.86	52.70	50.79	51.14
Max. dbr hp	30.91	28.67	48.64	46.23	46.20
Weight (lb)					
Unballasted	3,623	3,711	4,659	4,751	4,867
Ballasted	7,005	7,375	9,395	9,735	9,455
Engine	A-C		A-C		
Cylinders	4	4	4	4	6
Bore (in.)	3.50	3.50	4.0	4.0	3.5625
Stroke (in.)	3.875	3.875	4.50	4.50	4.375
Displ'mt (in.^{3})	149	149	226	226	262
Rated rpm	1,650	1,650	1,650	1,650	1,650
Comp. ratio	7.5:1	8.5:1	7.25:1	8.25:1	15.7:1
Carburetor (in.)	0.875	1	1	1	DNA*
Tires					
Front	5.50-16	5.50-16	6.00-16	6.00-16	6.00-16
Rear	12-26	12-26	14-28	14-28	14-28

* Does not apply.

Nebraska Test Results and Specifications Models D15 and D15II

	D15 Gas	D15 LPG	D15 Diesel	D15 II Gas	D15 II LPG
Test No.	795	797	796	837	838
Year	1961	1961	1961	1963	1963
Serial No.	3535	3623	2984D	13443	13442
Fuel	Gas	LPG	Diesel	Gas	LPG
Max. PTO hp	40.00	37.44	36.51	46.18	43.55
Max. dbr hp	35.33	33.22	33.32	38.33	36.02
Weight (lb)					
Unballasted	3,985	4,036	4,220	4,025	4,090
Ballasted	7,418	7,535	7,575	7,950	8,105
Engine	A-C	A-C	A-C	A-C	A-C
Cylinders	4	4	4	4	4
Bore (in.)	3.50	3.5	3.5625	3.625	3.625
Stroke (in.)	3.875	3.875	4.375	3.875	3.875
Displ'mt (in.^{3})	149	149	175	160	160
Rated rpm	2,000	2,000	2,000	2,000	2,000
Carburetor (in.)	0.875	0.875	DNA*	1.0	1.0
Tires:					
Front	6.00-16				
Rear	14.9-26				

* Does not apply.

Serial Numbers and Years Built for the D17 Series

D17 Series	Serial Nos.	Dates
(Series I)*		
Orange grill	1001-23363	1957-1959
Black bar grill	24001-31625	1959-1960
(Series II)*	32001-41540	1960-1962
Persian No. 2,		
cream wheels.		
Series III	42001-43358†	1962
	65001-72768	1963-1964
Series IV	75001-89213	1964-1967

* Unofficial designation used after Series III came out.
† No tractors between Serial No. 43358 and 65001.

Nebraska Test Results Models D10 and D12

	D10	D12	D10	D12
Test No.	724	723	812	813
Year	1959	1959	1962	1962
Serial No.	1164	1123	3501	3001
Max. PTO hp	28.51	28.56	33.46	33.32
Max. dbr hp	25.73	23.56	28.78	29.43
Weight (lb)				
Unballasted	2,860	2,945	3,001	3,051
Ballasted	4,801	5,449	5,560	5,630
Engine	A-C		A-C	
Bore (in.)	3.375		3.50	
Stroke (in.)	3.875		3.875	
Displ'mt (in.^{3})	138.7		149	
Rated rpm	1,650		1,650	
Comp. ratio	7.75:1		7.5:1	
Carburetor (in.)	.875		.875	
Tires:				
Front	5.00-15	5.00-15	5.00-15	5.00-15
Rear	10-24	11-24	11-24	11-24

Serial Numbers and Production Data for the Model D17 Tractor

Serial Numbers:

	Original (Series I*)	Black bar (Series I*)	Series II* Pers. No.2	Series III Fender Lts.	Series IV Hyd. Pump
1957	1001-4299	---	---	---	---
1958	4300-16499	---	---	---	---
1959	16500-23363	24001-28199	---	---	---
1960	---	28200-31625	32001-32099	---	---
1961	---	---	33100-38069	---	---
1962	---	---	38070-41540	42001-43358†	---
1963	---	---	---	65001-70610	---
1964	---	---	---	70611-72768	75001-77089
1965	---	---	---	---	77090-80532
1966	---	---	---	---	80533-86060
1967	---	---	---	---	86061-89213

Production:

	D17 Gas				D17 Diesel						LPG Tractors
	Dual	AFA	SFW	Cotton	Dual	AFA	SFW	Cotton	Cane	Total	in Gas Totals
1957	620	2,090	---	---	68	449	---	---	---	3,227	
1958	2,483	7,161	31	50	365	2,146	16	12	---	12,264	848
1959	2,135	6,697	15	80	230	1,853	22	20	---	11,052	120
1960	677	2,389	---	210	114	1,035	---	65	53	4,543	287
1961	762	2,937	---	271	75	750	---	81	88	4,964	216
1962	648	3,129	---	270	59	542	---	79	100	4,827	344
1963	651	3,899	---	114	45	650	---	40	180	5,579	
1964	DNA‡	DNA‡	DNA‡	DNA‡	DNA‡	DNA‡	DNA‡	DNA‡	DNA‡	4,237	
1965	DNA‡	DNA‡	DNA‡	DNA‡	DNA‡	DNA‡	DNA‡	DNA‡	DNA‡	3,414	
1966	DNA‡	DNA‡	DNA‡	DNA‡	DNA‡	DNA‡	DNA‡	DNA‡	DNA‡	5,264	
1967	DNA‡	DNA‡	DNA‡	DNA‡	DNA‡	DNA‡	DNA‡	DNA‡	DNA‡	3,169	
Total										62,540	

* Unofficial designation after the Series III came out.

† No tractors between Serial No. 43358 and Serial No. 65001.

‡ Details not available.

Serial Numbers and Production Data
Models D10 and D12

	D10 Black bar	Cream Grill Pers. Org 2	149 in.^{3}	Series II	Series III	Total
1959	1001-INA*	---	---	---	---	933
1960	INA-INA*	1950-INA*	---	---	---	752
1961	---	2801-3262	3501-INA*	---	---	478
1962	---	---	INA*-4511	---	---	1,009
1963	---	---	---	6801-7674	---	874
1964	---	---	---	7675-7850	9001-9203	377
1965	---	---	---	---	9204-9485	286
1966	---	---	---	---	9486-9794	281
1967	---	---	---	---	9795-9978	191
1968	---	---	---	---	9979-10100	123
Total						5,304
	D12					
1959	1001-INA*	---	---	---	---	734
1960	INA-INA*	1950-INA*	---	---	---	864
1961	---	2801-2919	3001-INA*	---	---	150
1962	---	---	INA*-3638	---	---	637
1963	---	---	---	5501-6011	---	411
1964	---	---	---	6012-6144	9001-9191	324
1965	---	---	---	---	9192-9509	315
1966	---	---	---	---	9510-9829	284
1967	---	---	---	---	9830-10051	229
1968	---	---	---	---	10052-10172	122
Total						4,070

* Information not available.

Nebraska Test Results and Specifications for the Models D10, D21, and D21 Series II

	D19 Gas	D19 Diesel	D19 LPG	D21	D21 Series II
Test No.	810	811	814	855	904
Year	1961	1962	1962	1963	1965
Serial No.	1199	1198	1200	1012	2109
Max. PTO hp	71.54	66.92	66.19	103.06	127.75
Max. dbr hp	63.91	62.05	58.29	93.09	116.41
Weight (lb)					
Unballasted	6,645	6,835	6,785	10,745	10,675
Ballasted	11,045	11,210	11,160	18,515	18,965
Engine: A-C	G262	D262	G262	3400	3500
Cylinders	6	6	6	6	6
Aspiration	Natural	Turbo.	Natural	Natural	Turbo.
Bore & stroke (in.)	3.625 × 4.375	3.625 × 4.375	3.625 × 4.375	4.25 × 5.00	4.25 × 5.00
Displacement (in.^{3})	262	262	262	426	426
Rated rpm	2,000	2,000	2,000	2,200	2,200
Compression ratio	8.0:1	14.0:1*	9.65:1	16.0:1	16.0:1
Tires:					
Front	7.50-16	7.50-16	7.50-16	11.00-16	11.00-16
Rear	18.4-34	18.4-34	18.4-34	24.5-32	24.5-32

* 15.0:1 Effective engine Serial No. 014419.

Serial Numbers and Production Data for Models D19, D21, and D21 Series II

Serial Numbers:

	D19	D21	D21 Series II
1961	1001-1249	---	---
1962	1250-7331	---	---
1963	12001-14944	1001-1416	---
1964	14945-16266	1417-2078	---
1965	---	2079-2129	2201-2407
1966	---	---	2408-2862
1967	---	---	2863-3776
1968	---	---	3777-4497
1969	---	---	4498-4609

Production:

	D19 Gas			D19 Diesel				Total		
	Dual*	RSFA†	HiCl‡	Dual*	RSFA†	HiCl‡	Beach§	D19	D21	D21 Ser. II
1961	1	63	---	---	1	---	---	65	---	---
1962	405	3,528	4	118	2,190	22	---	6,267	---	---
1963	117	1,921	5	36	816	48	1	2,944	414	---
1964	DNAII	DNAII	DNAII	DNAII	DNAII	DNAII	DNAII	1,315	664	---
1965	DNAII	DNAII	DNAII	DNAII	DNAII	DNAII	DNAII	DNAII	50	194
1966	DNAII	DNAII	DNAII	DNAII	DNAII	DNAII	DNAII	DNAII	---	421
1967	DNAII	DNAII	DNAII	DNAII	DNAII	DNAII	DNAII	DNAII	---	922#
1968	DNAII	DNAII	DNAII	DNAII	DNAII	DNAII	DNAII	DNAII	---	681
1969	DNAII	DNAII	DNAII	DNAII	DNAII	DNAII	DNAII	DNAII	---	111
Total								10,591	1,128	2,329

* Dual front wheels.
† Roll-Shift front axle.
‡ High clearance.
§ Beachmaster tractor for beach sanitizer.
II Does not apply
\# Includes 10 tractors with hydrostatic front-wheel assist, later converted to two-wheel drive.

Hundred Series Models and Production Dates	
One-Ninety gas	1964-1968
One-Ninety diesel	1964-1973
One-Ninety XT gas	1965-1970
One-Ninety XT LPG	1965-1968
One-Ninety XT diesel	1965-1971
Model 200	1972-1975
One-Seventy gas	1967-1971
One-Seventy diesel	1967-1971
Model 170 gas	1971-1973
Model 170 diesel	1971-1973
Model 175 gas	1973-1976
Model 175 diesel	1970-1980
One-Eighty gas	1969-1970
One-Eighty diesel	1967-1973
Model 185 diesel	1970-1981
Two-Twenty diesel	1969-1973
Two-Ten diesel	1970-1973

Calendar of Events One-Seventy, 170, 175, One-Eighty, and 185

Tractor	First Ser. No.	Date	Last Ser. No.	Date
One-Seventy Gas,dsl.	1005	8/67	7384*	12/71
Landhandler decals	6650	5/70		
170 Gas, diesel	7500	12/71	10300	12/73
175 diesel	1001	2/70	7509*	9/80
175 (248 engine)	1459	12/70		
175 (hyd. pump)	2035	6/73		
175 gas	2151	12/73		3/76
One-Eighty diesel	1007	5/67	12985	11/73
One-Eighty gas	6644	2/69		/70
Landhandler decals	9789	5/70		
185 diesel	1001	2/70	15961	9/81
Drop interchangeable front axle	12800	/78		

* Approximate

Nebraska Test Results for the One-Seventy, 170, 175, One-Eighty, and 185

	One-Seventy,170		175		One-Eighty		185
Fuel	Gas	Diesel	Gas	Diesel	Gas	Diesel	Diesel
Test No.	966	965	1156	1043	1005	964	1044
Year	1967	1967	1974	1970	1969	1967	1970
Serial No.	1134	1133	2151	1079	7058	1211	1082
Max. PTO hp	54.12	54.04	60.88	62.47	65.16	64.01	74.87
Max. Dbr hp	46.93	47.39	51.80	54.06	56.05	55.28	63.82
Weight (lb)							
Unballasted	5,720	5,950	6,390	5,925	6,420	6,510	6,710
Ballasted	9,010	9,160	8,400	8,400	10,270	10,300	9,400
Engine	A-C	Perkins	A-C	Perkins	A-C	A-C	A-C
Model	G226	4.236	G226	4.236*	G2500	2800	2800
Cylinders	4	4	4	4	6	6	6
Rated rpm	1,800	1,800	1,800	2,000	2,000	2,000	2,200
Bore & stroke	4.0 × 4.5	3.875 × 5.0	4.0 × 4.5	3.875 × 5.0	3.75 × 4.0	3.875 × 4.25	3.875 × 4.25
Displacement (in.^{3})	226	236	226	236	265	301	301
Comp. ratio	8.0:1	16.0:1	8.2:1	16.0:1	8.15:1	16.25:1	16.25:1
Carburetor (in.)	1.125	DNA†	1.25	DNA†	1.0	DNA†	DNA†
Tires-							
Front	7.50-16	7.50-16	7.50-16	6.50-16	7.50-16	7.50-16	9.5L-15
Rear	18.4-28	18.4-28	18.4-28	16.9-28	18.4-28	18.4-28	18.4-28

* Engine changed to Perkins 4.248 (3.975 × 5.0, 248 in.) at Serial No. 1459. Not retested.

† Does not apply.

Serial Numbers and Production Data for the One-Seventy, 170, 175, One-Eighty, and 185

Serial Numbers:

	One-Seventy	170	175	One-Eighty	185
1967	1005-2720	---	---	1007-2681	---
1968	2721-5373	---	---	2682-6093	---
1969	5374-6368	---	---	6094-9234	---
1970	6369-6987	---	1001-1476	9235-10564	1001-1951
1971	6988-7384*	7500-7796	1477-1623	10561-11728	1952-2934
1972	---	7797-8820	1624-1739	11729-12446	2935-3762
1973	---	8821-10300	1740-2152	12447-12985	3763-4960
1974	---	---	2153-3254	---	4961-6541
1975	---	---	3255-3739	---	6542-8366
1976	---	---	3740-4811	---	8367-10024
1977	---	---	4812-5662	---	10025-11625
1978	---	---	5663-6320	---	11626-13159
1979	---	---	6321-6998	---	13160-14671
1980	---	---	6999-7502*	---	14672-15647
1981	---	---	---	---	15648-15961

* Approximate serial number.

Production:

	One-Seventy/170			175			One-Eighty			185
	Gas	Diesel	Total	Gas	Diesel	Total	Gas	Diesel	Total	Diesel
1967	INA	INA	1,700	---	---	---	---	1,700	1,700	---
1968	INA	INA	2,673	---	---	---	---	3,157	3,157	---
1969	INA	INA	995	---	---	---	1,499	1,642	3,141	---
1970	463	170	633	---	475	475	88	1,232	1,320	941
1971	384	309	693	---	159	159	---	1,162	1,162	977
1972	741	356	1,097	---	104	104	---	731	731	840
1973	949	563	1,512	---	367	367	---	539	539	1,202
1974	---	---	---	722	424	1,146	---	---	---	1,582
1975	---	---	---	245	254	499	---	---	---	1,804
1976	---	---	---	42	1,013	1,055	---	---	---	1,679
1977	---	---	---	---	844	844	---	---	---	1,574
1978	---	---	---	---	656	656	---	---	---	1,553
1979	---	---	---	---	685	685	---	---	---	1,513
1980	---	---	---	---	504	504	---	---	---	975
1981	---	---	---	---	---	---	---	---	---	321
Total	---	---	9,303	---	---	6,494	---	---	11,750	14,961

Nebraska Test Results for the One-Ninety and One-Ninety XT

	One-Ninety		One-Ninety XT		
Fuel	Gas	Diesel	Gas	LPG	Diesel
Test No.	928	886	929	927	887
Year	1965	1965	1965	1965	1965
Serial No.	6955	1042	6973	7221	3454
Max. PTO hp	75.37	77.20	89.53	85.25	93.64
Max. Dbr hp	63.10	65.33	76.65	71.95	79.12
Weight (lb)					
Ballasted	11,295	11,255	11,355	11,295	11,300
Unballasted	7,665	7,845	7,675	7,715	7,945
Engine	A-C	A-C	A-C	A-C	A-C
Model	G2500	2800	G2800	G2800	2900
Aspiration	Natural	Natural	Natural	Natural	Turbocharged
Cylinders	6	6	6	6	6
Bore & stroke (in.)	3.75 × 4.00	3.875 × 4.25	3.875 × 4.25	3.875 × 4.25	3.875 × 4.25
Displacement (in.^{3})	265	301	301	301	301
Rated rpm	2,200	2,200	2,200	2,200	2,200
Compression ratio	8.0:1	16.25:1	8.0:1	9.0:1	16.25:1
Carburetor (in.)	1.5	DNA*	1.5	1.5	DNA*
Tires					
Front	7.50-16	7.50-16	7.50-16	7.50-16	7.50-16
Rear	18.4-34	18.4-34	18.4-34	18.4-34	18.4-34

* Does not apply.

Calendar of Events for the One-Ninety and One-Ninety XT

Tractor	First Serial No.	Date
One-Ninety gas, dsl.	1001	8/64
XT Diesel	2849	1/65
XT Gas	3209	2/65
XT LPG	4368	3/65
Series I (bar grill)	1001-8626	
No Tractors	8626-9000	
Transmission improvements	9004-17784	3/66
No Tractors	17785-19000	
Series II (two lights)	19001-22153	12/67
No tractors	22154-23000	
Series III (four-pin.)	23001-30938	12/68
One-Ninety Cane	10252	5/66
First air cond. cab	10962	5/66

Serial Numbers and Production Data for the One-Ninety and One-Ninety XT

Serial Numbers:

	One-Ninety, Series I Bar grill	One-Ninety XT Trans. imprv. Bar grill	Series II 2 lt.fender	Series III 4 pin. diff.
1964	1001-2484	---	---	---
1965	2485-8218	---	---	---
1966	8219-8626	9004-13272	---	---
1967	---	13273-17784	19001-19261	---
1968	---	---	19262-22153	23001-23233
1969	---	---	---	23234-25900
1970	---	---	---	25901-29135
1971	---	---	---	29136-31056
1972	---	---	---	31057-31117*
1973	---	---	---	31118-31140*

* All One-Ninety diesels.

Production:

	One-Ninety	One-Ninety XT	Total
1964	1,502	---	1,502
1965	1,384	4,293	5,677
1966	568	3,837	4,405
1967	564	4,221	4,785
1968	347	2,587	2,934
1969	115	2,552	2,667
1970	74	3,158	3,232
1971	76	1,771	1,847
1972	73	---	73
1973	23	---	23
Total	4,726	22,419	27,145

7000 Series Models and Production Data

Model	Chassis	Years
7000	Maroon	1975-1978
7000	Black	1978-1979
7010	Black	1979-1981
7020	Black	1977-1981
7030	Maroon	1973-1974
7040	Maroon	1974-1977
7045	Black	1977-1981
7050	Maroon	1973-1974
7060	Maroon	1974-1977
7060	Black	1978-1981
7080	Maroon	1974-1977
7080	Black	1978-1981
7580	Maroon	1976-1978
7580	Black	1978-1981
8550	Black	1977-1981

Serial Numbers and Productions Data for the 200, Two-Ten, and Two-Twenty

Serial Numbers:

	200	200*	Two-Ten	Two-Twenty
1969	---	---	---	1004-1937
1970	---	---	1004-1106	1938-2450
1971	---	---	1107-2081	2451-2625
1972	1004-3343	---	2082-2469	2626-2866
1973	3344-3558	4001-6293	---	---
1974	---	6294-9249	---	---
1975	---	9250-11521	---	---

Production:

	200	Two-Ten	Two-Twenty
1969	---	---	934
1970	---	103	530†
1971	---	961	161
1972	2,342	407	240
1973	2,507	---	---
1974	3,006	---	---
1975	2,271	---	---
Total	10,126	1,471	1,865

* Model 200 cost-reduced tractor.
† Includes 100 with front-wheel assist.
First Two-Twenty FWA — Serial No. 2223, June 1970.
Two-Twenty Landhandler decals, May 1970.

Nebraska Test Results for the Two-Ten and Two-Twenty

	Two-Twenty	Two-Ten
Test No.	1017	1065
Year	1969	1971
Serial No.	1138	1182
Max. PTO hp	135.95	122.40
Max. Dbr hp	117.21	104.95
Weight (lb)		
Ballasted	20,050	18,045
Unballasted	12,160	12,425
Engine	A-C 3500	A-C 3500
Cylinders	6	6
Aspiration	Turbocharged	Turbocharged
Bore & stroke (in.)	4.25 × 5.00	4.25 × 5.00
Displacement (in.^{3})	426	426
Rated rpm	2,200	2,200
Compression ratio	16.0:1	16.0:1
Tires		
Front	10.00-16	10.00-16
Rear	24.5-32	18.4-38

7000 Series Specifications

	7010	7020	7030	7040	7045	7050	7060	7080
Engine	Allis-Chalmers six-cylinder direct-injection diesel*							
Engine model	649T	649I	3500 II	3500 II	670T	3700	3700	3750HI
	---	---	---	---	---	---	3700 II	670HI
	---	---	---	---	---	---	670I	---
Aspiration	Turbo†	Turbo-int.‡	Turbo†	Turbo†	Turbo†	Turbo-int.‡	Turbo-int.‡	Turbo-int.‡
Bore × stroke	3.875 × 4.25	3.875 × 4.25	4.25 × 5.00	4.25 × 5.00	4.25 × 5.00	4.25 × 5.00	4.25 × 5.00	4.25 × 5.00
Displ'mt (in.^{3})	301	301	426	426	426	426	426	426
Rated rpm	2,300	2,300	2,300	2,300	2,300	2,300	2,300	2,550
Comp. ratio	16.25:1	16.25:1	16.0:1	16.0:1	16.0:1	16.0:1	16.0:1	16.0:1
Tires								
Front	10.00-16	11.00-16	11.00-16	11.00-16	11.00-16	11.00-16	14L-16.1	18.4-16.1
rear	18.4-38	20.8-38	20.8-38	20.8-38	20.8-38	20.8-38 duals	20.8-38 duals	20.8-38 duals
Power Trains:								
Power-Director								
Plates (low/high)	3/2	3/2	3/2	3/2	4/2	4/2	4/2	4/2 special
Power-Shift	Yes	Yes	No	Yes	Yes	No	Yes	No
Diff. ring gear								
width (in.)	2.30	2.30	2.30	2.30	2.55	2.55	2.55	2.55
Diff. lock plates	3	3	3	3	4	4	4	4
Diff. pinions	2	2	2	2	4	4	4	4
Final drive								
gear width (in.)	2.0	2.0	2.0	2.0	2.0	2.0	2.0	4.0
Final drive ratio	5.6:1	5.6:1	5.6:1	5.6:1	5.41:1	5.41:1	5.41:1	6.57:1
PTO (rpm)	540-1,000	540-1,000	540-1,000	540-1,000	1,000§	1,000§	1,000 §	1,000 ‖
Axle diameter								
Standard (in.)	3.625	3.625	3.625	3.75	3.75	3.625	3.75	4.0
Long/special	3.75	3.75	---	3.75/4.0	3.75/4.0	---	3.75/4.0	4.0

* Applies to all models in that category.
† Turbocharged.
‡ Turbocharged-intercooled.
§ 540-1000 optional
‖ 20 spline, 1.75-in. shaft.

7000 Series Nebraska Test Results

	7030	7050	7040PD	7040PS	7060PD	7060PS	7080	7020PD	7020PS	7045PD	7045PS	7010PD	7010PS
Test no.	1119	1120	1166	1196	1167	1197	1168	1260	1259	1261	1308	1345	1346
Year	1973	1973	1974	1975	1974	1975	1974	1977	1977	1977	1979	1979	1980
Max. PTO hp	130.98	156.49	136.49	136.30	161.51	161.42	181.51	123.85	123.79	146.18	146.88	106.53	106.72
Max. Dbr hp	110.29	131.54	114.23	114.34	136.12	141.67	148.89	102.09	101.89	122.90	123.35	87.83	89.10
Weight (lb)													
Ballasted	16,560	19,350	17,380	17,355	20,490	20,560	21,720	15,655	15,610	18,980	18,990	12,790	12,740
Unballasted	12,430	14,525	11,795	12,378	13,990	14,645	14,605	11,570	11,770	14,235	14,600	11,260	11,375

7000 Series Serial Numbers

Maroon Chassis

	7030	7050	7040	7060	7080	7580
1973	1001-2594	1001-1687	---	---	---	---
1974	2595-4398	1688-3300	1001-1302	1001-1298	1001-1006	---
1975	---	---	1303-4128	1299-2748	1007-1571	---
1976	---	---	4129-6839	2749-4581	1572-2500	1001-1287
1977	---	---	6840-8250	4582-5644	2501-2930	1288-1604
1978	---	---	---	---	---	1605-INA

Black Chassis

	7010	7020	7045	7060	7080	7580	8550
1977	---	1001-1316	1001-1233	---	---	---	1001-1082
1978	---	1317-2731	1234-2151	6001-6788	3001-3267	2001-2217	1083-1341
1979	1001-1924	2732-3841	2152-3398	6789-7692	3268-3647	2218-2485	1342-1552
1980	1925-2805	3842-4709	3399-4224	7693-8441	3648-3953	2486-2716	1553-1722
1981	2806-3433	4710-5209	4225-4888	8442-9142	3954-4225	2717-3018	1723-2021

7000 Series Production Records

	7030	7050	7040	7060	7080	7020	7045	7010	7580	8550	Total
1973	1,583	683	---	---	---	---	---	---	---	---	2,266
1974	1,813	1,619	286	289	3	---	---	---	---	---	4,010
1975	---	---	2,830	1,456	566	---	---	---	---	---	4,852
1976	---	---	2,711	1,833	480	---	---	---	284	---	5,308
1977	---	---	1,453	1,061	430	289	218	---	320	75	3,846
1978	---	---	---	793	267	1,438	930	---	216	265	3,909
1979	---	---	---	906	379	1,109	1,245	918	267	210	5,034
1980	---	---	---	749	305	868	827	880	232	170	4,031
1981	---	---	---	709	275	511	688	638	291	253	3,365
Total	3,396	2,302	7,280	7,796	2,705	4,215	3,908	2,436	1,610	973	36,621

Commonality

One big advantage of the 7000 Series was not apparent to their owners — their very high level of interchangeable parts. Obviously an important manufacturing advantage, this commonality also provided better availability of parts to dealers and customers. When the full 7000 Series was in production during 1980-81, the commonality was calculated as follows: (The engine as received from the Harvey plant was counted as one part.)

Tractor	Percent Common
7010	98.6
7020	97.4
7045	97.9
7060	98.2
7080	96.3
7580	90.6
8550	58.0

Model 7000 Nebraska Test Results, Serial Numbers and Production Data

Model	7000
Test No.	1195
Year	1975
Max. PTO hp	106.44
Max. Dbr hp	89.46
Weight (lb)	
Ballasted	12,580
Unballasted	9,550
Engine	A-C 2900
Cylinders	6
Aspiration	Turbo
Rated rpm	2,200
Bore × stroke	3.875 × 4.25
Displ'mt (in.^{3})	301
Comp. ratio	16.25:1

Serial Numbers:

Year	Maroon Chassis	Black Chassis
1975	1001-1647	---
1976	1648-5035	---
1977	5036-6372	---
1978	6373-6760	8000-8962
1979	---	8963-9503

Production:

Year	Maroon	Black	Total
1975	639	---	638
1976	3,385	---	3,385
1977	1,346	---	1,346
1978	393	950	1,343
1979	---	551	551
Total	5,763	1,501	7,264

6000 Series Nebraska Test Results, Serial Numbers, and Production Data

Nebraska Test Results:

	6060	6070	6080
Test No.	1397	1545	1398
Year	1981	1984	1981
Max. PTO hp	63.83	70.78	83.66
Max. Dbr hp	50.87	58.35	70.50
Weight (lb)			
Ballasted	8,945	10,515	11,790
Unballasted	6,465	7,460	7,590
Engine (A-C)	433T	433T	433I
Aspiration	Turbo*	Turbo*	Turbo-Int†
Bore	3.875	3.875	3.875
Stroke	4.25	4.25	4.25
Displ'mt (in.^{3})	200	200	200
Comp. ratio	14.1:1‡	15.1:1	14.1:1‡
Tires			
Front	7.50-16	7.50-16	7.50-16
Rear:	15.5-38	16.9-34	18.4-34

* Turbocharged.
† Turbocharged-intercooled.
‡ Later 15.0:1, see below.

Serial Numbers:

	6060	6070	6080
1980	1001-1296	---	1001-1151
1981	1297-2462	---	1152-3001
1982	2463-3893	---	3002-4566
1983	3894-4571	---	4567-5779
1984	4572-5027*	1001-1608	5780-6852
1985	---	1609-1972*	6853-7698*

* Last tractor.

Production:

	6060	6070	6080	Total
1980	296	---	151	447
1981	1,166	---	1,850	3,016
1982	1,431	---	1,565	2,996
1983	678	---	1,213	1,891
1984	456	608	1,073	2,137
1985	---	364	846	1,210
Total	4,027	972	6,698	11,697

Running changes — effective serial numbers:

Reduced front-wheel assist turning radius 6060 none, 6070 all, 6080 Serial No. 6325.

Compression ration 15.0:1: 6060 Engine Serial No. 17813, 6070 all, 6080 engine Serial No. 20833.

Manifold air heaters — (12-83) 6060 Serial No. 4533, 6070 all, 6080 Serial No. 5639.

Running Changes — 8000 and 4W Series*

Change	8010	8030	8050	8070	4W220	4W305
Ambac injection pump	†	---	---	---	---	---
Throttle in console (August, 1983)	1925	2307	2197	1711	1106	1133
Lever type hydraulic couplers (December, 1983)	2104	2497	2383	1897	1130	1160
Faster low range, +6.5% (April, 1984)	---	---	2814	2359	---	---
Power shift gear, 274153	2430	2924	2905	2453	---	---
Glide-ride seat attenuator (April, 1984)	2391	2885	2831	2374	---	----
8070 2WD narrow final drive	---	---	---	2663	---	---

* Changes are noted by the effective serial numbers.
† Engine Serial No. 49K-21508.

Serial Number Information and Production Data for the 8000 and 4W Series

Serial Numbers:

	8010	8030	8050	8070	4W220	4W305
1981	1001-1019	1001-1008	1001-1015	1001-1003	1001-1002	1001-1003
1982	1020-1711	1009-2092	1016-1923	1004-1429	1003-1080	1004-1111
1983	1712-2265	2093-2700	1924-2495	1430-2089	1081-1144	1112-1175
1984	2266-2608	2701-3145	2496-3186	2090-2902	1145-1175*	1176-1337
1985	2609-2832*	3146-3328*	3187-3336*	2903-3354†	---	1338-1412*

Production:

	8010	8030	8050	8070	4W220	4W305	Total
1981	19	8	15	3	2	3	50
1982	692	1,084	908	426	78	108	3,296
1983	554	608	572	660	64	64	2,522
1984	343	445	691	813	31	162	2,485
1985	224	183	150	452	---	75	1,084
Total	1,832	2,328	2,336	2,354	175	412	9,437

* Last tractor of model built.
† Last 8000 built.

8000 Series Nebraska Test Results

	8010	8030	8050	8070
Test No.	1446	1447	1448	1449
Year	1982	1982	1982	1982
Max. PTO hp	107.38	133.75	152.40	170.72
Max. Dbr hp	91.46	115.22	131.44	145.36
Weight (lb)				
Ballasted	13,185	17,100	19,380	22,370
Unballasted	12,630	15,015	15,305	16,850
Engine	AC	AC	AC	AC
Model	649T	670T	670I	670I
Cylinders	6	6	6	6
Aspiration	Turbo†	Turbo†	Turbo-int.‡	Turbo-int.‡
Bore (in.)	3.875	4.25	4.25	4.25
Stroke (in.)	4.25	5.0	5.0	5.0
Displ'mt (in.^{3})	301	426	426	426
Comp. ratio	15.5:1	15.5:1	15.5:1	15.5:1
Tires				
Front	10.00-16	11.00-16	14L-16.1	18.4-16.1
Rear	18.4-38	18.4-38	20.8-38	20.8-38
	---	Duals	Duals	Duals

* Power-shift only.
† Turbocharged.
‡ Turbocharged-intercooled.

Serial Numbers and Production Data for Four-Wheel Drive Tractors

Serial Numbers:

	440 Maroon	7580 Maroon	7580 Black	8550	4W220	4W305
1972	1001-1139	---	---	---	---	---
1973	1140-1339	---	---	---	---	---
1974	1440-1650	---	---	---	---	---
1975	1651-INA*	---	---	---	---	---
1976	INA*-2010	1001-1286	---	---	---	---
1977	---	1287-1604	---	1001-1082	---	--- ---
1978	---	1605-INA	2001-2217	1083-1341	---	---
1979	---	---	2218-2485	1342-1552	---	---
1978	---	---	2486-2716	1553-1722	---	---
1981	---	---	2717-3018	1723-2021	1001-1002	1001-1003
1982	---	---	---	---	1003-1080	1004-1111
1983	---	---	---	---	1081-1144	1112-1175
1984	---	---	---	---	1145-1175	1176-1337
1985	---	---	---	---	---	1338-1412

* Information not available.

Production:

	440	7580	8550	4W220	4W305	Total
1972	138	---	---	---	---	138
1973	199	---	---	---	---	199
1974	300	---	---	---	---	300
1975	302	---	---	---	---	302
1976	94	284	---	---	---	378
1977	---	320	75	---	---	395
1978	---	216	265	---	---	481
1979	---	267	210	---	---	477
1980	---	232	170	---	---	402
1981	---	291	253	2	3	549
1982	---	---	---	78	108	186
1983	---	---	---	64	64	128
1984	---	---	---	31	162	193
1985	---	---	---	---	75	75
Total	1,033	1,610	973	175	412	4,203

Nebraska Test Results and Specifications for the Allis-Chalmers Four-Wheel Drive Tractors

	T16	440	7580	8550	4W220	4W305
Test No.	None	None	1229	1262	None	None
Year			1976	1977		
Max. engine hp	150*	218*	220*	300*	220*	305*
Max. PTO hp	†	†	186.35	253.88	180*	250*
Max. Dbr hp	---	165*	153.70	217.92	---	---
Weight (lb)						
Ballasted	---	---	25,000	29,360	---	---
Unballasted	16,300	17,500	23,510	27,040	21,475	27,100
Engine make	A-C	Cummins	A-C	A-C	A-C	A-C
Model	7000‡	555C	3750 MKII	6120T	670HI	6120T
Cylinders	4	V8	6	6	6	6
Aspiration	Turbo§	Natural	Turbo-int.II	Turbo§	Turbo-int.II	Turbo§
Bore (in.)	4.4375	4.625	4.25	5.25	4.25	5.25
Stroke (in.)	5.5625	4.125	5.00	5.625	5.00	5.625
Displ'mt (in.^{3})	344	555	426	731	426	731
Rated rpm	2,200	2,800	2,550	2,550	2,400	2,400
Comp. ratio	INA#	17.0:1	16.0:1	16.25:1	15.5:1	16.25:1
Speeds						
Forward	6 PS	10	20	20	20	20
Reverse	1	2	4	4	4	4
Tires:	20.5-25	23.1-26	18.4-38	18.4-38	18.4-38	18.4-38
	---	---	Duals	Duals	Duals	Duals

* Manufacturer's estimate.
† No power take-off available.
‡ Effective Serial No. 1910 the A-C 3500 engine was installed. Similar to 7580 except 2,200 rated rpm.
§ Turbocharged.
II Turbocharged-intercooled.
Information not available.

Field Performance — Two-Wheel Drive Versus Front-Wheel Assist

Unplowed ground:

Test	Tractor	Time to Use 1 gal (min)	Distance (ft)	acre/gal	gal/acre	gal/h	acre/h	Speed (mph)	Wheel Revs. Loaded	Wheel Revs. No Load	Slip (%)
A	8050 FWA SGL	5.75	1,968	0.62	1.61	10.43	6.48	3.89	133	116	14.9
B	8050 FWA DLS	6.20	1,983	0.63	1.59	9.74	6.05	3.63	130	116	12.1
C	8050 2WD DLS	7.17	1,931	0.58	1.72	8.27	4.83	2.90	136	107	26.5
Plowed ground (Area used in tests A,B, and C):											
D	8050 2WD DLS	6.58	2,171	0.685	1.46	9.12	6.25	3.75	149	127	17.3
	8050 FWA DLS	6.55	2,426	0.765	1.31	9.16	7.01	4.21	152	142	7.0
E	8050 2WD DLS	6.49	2,200	*	*	*	6.42	3.85	150	125.5	19.1
	8050 FWA DLS	6.1	2,200	*	*	*	6.83	4.10	140	125.5	11.5

* Not tested.

Summary:

Test A compared to B: Duals on the front-wheel assist reduced slippage slightly, but not enough to affect performance and economy very much.

Test B compared to C: Front-wheel assist with duals covered 25 percent more acres per hour that 2WD with duals in unplowed round.

Comparing the averages of tests A, B, and C: The 8050 FWA covered 25 to 34 percent more acres per hour on the same amount of fuel, and reduced slippage about 50 percent.

Test D: Front-wheel assist with duals covered 12 percent more acres per hour that 2WD with duals in plowed ground.

Test E: 8050 FWA covered 6 percent more acres per hour than 2WD.

Nebraska Test Results and Specifications for the Allis-Chalmers Imported Tractors

	ED40	One-Sixty	6040	5040	5045	5050	5015	5020	5030	6140
Test No.	None	1028	None	1230	None	1240	None	1270	1271	1445
Year	---	1969	---	1976	---	1977	---	1978	1978	1982
Engine hp	41*	---	---	---	---	---	18*	25*	31*	---
Max. PTO hp	33.7*	40.36	40.36*	40.05	44.0*	51.46	15*	21.79	26.42	41.08
Max. Dbr hp	---	36.71	36.71	35.23	---	43.29	---	17.37	22.47	33.31
Weight (lb)										
Ballasted	---	6,070	---	6,240	---	7,690	---	3,358	3,876	6,195
Unballasted	---	4,680	---	4,060	---	4,410	---	2,060	2,530	4,460
Shipping wt.										
Two-wheel	3,625	4,250	4,250	3,850	3,850	4,150	1,387	1,900	2,280	4,228
Front assist	---	---	---	---	---	4,740	1,557	2,150	---	4,628
Engine make	Standard	Perkins	Perkins	UTB†	Fiat	Fiat	Toyosha	Toyosha	Toyosha	Toyosha
Cylinders	4	3	3	3	3	3	3	2	2	3
Displ'mt (in.^{3})	137.9	152.7	152.7	142.8	158.6	167.8‡	61.3	77.0	90.2	142.0
Rated rpm	2,250	2,250	2,250	2,400	2,600	2,700	2,500	2,500	2,500	2,500
Comp. ratio	INA§	17.4:1	18.5:1	17.0:1	17.0:1	17.0:1	22.5:1	23.0:1	23.0:1	23.0:1
Speeds std/opt.										
Forward	8	10/6	10	6/9ll	8	8/12	9	12	12	10
Reverse	2	2/1	2	2/3	2	2/3	3	3	3	2
FWA avail.	No	No	No	No	No	Yes	Yes	Yes	No	Yes

* Manufacturer's estimate.

† Universal Tractor Brasov.

‡ Early tractors — 158.6 in.^{3}

§ Information not available.

ll 1979 and later — 8/12 speeds.

Serial Numbers and Import Quantities for Allis-Chalmers Imported Tractors

Serial Numbers*

	5040	5045	5050	5020	5030	5015	6140	
1976	408445-	---	---	---	---	---	---	
1977	462148-	---	573461-	1001-2219	---	---	---	
1978	410384-	---	579832-	2220-3090	1001-2004	---	---	
1979	473000-	---	584000-	3091-4114	2005-2254	---	---	
1980	474000-	---	591000-	4115-5789	2255-2975	---	---	
1981	---	988501-	596014-	5790-7033	2976-3519	---	---	
1982	---	---	597730-	7034-8387	3520-4065	1001-1726	1001-1446	
1983	---	---	599290-	8388-8733	4066-4213	1727-3276	1447-1850	
1984	---	---	---	8734-9216	4214-4368	3277-4231	1851-2710	(6140)
1985	---	---	---	9217-INA	4369-INA	4232-INA	2711-2714	3545-3732

* No records of ED40, One-Sixty, 160, or 6040.

Quantities Imported:

	ED40	One-Sixty	160	6040	5040	5045	5050	5020	5030	5015	6140
1964	450	---	---	---	---	---	---	---	---	---	---
1969	---	1,655	---	---	---	---	---	---	---	---	---
1970	---	1,460	---	---	---	---	---	---	---	---	---
1971	---	995	---	---	---	---	---	---	---	---	---
1972	---	---	587	---	---	---	---	---	---	---	---
1973	---	---	900	---	---	---	---	---	---	---	---
1974	---	---	---	499	---	---	---	---	---	---	---
1975	---	---	---	---	411	---	---	---	---	---	---
1976	---	---	---	---	1,630	---	405	---	---	---	---
1977	---	---	---	---	1,772	---	1,505	1,218	---	---	---
1978	---	---	---	---	INA*	---	INA*	870	1,004	---	---
1979	---	---	---	---	INA*	---	INA*	1,024	250	---	---
1980	---	---	---	---	INA*	---	1,023	1,675	721	---	---
1981	---	---	---	---	---	599	526	1,244	544	---	---
1982	---	---	---	---	---	---	520	1,354	546	726	446
1983	---	---	---	---	---	---	73	346	148	1,550	402
1984	---	---	---	---	---	---	---	483	155	955	860
1985	---	---	---	---	---	---	---	INA*	INA*	INA*	190
Total	450	4,110	1,487	499	6,000†	599	6,500†	8,500†	3,500†	3,750†	1,898

* Information not available.

† Total is estimated due to lack of information.

Springfield Crawler Tractors

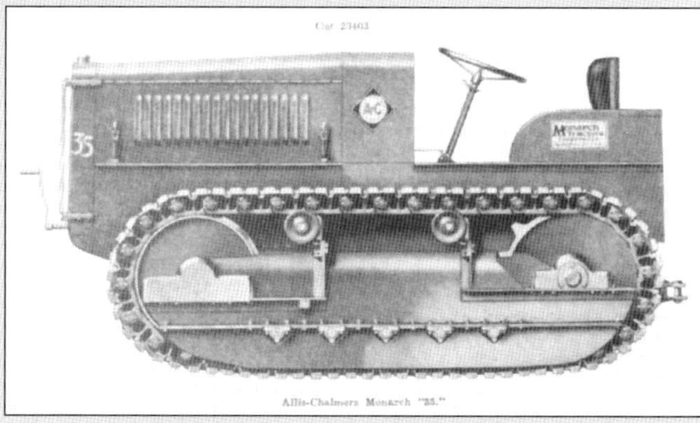

The Monarch 35 was the first new crawler to carry the Allis-Chalmers name. Its A-C engine had 461 in.^{3} and developed 41-drawbar hp.

This Monarch 50 with bulldozer was working on a new drive along the Hudson River (dated January 1929).

There was a time, before wheel tractors grew up, when nothing but a good crawler could handle the big tough jobs. One of those tough jobs involved sticking to the steep slopes of the Palouse country in eastern Washington and Idaho. Where the fields in Montana were as big as the sky they were under was another. Where West Coast soils were so heavily compacted by constant irrigation that they would yield to no other force, except for perhaps an occasional earthquake. Where the fields were pock-marked with hundreds of sinkholes that would not dry up, and the season was short. Where a farmer wanted to pull something larger than a four-bottom moldboard plow. Where delicate soil structures insisted upon a light step. In these conditions the farm crawler ruled the fields.

Sure, crawlers had a few shortcomings, too. The highway folks detested them. They weren't easy to move from farm to farm. Abrasive soils ate their track rails, pins, bushings, rollers, and sprockets, none of which sold for pocket change. They were slow, but they could also work wide and deep. But until someone figured out how to make the four-wheelers work, crawlers were the best way to get the tough jobs done.

During the development years, about as many tracklayers as wheel tractors were offered to farmers. Paging through R. B. Gray's *The Agricultural Tractor 1855-1950* reveals an astonishing number of methods and arrangements for transmitting power to the ground. Until Allis-Chalmers pioneered the low-pressure pneumatic tractor tire in 1932, tracks probably were a better solution than wheels. Steel wheels won over tracks primarily on cost and simplicity, certainly not on superior performance. Perhaps the best of both worlds would be attained as rubber-tracked crawlers found broad acceptance in agriculture.

The Monarch Tractor

In 1926, the Company formed the Tractor Division with its own sales and distribution system. The new division needed growth opportunities. A related product line, such as crawler tractors, could lead to participation in both the earthmoving equipment and agricultural markets. The Monarch Tractor Corporation of Springfield, Illinois, was in financial straits and A-C was able to buy it for half a million dollars in 1928 — a real bargain. Harry Merritt, the Tractor Division manager, was instrumental in this acquisition.

The product line consisted of three models, the C25-35, which was replaced by the 35; the 10-ton F, later referred to as the F75 or 75; and the 6 ton, which would also be known as the 50. The 6-ton (50) Monarch tractor was powered by a Stearns engine (A-C acquired Stearns in 1930).

The annual Road Show for construction machinery was held in Cleveland, Ohio, in January 1929, and the three Monarch crawlers were placed on display. According to Harry Hoffman, advertising manager at the time, the feedback was very negative. He explained that Harry Merritt was disheartened to learn that the Monarch tractors were too expensive,

A Monarch 50 tractor clearing the way for a new highway at Yonkers, N.Y. (from a January 1929, company sales bulletin).

If Paul Bunyan were here he would trade his big blue ox for a Monarch 75 logger. This Monarch is going after a "Monarch of the woods" in the New Mexico Rockies (from a January 1929, sales bulletin).

The Model L was the big one. The six-cylinder engine had 844 in.^{3} First tested at 92, it later developed nearly 109-belt hp. It weighed 22,400 lb and sold for $4,835 with lights and starter in 1934.

too slow, and not competitive. He then withdrew the whole line from sales and started on a project to redesign the tractors to be competitive. The tractors had open or partially-enclosed chain and sprocket final drives and differential and brake steering, neither of which were competitive with tractors having enclosed final drives and clutch steering. Monarch tractors had steering wheels which also separated them from major competitors with lever steering.

The first new tractor based on the redesign was the Model 35. It was described as being lighter, faster, stronger, and lower-priced than competitors. Two larger redesigned models quickly followed. With these important changes the tracklayers were back in business. The Monarch tractors laid the foundation for the Allis-Chalmers line of crawlers which anchored the Construction Machinery Division line and were a valuable addition to the Farm Equipment Division's wheel tractors.

Allis-Chalmers Crawlers with Spark-Ignition Engines

New models bearing the Allis-Chalmers name started flowing out of Springfield. The Model L replaced the 75, and was tested at Nebraska in 1932. The Special K replaced the Model 35 and was tested in 1933. The Model M was a new size at 35.05-belt horsepower. The name Monarch would never be used again.

The Models LO and KO tractors appeared in a 1934 price list, but didn't make an appearance at the Nebraska tests until 1937. By that time the Model SO had been added to the line. The "O" tractors had the "Oil" engine, which had spark ignition with fuel injection, a relatively high (for that time) 6.5:1 compression ratio, and burned diesel fuel. Only air entered the combustion chamber on the intake stroke. A Bosch fuel pump injected a measured charge of fuel into the cylinder, and a high-tension magneto provided electricity to the spark plug for ignition. Easier starting and reduced maintenance were claimed advantages. The same engines were supplied as the LO and EO power units.

The LO developed about the same power as the previously tested L, with 91.56 and 91.93 horsepower, respectively. The WKO had 4 more belt power than its gasoline-powered counterpart, the Special K. The "O" powered tractors delivered about 20 percent more horsepower hours per gallon than their gasoline or distillate-fueled counterparts, so their 27- to

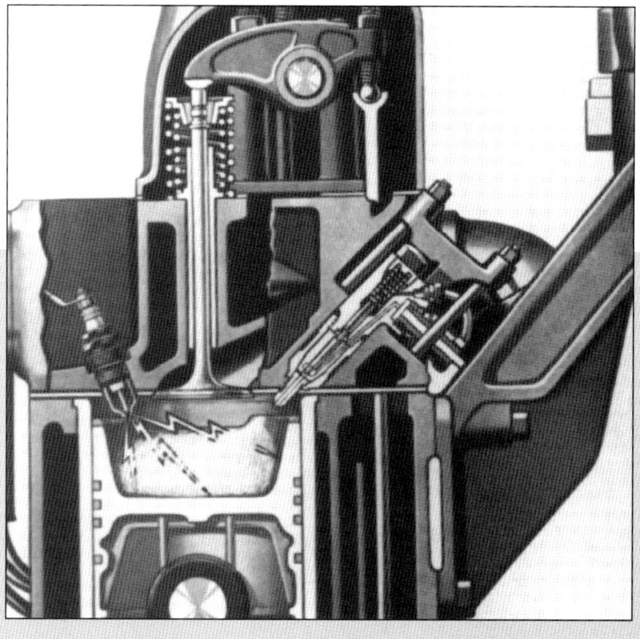

The oil engines were fuel-injected with spark ignition and burned diesel fuel. The compression ratio was 6.5:1. Easier starting was a claimed advantage. The oil engines were soon replaced by diesels.

A Model M with a "side-saddle" operator's station. Application was probably for a pipe-laying boom.

The Model SO had the same displacement as the Model S gas-burner but had 10 fewer belt horsepower. More economical, it delivered 20 percent more horsepower-hours per gallon of fuel.

This Model KO is pulling a Heil Dig-n-Carry RS4-40 scraper near Springfield, Ill., in 1937. The KO oil engine had 563 in.^{3} and delivered 49-drawbar hp. The KO had about 5-percent less power than the K, but burned 26 percent less fuel. It sold for \$3,150 which was \$750 more than the K.

30-percent price premium could be justified. At best the oil engines proved to be only an interim step on the way to diesels.

There is no indication that Allis-Chalmers attempted to manufacture diesel engines, but if Springfield tracklayers were to become a major competitor, diesel power was inevitable. In 1938, 10 Model LD tractors were built at Serial Nos. 8-17, and appear to have been a pilot run of diesel-powered tractors. One account tells that in 1938 and 1939, 16 experimental General Motors diesel-powered tractors were tested on earthmoving jobs in Harlingen, Texas, and on the Pennsylvania turnpike, as well as in West Coast orchards.

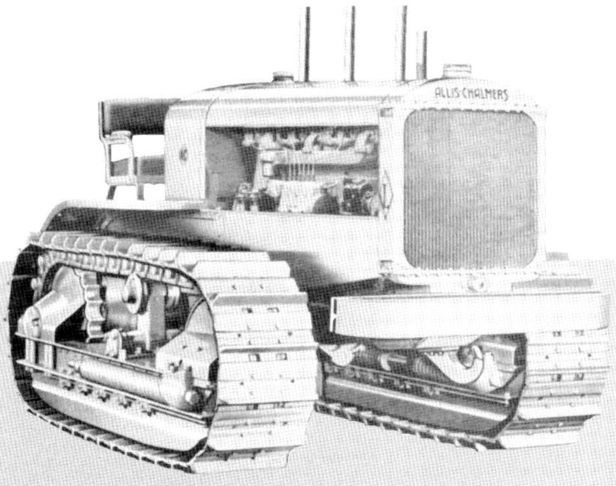

The Model LO (oil engine) delivered 91-belt and 76-drawbar hp. The 844-in.^{3} six turned only 1,050 rpm.

The Model M had a four-cylinder 300-in.^{3} engine and it developed 35-belt hp on either gasoline or distillate. In 1936 it received the 318-in.^{3} engine but was not retested. It sold for $1,650 and the wide tread for hillsides was $100 more.

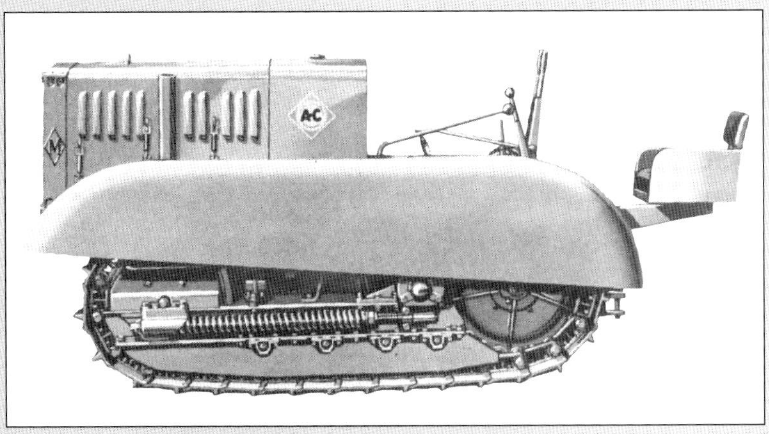

The Model M orchard tractor had a low operator's station and full fenders to protect low branches and fruit, but was mechanically similar to the standard model.

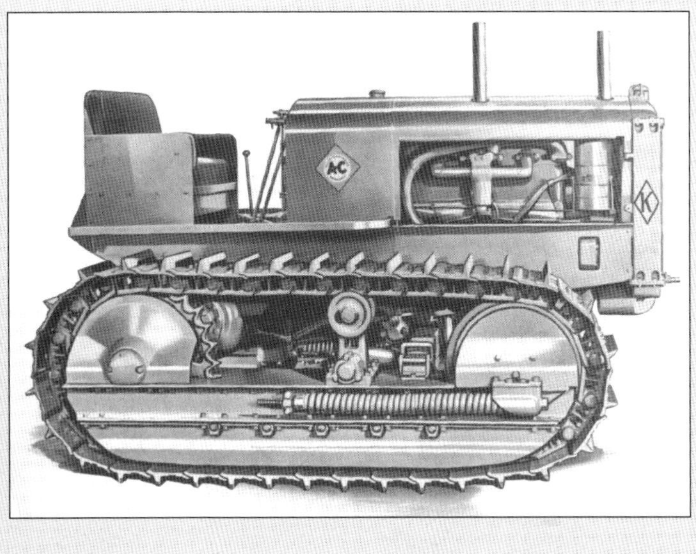

The 62-hp Model K had a 510-in.^{3} four-cylinder engine. It weighed 11,000 lb and had a 1934 list price of $2,400.

The HD14 was the first of the GM-powered crawlers. The 6-71 engine developed 145-belt hp. It pulled 97 percent of its own weight of 28,750 lb. Its construction counterpart, the HD14C, had a torque-converter drive.

The HD10 used the 4-71 engine and developed 98-belt and 82-drawbar hp. It had a six-forward, two-reverse transmission and weighed 21,630 lb. The HD10W had a 74-in. tread or gauge.

The HD7 had the 3-71 engine and produced 68-belt and 57-drawbar hp. The transmission was a four-forward, one-reverse sliding-gear type, and the tractor weighed 14,175 lb. The HD7W was a wide-gauge version used on hillsides.

Equipped with the 2-71 engine, the HD5 had 48-belt and 38-drawbar hp. It had a new power train with a five-speed transmission and double-reduction final drives.

The HD9 had a 4-71 engine that developed 79-belt and 67-drawbar hp. The transmission was a new constant-mesh six-forward, three-reverse unit. The HD9 weighed 19,945 lb.

The HD15 with the 6-71 engine had 117-belt and 104-drawbar hp and shared the six-speed transmission with the HD9. It weighed 30,895 lb.

The HD20 cross-section shows the double-reduction final drive used by all new models after 1946. Final drive gears were smaller and stronger. Clearance under the tractor centerline was increased.

The HD19H with 129-belt hp replaced the HD14. It had torque-converter drive with a two-forward, one-reverse transmission and the double-reduction final drives. It weighed 40,395 lb.

The HD Series with GM Engines

Production of the HD line with GM two-cycle diesels began in 1939. The GM diesels were two-stroke cycle engines, which meant that each cylinder had a power stroke on every revolution of the crankshaft, as compared to the four-stroke cycle which fires each cylinder on every other revolution. Since there was no intake stroke in a two-stroke cycle, an inherent shortcoming was the lack of a full charge of intake air which would greatly reduce efficiency and power if it were not compensated in some manner. The GM engines were equipped with Roots blowers which supplied a large volume of air to the combustion chamber through the intake port as the piston neared the bottom of its power stroke. This also helped purge the combustion chamber of exhaust gases. The Roots blower was geared to the engine crankshaft, similar to a supercharger. Performance of the GM-powered tractors was very good and fuel efficiency in general was slightly better than the oil engines and naturally-aspirated diesels, but somewhat less than later turbocharged diesels.

The HD14 went into production in late 1939, followed shortly by the HD10 and HD7. In World War II the military versions of the HD10 and HD7 went to war. The HD7, HD10, and HD14 tractors served the middle of the market, but both larger and smaller tractors were needed. Delayed by the war, the HD5 and HD19 became part of the line in 1946.

In 1950, the HD15 replaced the HD14, while the HD9 displaced the HD7, and the HD19 was upgraded to the HD20. These, and the HD5, made up the Allis-Chalmers crawler line until 1955 when they were all replaced by a new line — completely powered by A-C's own engines. The GM-powered tractors were all used extensively in farming. The "W" wide gauge models were especially well adapted to working on the hills of eastern Washington and Idaho. The HD19 and HD20 torque-converter models were not great open-field tractors except for very deep subsoiling

The HD20 replaced the HD19H and was powered by the GM 6-110 660-in.^{3} engine. It had torque-converter drive and a two-forward, one-reverse gearbox. At Nebraska it pulled 97 percent of its 42,625 lb.

"During the War They Even Fought on Tractors"

This letter appeared in the February 9, 1993, issue of Izvestia, the prominent Russian newspaper, and is quoted as translated:

Our magazines are publishing an epic called "The Red Wheel" by A. Solzhenitzen. In it I read: Arriving from the front to Petrograd in 1916, the Colonel Varatinzev is being asked many questions about the war, Such as, "How do the tractors 'Allis-Chalmerse' travel the roads?" It's interesting that those tractors were also used in WWII.*

In February, 1943, I was named the commander of the division of the Heavy Artillery Unit. In order to pull canons the division received fifteen American "Allis-Chalmerse" which are reliable, easily operated, and attractive; maneuverable HD7 and HD10.

In June we already departed for the front of the North of Kursk, and until the end of the war these tractors served us without fail through winter and summer, on the roads and off, of Russia, Ukrain, Belarussia, Poland, and Germany. Due to the reliable construction, the tractors withstood long night marches, overlooking the spring and fall thaws. As well as the unusual winter and spring storms of 1944 when the snow drift peaked the tops of houses.

These tractors had one disadvantage, they were made for agricultural purposes and lacked a cabin. This wasn't so upsetting, many of us had started war on horse wagons and simply switched from saddle to the seat of a tractor.

This has all been brought to mind due to the fact that it is the upcoming fiftieth anniversary of the Kurskaya Dooga battle, and if the company "Allis-Chalmers" exists we wish it good luck and success.

M. Fooks, World War II veteran, Perm City, Russia.

* Writer's note: There were no Allis-Chalmers crawlers in 1916, but the company sold a number of the Tractor-Trucks to Russia during World War I.

and reclamation projects which called for turning the soil with huge single-bottom plows capable of working several feet deep.

The HD Tractors with A-C Engines

The acquisition of the Buda Company in 1953 brought the diesel engine technology necessary if A-C was to have complete control of its construction machinery and farm equipment lines. The first available large diesels were the four-cylinder D344, the six-cylinder D516, and the six-cylinder D844 and DS844 engines, all with the Lanova (energy cell) combustion system. They were installed in existing crawlers to provide the HD6B, HD11B, HD16A, HD16AC, and HD21AC tractors. Models with "C" had a torque-converter drive and were built specifically for bulldozing and other earthmoving applications.

In the early 1960s the A-C engines were redesigned to direct-injection open-chamber type, a feature that provided easier starting and improved fuel economy. They were known as the Thousand Series and the 6000 was installed in the HD6, the 10000 in the HD11, the 16000 in the HD16, and the 21000 in the HD21.

This HD6E is equipped with an Oxnard toolbar and 900 three-bottom two-way plow. The tractor had a "hard nose" with built-in dozer lift-cylinder mounts.

The HD16 farm crawler had an 844-in.^{3} naturally-aspirated six-cylinder engine to crank out 160-engine and 137-drawbar hp. Maximum pull was 40,700 lb.

◀ *An HD11 Series B farm crawler working with an Oxnard 600 tool carrier and five subsoilers. The tractor had a 516 in.^{3} six that produced 126-engine and 106-drawbar hp.*

▲ *An HD16 Series B farm crawler had a turbocharged 844-in.^{3} engine with 172-engine and 148-drawbar hp. The HD11 and HD16 Series B tractors were the last farm crawlers, about 1972.*

▶ *The turbocharged HD6AG had 90-engine hp, about 20 percent more than the HD6B, which improved performance in the hills and higher elevations of the Palouse. The A-C 7000 engine had 344 in.^{3}*

The AG Models

Up until around 1965, A-C crawlers used for farming were little different from the standard models. Most farmers neither needed nor wanted all the heavy duty track equipment or extra-heavy radiator grills normally supplied for dozer tractors. Many of them, especially those in the hills, needed more power to top the crests with a wide hitch of rod weeders or treaders. To meet this need the AG models with special specifications were introduced. The HD6AG had a turbocharger added and engine horsepower jumped from 75 to 90, a 20 percent boost. The HD11AG also got a turbocharger and engine power leaped from 99 to 130, a 30 percent improvement. Heavy duty track guards were replaced with light guards. Hard noses with dozer attachment points gave way to much lighter, but still very durable radiator grills.

From the early days of farm crawlers, BeGe hydraulic systems were used for implement control. With a PTO-driven hydraulic pump, these relatively low-cost units were self-contained with their own oil tank and controls. There were two units, the 1000 and the 1400. Prices ranged from $300 to $450, depending the tractor. BeGe equipment was sold through the Oxnard plant.

As everyone became more cognizant of the need for improved safety in mobile equipment, the tracklayers were equipped with massive Roll Over Protective Structures (ROPS) as standard equipment. The ROPS also provided a good substructure for a cab, thus providing more operator comfort as well as safety. All models were sold for farm applications. Some sales history follows:

Year	HD6	HD11	HD16	HD21
1964	101	81	31	1
1965	112	82	36	3
1966	126	101	35	2
1967	92	103	25	0
1968	70	100	28	0
1969	70	61	20	0
1970	35	26	3	0

The downward trend highlights the impact of the large four-wheel drive tractors on crawler sales. Until the late 1960s, the company could afford to be somewhat complacent about the four-wheelers because of its strong Springfield sales but the trend was impossible to ignore. Significant farm crawler sales would soon be history. As the company reorganized into producing-plant profit centers, the Construction Machinery Division took over crawler sales, resulting in the Farm Equipment Division losing that income. After the formation of Fiat-Allis in 1974, dealers had to have a Fiat-Allis contract to get tractors. These changes were enough to discourage some dealers from trying to stay in the crawler business, and as they watched the big four-wheelers move in, many turned to other companies who had four-wheel drive tractors for something to sell. It was enough to make the farm division get serious about four-wheel drive tractors — not a minute too soon.

Nebraska Test Results and Specifications for Monarch Tractors						
Make	Monarch	Monarch	Monarch	Monarch	Monarch	Monarch
Model	50	35	6 Ton	10 Ton, F	C 25-35	D 6-60
Test No.	179	171	147	139	113	108
Year	1930	1929	1927	1927	1925	1924
Serial No.	61999	39277	60172	10191	30 20	6044
Fuel	Gas	Gas	Gas	Gas	Gas	Gas
Max. belt hp	62.18	No test	No test	No test	43.67	60.21
Max dbr. hp	53.28	40.99	50.55	78.17	37.59	53.25
Test weight (lb)	15,100	10,680	13,755	21,700	10,630	16,789
Max. pull (lb)	10,573	8,450	10,537	20,050	6,680	12,922
Percent pull	70	79	76	92	63	77
Speed at max. pull (mph)	1.78	1.82	1.80	1.40	2.11	1.50
Engine make	A-C	A-C	Stearns	Beaver-LeRoi	Beaver	Beaver
Cylinders	4	4	4	4	4	6
Bore(in.)	5.25	4.7	5.12	6.50	4.75	4.7
Stroke (in.)	6.50	6.50	6.50	7.00	6.00	6.00
Displ'mt (in.^{3})	563	460	536	929	425	637
Rated rpm	1,000	930	1,000	850	1,200	1,200
Speeds						
Forward	2	3	3	3	3	3
Reverse	1	1	1	1	1	1

Serial Numbers for Monarch Crawlers					
	F-10 Ton	F-75	G-5 Ton	H-6 Ton	H-50
1926	10001-10081	---	7001-7090	---	---
1927	10082-10240	---	7091-7117	60001-60206	---
1928	10241-10250	70001-70207	---	60207-60297	60298-60610
1929	---	70208-70654	---	---	60611-61558
1930	---	70655-70984	---	---	61559-62146
1931	---	70985-71066	---	---	62147-62297

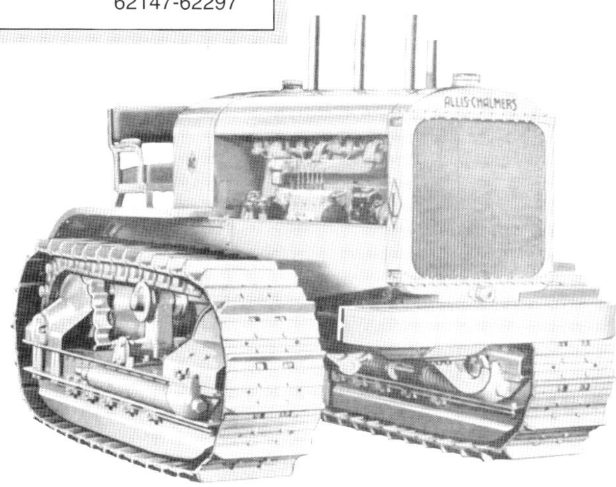

Nebraska Test Results and Specifications for Allis-Chalmers Crawlers with Spark-Ignition Engines

	L	WS	WK	LO	SO	WKO	M	M	SPL.K	L
Test No.	338	337	336	287	286	285	239	216	215	200
Year	1939	1939	1939	1937	1937	1937	1935	1933	1933	1932
Serial No.	L3133	WS901	WK8276	LO2459	S0218	WKO6989	M3120	M255	2848	L69
Fuel	Gas	Gas	Gas	Diesel	Diesel	Diesel	Dist.	Gas	Gas	Gas
Max. belt hp	108.84	84.34	62.22	91.56	74.82	59.06	35.05	35.43	55.24	91.93
Max. dbr. hp	91.99	68.86	53.60	76.75	62.39	49.26	28.66	29.65	47.87	76.01
Test weight (lb)	26,105	20,330	13,340	24,925	20,100	13,000	6,855	6,620	11,670	22,027
Max. pull (lb)	23,851	17,843	11,785	20,273	16,732	11,685	5,576	5,166	8,865	15,086
Percent pull	91	88	88	81	83	90	81	78	76	68
Speed at max. pull (mph)	1.42	1.44	1.65	1.41	1.41	1.62	1.56	2.15	2.03	1.88
Engine make	A-C	A-C	A-C	A-C	A-C	A-C	A-C	A-C	A-C	A-C
Cylinders	6	4	4	6	4	4	4	4	4	6
Bore (in.)	5.25	5.75	5.00	5.25	5.75	5.25	4.38*	4.38*	5.00	5.25
Stroke (in.)	6.50	6.50	6.50	6.50	6.50	6.50	5.00	5.00	6.50	6.50
Displ'mt ($in.^3$)	844	675	510	844	675	563	300	300	510	844
Rated rpm	1,050	1,050	1,050	1,050	1,050	1,050	1,200	1,200	1,050	1,050
Comp. ratio	5.2:1	5.0:1	5.2:1	6.5:1	6.5:1	6.5:1	4.2:1	4.7:1	5.2:1	5.2:1
Carburetor (in.)	(2)1.125	1.5	1.5	DNA†	DNA†	DNA†	1.25	1.25	1.5	(2)1.5
Speeds										
Forward	6	5	4	6	5	4	4	4	3	6
Reverse	2	1	1	2	1	1	1	1	1	2
Tracks										
Links	35	32	34	35	32	34	31	31	34	INA‡
Shoes (in.)	20	18	18	20	18	18	12	12	18	16
Gauge (in.)§	68	62/74	48/63	68	62/74	48/63	40/50	40/50	INA‡	68
Rollers	5	5	5	5	5	5	4	4	5	5

* Model M bore increased to 4.50 in. at Serial No. 2942 in 1936.

† Does not apply.

‡ Information not available.

§ Standard/wide gauge — "W" tractors have wide gauge.

Serial Numbers for Spark-Ignition Crawlers

	K/KO		L/LO		M	S/SO
1929	1-48	---	---	---	---	---
1930	49-1372	---	---	---	---	---
1931	1373-2333	---	1-34	---	---	---
1932	2334-2653	---	35-498	---	1-41	---
1933	2654-3045	---	499-660	---	42-401	---
1934	3046-3593	6001-6018	661-887	4001-4021	402-841	---
1935	3594-4792	---	888-1438	---	842-1941	---
1936	4793-	6019-6335	1439-2136	---	1942-3841*	---
1937	---	6336-7707	2137-2705	---	3842-7066	3-412
1938	---	7708-8017	2706-2943	LD	7067-8126	413-584
1939	---	8018-8522	2944-3232	8-17	8127-9539	585-1084
1940	---	8523-8956	3233-3251	---	9540-11379	1085-1127
1941	---	8957-9268	3252-3272	---	11380-12946	1128-1211
1942	---	9269-9393	3273-3357	---	12947-14524	1212-1227
1943	---	9394-9468				

* Model M cylinder bore increased to 4.50 in. at Serial No. 2942.

Serial Number Information for Allis-Chalmers Crawlers with GM Engines

	HD14,HD14C	HD7,HD7W, Military	HD10,HD10W, Military	HD10 Continued (Showing Overlapped Serial Numbers)		
1939	18-25	---	---	---	---	---
1940	26-548	3-502	2-641	---	---	---
1941	549-1165	503-1136	642-1445	---	---	---
1942	1166-2112	1137-2980	1446-2081	2452-2855	---	---
1943	2113-3137	5403-5752	2082-2338	2856-3651	---	---
1944	3138-4258	5753-8948	2339-2451	3652-4094	4152-4336	4452-4896
1945	4259-5454	8949-12090	---	4095-4151	4337-4451	4897-5963
1946	5455-5814	12091-13077	---	---	---	5964-6462
1947	5815-6422	13078-15121	---	---	---	6463-7601
1948	---	15122-16751	---	---	---	7602-8675
1949	---	16752-18085	---	---	---	8676-9630
1950	---	18086-18505	---	---	---	9631-10198

	HD5	HD15	HD9	HD19	HD20
1946	1-6	---	---	---	---
1947	7-1357	---	---	4-120	---
1948	1358-4315	---	---	121-1195	---
1949	4316-7498	---	---	1196-2001	---
1950	7499-11070	-1	-1	2002-2653	---
1951	11071-14289	2-810	2-737	---	3001-3827
1952	14290-17557	811-1857	738-1882	---	3828-4922
1953	17558-21836	1858-2855	1883-3590	---	4923-5736
1954	21837-25563	2856-3684	3591-5208	---	5737-6100
1955	25564-29255	3685-3909	5209-5850	---	---

Nebraska Test Results and Specifications for Allis-Chalmers Crawlers with GM Two-Cycle Engines

Model	HD20	HD15	HD9	HD19H	HD5B	HD14	HD10W	HD7W
Test No.	465	464	463	397	396	362	361	360
Year	1951	1951	1951	1948	1948	1940	1940	1940
Serial No.	3275	90	196	452	2104	265	201	176
Fuel	Diesel	Diesel	Diesel	Diesel	Diesel	Diesel	Diesel	Diesel
Max. belt hp	No test	117.68	79.10	129.08	47.85	145.39	98.47	68.68
Max. dbr. hp	116.69	104.37	67.39	110.73	38.24	126.98	82.19	57.29
Weight (lb)	42,625	30,895	19,945	40,395	11,815	28,750	21,630	14,175
Max. pull	41,321	29,400	19,035	37,536	10,059	28,019	19,002	12,171
Speed at max.								
pull (mph)	0.96	1.34	1.30	1.08	1.41	1.63	1.60	1.77
Percent pull	97	95	95	93	85	97	88	86
Engine make	GM	GM	GM	GM	GM	GM	GM	GM
Model	6-110	6-71	4-71	6-71	2-71	6-71	4-71	3-71
Cylinders	6	6	4	6	2	6	4	3
Bore (in.)	5.00	4.25	4.25	4.25	4.25	4.25	4.25	4.25
Stroke (in.)	5.60	5.00	5.00	5.00	5.00	5.00	5.00	5.00
Displ'mt (in.^{3})	660	426	284	426	142	426	284	213
Rated rpm	1,700	1,600	1,600	1,750	1,800	1,500	1,600	1,500
Comp. ratio	18.0:1	16.0:1	16.0:1	16.0:1	16.0:1	16.0:1	16.0:1	16.0:1
Trans. type	TC*	CM†	CM†	TC*	SG‡	SG‡§	SG‡	SG‡
Speeds:								
Forward	2	6	6	2	5	6	6	4
Reverse	1	3	3	1	1	2	2	1
Tracks								
Links	37	38	38	INA‖	33	35	32	34
Shoes (in.)	28	24	20	INA‖	16	22	20	18
Gauge (in.)#	84	74	74	84	60	68	62/74	48/63
Track length (ft)	27.8	25.0	22.2	27.8	19.0	23.0	21.0	18.5
Rollers	6	6	6	6	4	5	5/6	5

* Torque converter.
† Constant mesh.
‡ Sliding gear.
§ Also HD14C — Torque converter, 2 forward, 1 reverse, no test..
‖ Information not available.
Standard/wide — "W" tractors have wide gauge.

The Model S had an enormous four-cylinder 675-in.^{3} engine and pulled 84-belt hp.

Only 30 HD3 (1942) tractors were built. Serial numbers were 3 to 32, with no Serial No. 20 or 30. Specifications are not known — a neat little crawler.

Nebraska Test Results and Specifications for Allis-Chalmers Crawlers with A-C Engines

	HD21A	HD11B	HD6B	HD16A	HD16AC	HD21AC
Test No.	664	581	580	552	551	550
Year	1958	1956	1956	1955	1955	1955
Serial No.	1002	1675	2851	282	280	7302
Max. belt hp	No test	89.75	60.51	133.83	No test	No test
Max. dbr hp	147.18	73.80	49.95	118.69	101.33	135.12
Test weight (lb)	47,455	22,375	13,580	32,375	32,135	44,725
Max. pull (lb)	43,250	20,468	12,636	28,743	29,130	40,563
Percent pull	91	91	93	89	91	91
Speed at max. pull (mph)	1.26	1.33	1.37	1.32	1.35	1.22
Engine make	A-C	A-C	A-C	A-C	A-C	A-C
Cylinders/aspiration	6/turbo	6/nat.	4/nat.	6/nat.	6/nat.	6/turbo
Bore (in.)	5.25	4.44	4.44	5.25	5.25	5.25
Stroke (in.)	6.50	5.56	5.56	6.50	6.50	6.50
Displacement (in.^{3})	844	516	344	844	844	844
Rated rpm	1,825	1,800	1,800	1,600	1,800	1,800
Comp. ratio	14.5:1	15.0:1	15.0:1	14.15:1	14.15:1	13.3:1
Transmission type	TC*	SG†	SG†	SG†	TC*	TC*
Speeds						
Forward	2	6	5	6	3	2
Reverse	1	3	1	3	1	1
Tracks						
Links	40	38	34	38	38	37
Shoes (in.)	26	20	16	22	22	28
Gauge (in.)	84	74	60	74	74	84
Track length (ft)	30.0	22.2	18.5	25.0	25.0	27.8
Rollers (std/opt)	7	6	4/5	6/7	6/7	7

* Torque converter.
† Sliding gear.

The HD21A was an improved version of the HD21AC with 147-drawbar hp. It had the same 844-in.^{3} engine with a higher compression ratio. Maximum pull was 91 percent of its 47,455 test weight. The HD21A was the last A-C crawler tested in Nebraska in 1958.

The HD21AC had 135-drawbar hp produced by an 844-in.^{3} turbocharged six. It pulled 91 percent of its 44,725 lb and had a torque-converter drive.

Specifications for Agricultural Crawlers

	HD6B HD6E†	HD6AG*	HD11B	HD11E HD11AG*	HD11AG* Series B	HD16D	HD16AG* Series B
Max. engine hp	75	90	110	130	126	160	172
Max. belt hp	63	---	---	122	---	---	---
Max. dbr. hp	52	---	77	95	106	137	148
Max. pull (lb)	15,485	---	21,980	29,050	28,750	40,700	39,700
Engine make	A-C	A-C	A-C	A-C	A-C	A-C	A-C
Model	6000	7000	10000	11000	11000	16000H	16000T
Cylinders-Aspiration	4-Nat.	4-Turbo	6-Nat.	6-Turbo	6-Turbo	6-Nat.	6-Turbo
Bore (in.)	4.44	4.44	4.44	4.44	4.44	5.25	5.25
Stroke (in.)	5.56	5.56	5.56	5.56	5.56	6.50	6.50
Displacement (in.^{3})	344	344	516	516	516	844	844
Rated rpm	1,800	1,800	1,800	1,800	1,800	1,600	1,600
Comp. ratio	16.25:1	16.25:1	16.25:1	16.25:1	16.25:1	14.5:1	14.5:1
Speeds:							
Forward	5	5	6	6	6	6	6
Reverse	1	1	3	3	3	3	3
Tracks							
Links std/opt.	34/37	34/37	40	40	40	41	41
Shoes (in.)	13	13	16	16	18	20	20
Gauge (in.)	60	60	74	74	74	74	74
Weight (lb)							
Shipping	12,600	---	21,000	21,700	---	---	---
Field	---	14,000	---	23,000	25,000	---	38,500

* AG versions were equipped for agricultural applications with turbochargers, lighter weight grills and bumpers, and reduced track guard equipment.

† HD6E had power steering, power brakes, raised seat, multi-disk oil clutch.

The turbocharged HD11AG had a six-cylinder 516 in.^{3} that produced 130-engine, 122-belt, and 95-drawbar hp. The transmission was a six-forward, three-reverse constant mesh unit.

An HD16D equipped with a hard nose for dozer installation is compared to an HD16 farm crawler with the light grill. Construction tractors were painted industrial yellow while farm crawlers were painted Persian orange.

Engines and Power Units

The B15 and B125 power units ran the rides at fairs and carnivals. Their quiet, reliable operation made them the favorites of the traveling shows.

Remember when you were young and the carnival came to town? Along with all the games you could play to win a teddy bear for your best girl, there were also rides. A carousel (merry-go-round in Indiana), a Ferris wheel, and maybe a Tilt-a-Whirl, just waiting to take you and your special friend on a thrilling, but brief adventure. When we didn't have a date (which was usually the case) and had to ride with one of the guys, there would be time while waiting for the action to start to notice that there was a little gasoline engine down there amongst the works. It would be just setting there chuckling to itself, waiting. When the ride finally filled up, the greasy handed operator would engage the clutch, the governor would pop the throttle open, and the little engine would bark a few times before settling down to a healthy cackle as it carried us away to another world.

If you had looked closely, you would probably have seen Allis-Chalmers on the hood of the little worker. A-C engines were favorites of the carnivals and perhaps as many as 90 percent of the rides were put through their paces by B15 and B125 power units. Where the show grounds have not yet been electrified, these dependable little motors are still making the wheels go 'round.

Not all Allis power units had such romantic jobs. There was water to pump, grain to grind, logs to saw, and winches to wind. Until the power lines spread like huge spider webs out from the towns and electric motors took over those jobs, the stationary engine was the best way to turn the wheels. Short term power requirements were met by tractor belt power, but for the long term applications, a dedicated power unit provided ready and reliable operation.

The auxiliary engine allowed a farmer to use a larger machine than his tractor could handle on the power take-off, and A-C provided auxiliary engines for All-Crop harvesters, Roto-Balers, forage harvesters, and blowers. The engine product line was always directly related to the tractors. Every tractor engine

The West Allis carbureted power units were adaptable to gasoline, LP gas, natural gas, and low-grade liquid fuels. Enclosed and open power units with cooling systems or fan-to-flywheel engines were furnished to the customer's specifications. The well-known Murphy safety switches were available to stop an engine in case of low oil pressure or high coolant temperature.

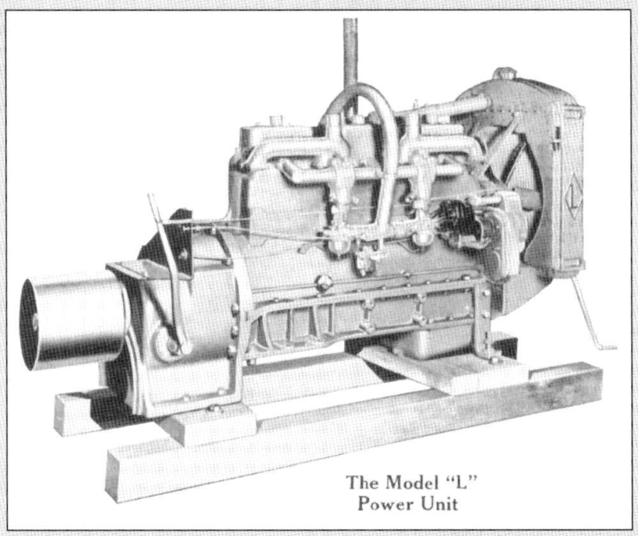

The L and L90 power units were rated at 90-maximum hp, which grew to 110 hp in the L844. These 844-in.^{3} six-cylinder engines were rated at 1,000 and 1,050 rpm.

The U, U40, and U318 displacement was 284, 300, or 318 in.^{3} as the bore grew from 4.25 to 4.375 or 4.50 in., the same as in the tractors. Nominal rating was 40 hp at 1,200 rpm for the U and U40. The U318 rating was 45.2 at 1,200 or 50.9 at 1,400 rpm.

was also offered as a power unit. The company had a long list of OEM customers who had engines modified to their specifications.

The Models L, E, and U Power Units

The Model L power unit was the largest engine in the line. A six-cylinder unit with a bore and stroke of 5.25 × 6.50 inches, it had a displacement of 844 cubic inches. It could develop 90-maximum horsepower, and was later called the the L90. The further upgraded L844 had 110 horsepower and the governed speed was 1,050 rpm, which was relatively slow by modern criteria. The L90 sold for $1,675.

The Model E power unit was the four-cylinder counterpart and had a displacement of 563 cubic inches and was rated at 1,000 rpm. With the 5.25-inch bore, it was called the E60 power unit. With a smaller 4.75-inch bore, displacement was reduced to 460 cubic inches and it was designated the E50. They were rated at 60- and 50-maximum horsepower, respectively. The E60 was priced at $820 and the E50 sold for $750. A later version was called the E563 and developed 74 horsepower. Compression ratios of 4.0:1, 4.4:1, 5.2:1, and 6.0:1 were available.

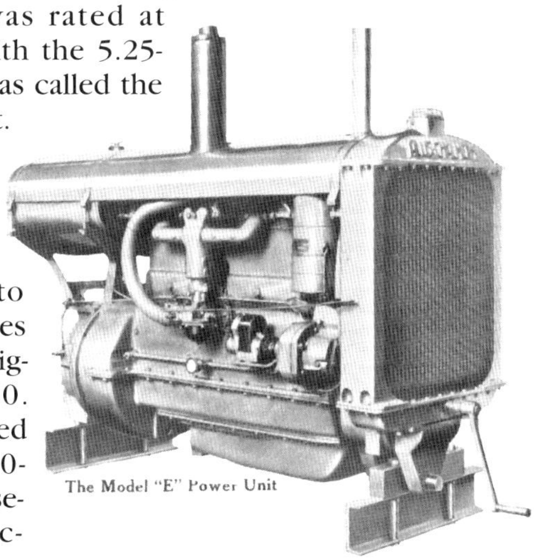

The big four-cylinder power units had a 6.5-in. stroke. The E50 had a 4.75-in. bore, 460 in.^{3}, and 50 hp. The E60 and E563 had a 5.25-in. bore, 563 in.^{3}, and 60 rated hp. They turned 1,000 rpm. Later the E563 rating was increased to 74 hp at 1,050 rpm.

Models LO and EO power units were also available. The "O" (oil) engines were spark ignition types which burned commercial diesel fuel. Fuel was delivered to the combustion chamber by an injector, somewhat like a diesel, but the compression ratio of 6.5:1 was far too low to fire the fuel charge. With a bore and stroke of 5.25 × 6.50 inches, the displacement was 844 cubic inches in the LO and 563 cubic inches in the EO.

Model U Power Unit

The Model U power unit was the same as the Model U tractor engine, therefore the first Model U had the 4.25-inch bore like the Continental engine. Later, when the bore was increased to 4.375 inches, it was called the U40. Finally, with the 4.50-inch bore it was designated the U318. The stroke remained constant at 5.0 inches, so displacement was 284, 300, or 318 cubic inches as the engine grew with the tractor. It was rated at 1,200 or 1,400 rpm. Horsepower grew from 40 to 45, and then to 51 as the displacement increased. Compression ratios of 4.2:1, 4.7:1, 5.2:1, and 5.5:1 were available. The U40 power unit sold for $555 in 1934.

The B15 and B125 Power Units

The B15 and the B125 were the dependable little engines that made the wheels go around at the fairs and carnivals. No doubt a reasonable price got them their jobs, but economical operation and reliable long-lived service helped them keep their positions.

The B15 power unit joined the line about the same time as the Model B tractor in 1937, and shared its specifications. With a 3.25 × 3.50-inch bore and stroke, displacement was 116 cubic inches. The B15 was upgraded to the B125 in the early 1940s with a 3.375-inch bore so displacement increased to 125 cubic inches. The gasoline powered model could develop 28.3-maximum brake horsepower at 1,900 rpm. Intermittent brake horsepower was 25.5 and continuous brake horsepower dropped to 22.6. No doubt those carnival rides required far less power than that, which explains the engine's longevity in these applications.

The B15 and B125 were also used as auxiliary engines on All-Crop harvesters and forage harvesters. They were offered on Roto-Balers, but found little use there since the round baler did not consume much power. Many A-C forage blowers were equipped with auxiliary engines to keep a tractor free for other work.

The B15 had 116 in.^{3} and the B125 was increased to 125 in.^{3} They not only made the wheels go 'round but were used as auxiliary engines on combines and forage harvesters.

The G-138, G-149, and G-160 Power Units

The G-138, G-149, and G-160 power units were all based on the 3.875-inch stroke crankshaft and block. Cylinder bores were: G-138, 3.375 inches; G-149, 3.50 inches; and G-160, 3.625 inches. Displacement was the same as the model designation and they all had the Power-Crater combustion chamber for improved power and economy.

The first use for this engine series was the 149-cubic inch version in the D14 in early 1957. The D14 auxiliary engine followed shortly, and could be mounted on the 60, 66, 72, and 90 All-Crop harvesters, standard and Model 50 forage harvesters, forage blower, and standard and No. 10 Roto-Balers, taking the place of the B125 auxiliary engines. The D14 auxiliary engine had a list price of $685, mounting parts sold for $95 to $195, and the electric starter cost $98.

The G-149 power unit was also offered in 1957. The G-138 power unit was added as the D10 and D12 tractors were launched in 1959. The G-160 came along with the D15 Series II in 1963, and replaced the G-149 at that time. The 160-cubic inch auxiliary engine was listed for the A-C potato harvester. Maximum brake horsepower ratings were: G-138, 39; G-149, 45; and G-160, 52. With a total spread of only 13-maximum brake horsepower, the G-149 was not needed.

Besides the gasoline powered version, models for natural gas, LP gas, and low-grade fuels with a range of compression ratios from 5.2:1 to as high as 9.0:1 were cataloged. Operating speed range was 800 to 1,000 rpm at the low end and 1,800 to 2,000 rpm at the high end, depending on the engine. The G Series power units were phased out as the production of tractors using these engines (the D10, D12, and D15) was discontinued in 1968.

The G138, G149, and G160 power units all had the 3.875-in. stroke crankshaft and block, with ratings of 39, 45, and 52 hp. They were adaptable to gasoline, LP gas, or natural gas fuels.

The engine test line for West Allis engines was a busy place. Engine runs included part-load break-in time and a full-load dynamometer test.

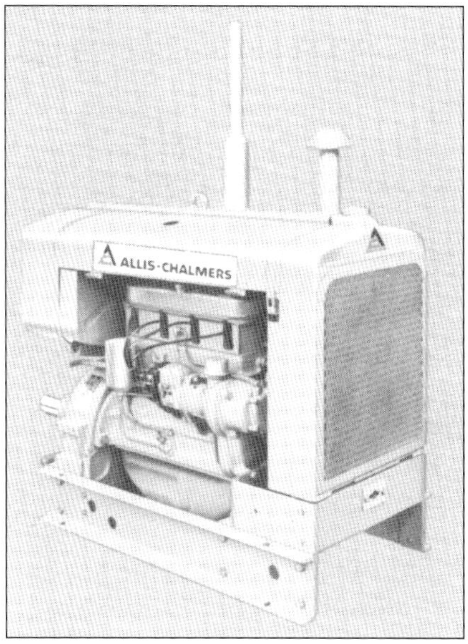

The G226 power unit was the end of the long line of 4-in. bore West Allis engines. Its last rating had grown to 67-maximum brake hp.

The W-25, W-201, W-226, and G-226 Power Units

The W- Series and G-226 power units were based on the 4-inch bore block which began with the Model WC in 1934 and continued until the end of Model 175 gas tractor production in March 1976. During that 42-year period, the engine grew from a modest 21-belt horsepower in the WC to a muscular 67-brake horsepower in the G-226 power unit. Perhaps the only common feature between the two was the 4.0-inch bore, since there had been literally hundreds of changes and improvements incorporated into the engine.

The W-25 power unit originated with the W engine which also powered the WC and WF tractors. The square 4.0- × 4.0-inch bore and stroke displaced 201 cubic inches. An engineering specification listed a W-25 power unit and a W-25 Hi-speed power unit, but neither were defined. This engine was also sold to LeRoi on an OEM basis.

The W-201 power unit shared its background with the WD engine and all its improvements. Still 201 cubic inches, but it had slightly higher rpm ratings.

When the "45" engine, with its 4.5-inch stroke and 226-cubic inch displacement, was developed for the WD45 tractor, the W-226 power unit became available. This brought the new Power-Crater combustion chamber which improved fuel economy about 6 percent and it developed 60-maximum brake horsepower.

The "17" engine provided the basis of the G-226 engine and power moved up to 67-maximum brake horsepower at 1,800 rpm, an increase of over 11 per-

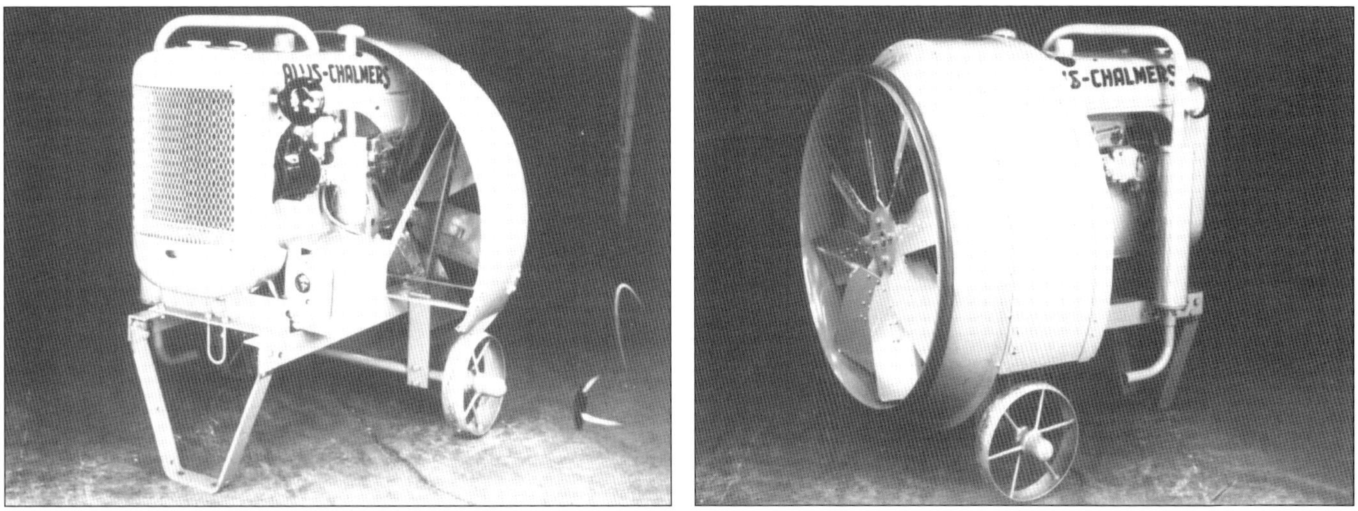

The B15 blower was intended for hay and crop drying but could be used as a power unit by removing the fan. This 1945 experimental unit was never produced.

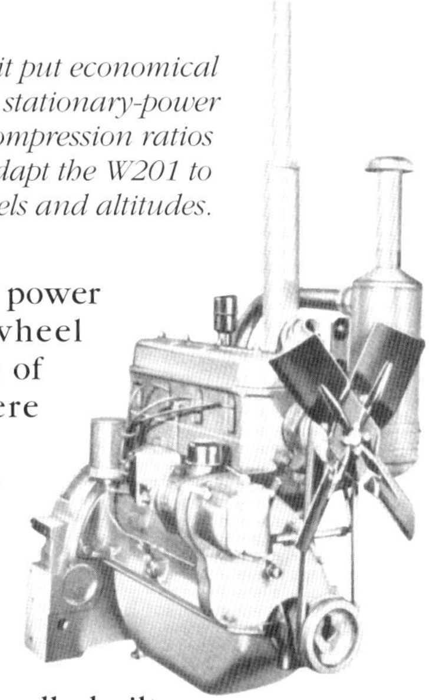

The W201 power unit put economical WC-WD power into stationary-power applications. Several compression ratios were available to adapt the W201 to different fuels and altitudes.

cent. Closed and open power units and fan-to-flywheel engines with a choice of fuel equipment were offered.

When all the tractors, power units, OEM engines, combines, cotton pickers, etc., powered by some version of this well-designed and ruggedly built engine were totaled, approximately 500,000 engines capable of a total of 22,500,000 horsepower were built. At 700 pounds each, that comes to 350 million pounds of engines.

Allis-Chalmers Diesel Engines

A railroad mechanic, George Chalendar, conceived an idea which grew to a leading manufacturer of material-handling equipment and diesel engines. He had worked for the Chicago, Burlington, and Quincy Railroad and knew that section hands travelled to work sites by foot or wagon. His idea was a handcar which allowed the workers to use the tracks to get to their work. He obtained some orders, but needed a place to build them. He bought a plant vacated by the Babcock Reaper Company in Buda, Illinois, which is 97 miles southwest of Chicago. In 1881 he founded the Buda Foundry and Manufacturing Company to produce handcars and other railroad-related equipment. Business was good and the company soon outgrew its 3-acre location in Buda. He found a good place in Harvey, Illinois, to relocate.

The Harvey Land Association had just the right spot. The 6-acre triangular tract was bounded by three railroads: the Illinois Central, the Grand Trunk, and the Baltimore and Ohio. That location could not have been more appropriate. More railroad products followed, but as the automobile industry was rapidly expanding, Buda saw an opportunity. In 1919, Buda jumped into the market with a small two-cylinder vehicle. This was followed by a four-cylinder automotive engine that was used by Hudson, National, and the Apperson Jack Rabbit automobiles.

In 1926 Buda began production of diesel engines and generator sets for industrial, marine, and agricultural applications. In 1938 they introduced the Chore Boy, a small industrial platform-type truck, and this led to the manufacture of lift trucks. When Allis-Chalmers purchased the Buda Company in 1953, it was an important producer of diesel engines, lift trucks, generator sets, and some specialized products related to the railroad industry. By then Buda diesel engines were well established.

Allis-Chalmers needed a captive source of diesel engines for the Springfield crawlers, Cedar Rapids motor scrapers, and a growing requirement in farm equipment. Buda fulfilled that need. Buda engines available at the time of the buyout included the four-cylinder 153 and 182 series engines and the six-cylinder 230 and 273 series engines. Since the company had been doing considerable business in the Southwest, they also offered natural gas and LP gas models. Their farm equipment OEM accounts included Cockshutt, Mercer-Robinson, and Intercontinental. Only Cockshutt was of any consequence, the others did not survive the big tractor-maker shakeout of the early 1950s.

The larger Buda engines were the D344, D516, and D844, the numbers indicating their displacement in cubic inches. These diesels had the Lanova (energy cell) combustion system, which was rapidly becoming obsolete. By the 1960s the engines had been redesigned to the more modern direct-injection type and these became known as the Thousand Series.

The Chore Boy industrial truck of 1938 could do the work of three men with hand trucks. It was the start of a full line of lift trucks.

The ink on the contract was probably still damp when tractor engineering moved to install a 6BD-230 diesel engine in a WD tractor. After the acquisition by A-C, the new 262-cubic inch six-cylinder engine was developed. With a 3.562-inch bore and 4.375-inch stroke, it shared neither with the existing 230 or 273 engines. The diesel version was first used in the D17D in the fall of 1957. The gas and turbocharged diesel powered the D19 tractors, and all three powered various combines and cotton pickers.

In the early 1960s two engine lines were developed:

Series	**Bore × stroke**	**Cylinders-displ'mt**
2000	3.75 × 4.00	4-177, 6-265 in.^{3}
	3.875 × 4.00	4-190, 6-283 in.^{3}
	3.875 × 4.25	4-200, 6-301 in.^{3}
3000	4.25 × 5.00	6-426 in.^{3}

The 2000's were to be offered for gasoline, LPG, and diesel fuels. Larger 3000 diesels were available with natural aspiration or turbocharging. It was soon evident that all these configurations were not required to serve the engine market. As a result, some of the smaller four-cylinder engines were never produced. For instance, if diesel engines smaller than 200- or 301-cubic inches were needed, it proved to be more economical to supply the larger displacement with reduced fuel input. For this reason the 301-cubic inch diesel was used in the One-Eighty, instead of the originally planned 283-cubic inch engine.

All the gasoline and LPG engines were based on the diesel block and were more substantially built than necessary for these fuels. This raised their cost and they were not competitively priced. The gas-burners were the first to be phased out when lower-priced automotive engines filled the market needs. As a result the Gleaner F, G, and K combines were converted to General Motors automotive gasoline engines.

By 1978, most naturally-aspirated diesels had been dropped and metric model designations adopted. For example, the 649I had six cylinders, 4.9 liters, was turbocharged and intercooled. A suffix on the model number indicated the aspiration; no suffix, naturally aspirated; T, turbocharged; LT, lightly turbocharged;

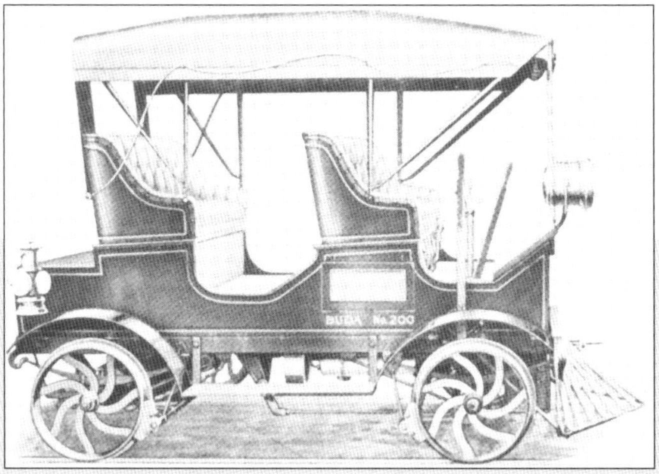

The engine-powered inspection car was among the many Buda railroad products. It all started with handcars for section hands.

This 6BD230 is typical of the engines Buda was building when A-C acquired the company in 1953. There were five sizes ranging from 153 to 844 in.^{3} that could develop from 30 to 197 hp.

After building automotive engines for 16 years, Buda built its first diesel in 1926. The line grew to 42 displacement sizes of gasoline and diesel engines and 193 engine-generator set models by 1940.

The Farm Equipment Division had a pair of cutaway 426-in.^{3} engines which told the A-C engine story at fairs and farm shows. They attracted much attention and were effective sales tools.

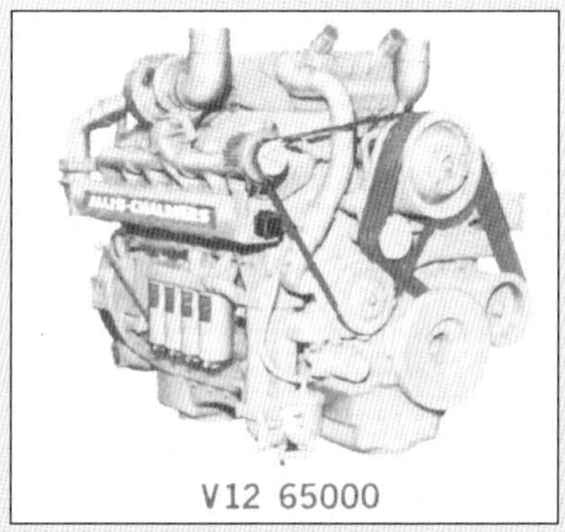

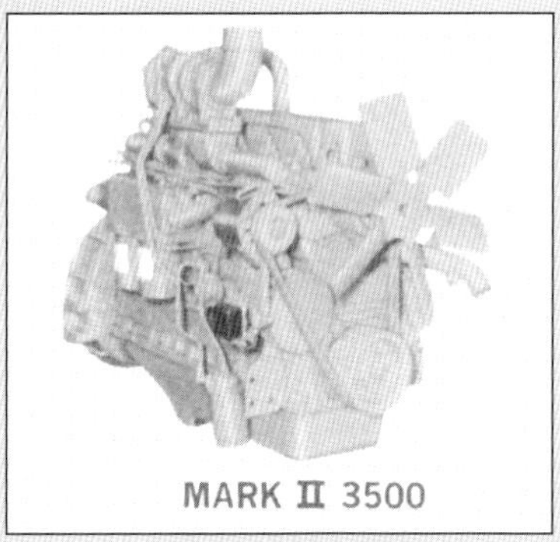

The 2900 Mark II (649T) and 3500 Mark II (670T, 670I) engines were used in 100-plus hp tractors, large conventional combines, and rotary Gleaner combines. They were the workhorses of the ag products.

The 25000 Mark II (6138I) developed 450-maximum brake hp and the huge 65000 turbocharged and intercooled V12 pulled 900-brake hp. There were built for construction machinery and truck applications.

I, turbocharged and intercooled. The engine line that remained was as follows:

Model	Liters	Displ'mt (in.^{3})	Maximum Horsepower	At Rpm
433T	3.3	200	90	2,400
433I	3.3	200	100	2,400
649	4.9	301	85	2,600
649T	4.9	301	135	2,600
649I	4.9	301	155	2,600
670T	7.0	426	175	2,400
670I	7.0	426	200	2,400
685T	8.5	516	220	2,200
685I	8.5	516	275	2,200
6120T	12.0	731	325	2,500
6138LT	13.8	844	325	2,100
6138T	13.8	844	375	2,100
6138I	13.8	844	450	2,100

A bleak future faced these good engines. When the Allis-Chalmers Agricultural Equipment Company was sold to Klockner-Humboldt-Deutz in early 1985, KHD did not purchase the Engine Division. The Harvey plant produced engines for the West Allis tractor plant until late 1985, and for the Independence plant until mid-1986, when it was closed.

The Engine Division had many OEM customers, among them New Idea who used the 3500 Mark II 426-in.^{3} turbocharged diesel in their Model 709 Uni-System power unit. It delivered 160-PTO hp.

Allis-Chalmers Engine Division Products							
Bore & Stroke (in.)	**Cyl.-Asp.***	**Cubic inches**	**Fuel†**	**Horse-power‡**	**Model (s)**		**Farm Equipment Applications Tractors/Combines/Cotton Harvesters**
3.437 × 4.125	4	153	G	---	G153		None
	4	153	D	---	D153		None
	6	230	G	---	6B230		-/ A,R,B
	6	230	D	---	6BD230		WD45D
3.562 × 4.375	4	175	D	49	D175		D15D,HD3,I60D,615LBH,500FL
	6	262	G,LP	---	G262		D19/ C,CH,F/ 616,801,802
	6	262	D	---	D262		D17D/—/ 622,616,500
	6T	262	D	77	D262		D19D/—/ 622,801,802
3.75 × 4.25	4	182	G	---	G182		None
	4	182	D	---	D182		None
	6	273	G	---	G273		None
	6	273	D	---	D273		None
3.875 × 4.25	4	200	G	---	G2200		H4,
	4	200	D	---	2200		HD4,500FL,615LBH
	4T	200	D	90	2300	433T	6060,6070/ K,K2,
	4TI	200	D	100	2350	433I	6080/ F,F2,F3/ 880
3.75 × 4.0	6	265	G	---	G2500		190,180
3.875 × 4.25	6	301	G,LP	---	G2800		190XT/ F,G,GH
	6	301	D	85	2800	649	190,185,180/ F/ 860
	6T	301	D	135	2900	649T	190XT,200,7000,7010,8010/ G,GH,L,L2,L3,M,M2,M3,MH,MH2,MH3
	6TI	301	D	155	2950	649I	7020/ M3
4.25 × 5.00	6	426	D	125	3400		D21
	6T	426	D	175	3500	670T	D21II,210,220,7030,7040,7045,8030/ L2,L3,MH2,MH3
	6TI	426	D	200	3700	670I	7050,7060,8050,8070/ N5,N6
	6TI	426	D	220	3750	670HI	7080,7580,4W220/ N6
4.437 × 5.562	4	344	D	---	HD344	6000	HD6
	4T	344	D	---	7000		HD6E,T16
	6	516	D	---	HD516	10000	HD11B
	6T	516	D	220	11000	685T	HD11E,HD11AG,HD11 SER. B/ N7
5.25 × 5.625	6T	731	D	305	6120T		8550,4W305
5.25 × 6.50	6	844	D	216	HD844	16000	HD16,HD16AC
	6T	844	D	---	16000T		HD16
	6T	844	D	---	16000H		HD16D
	6LT	844	D	325	17000	6138LT	HD16 SER.B
	6T	844	D	---	HDS844	21000	HD21AC
	6T	844	D	375	21000H	6138T	None
	6TI	844	D	450	25000	6138I	None
6.0 × 6.0	V8T	1358	D	552	V8T		None
	V8TI	1358	D	600	V8TI		None
	V12T	2035	D	830	V12T	61000	None
	V12TI	2035	D	900	V12TI	65000	None

* Aspiration: No suffix — natural, T — turbocharged, I — intercooled, LT — lightly turbocharged.

† G — gasoline, D — diesel, LP — LP gas. Most gas engines were available as natural gas power units.

‡ Horsepower varies, horsepower shown is highest for any application.

West Allis Engines with 6.5-inch Stroke

	A	K	E	L	S	EO/KO*	LO*	SO*
Cylinders	4	4	4	6	4	4	6	4
Bore (in.)	4.75	5.0	5.25	5.25	5.75	5.25	5.25	5.75
Stroke (in.)	6.50	6.50	6.50	6.50	6.50	6.50	6.50	6.50
Displ'mt (in.^{3})	460	510	563	844	675	563	844	675
Rpm (normal)	1,000	1,050	1,000	1,050	1,050	1,000/1,050	1,050	1,050
Compression ratio-mark:†								
Distillate	---	4.0-K	4.0-K	4.0-K	---	---	---	---
Gasoline	4.7-G	4.4-G	4.4-G	4.4-G	6.0-G	---	---	---
Dist. (high alt.)	---	5.2-A	5.2-A	5.2-A	5.0-A	---	---	---
Gas (high alt.)	---	5.8-B	6.0-B	6.0-B	---	---	---	---
Diesel fuel	---	---	---	---	---	6.5-EO	6.5-LO	6.5-SO
Applications:								
Tractors	15-30	E25-40	E25-40	L	S	KO	LO	SO
	18-30	K	TW	---	---	---	---	---
	20-35	A	---	---	---	---	---	---
	A	54 Patrol	---	---	---	54KO Patrol	---	---
Power units	---	UE Cooper	UE Cooper	---	---	---	---	---
	E50	---	E,E60	L,L90	---	EO	LO	---
	---	---	E563	L844	---	---	---	---

* Engines with "O" have spark ignition and burn diesel fuel.

† Letter on the engine serial number identified the compression ratio.

West Allis Engines with 5.0-inch Stroke

	Cont'l	UM	UMA
Cylinders	4	4	4
Bore (in.)	4.25	4.375	4.5
Stroke (in.)	5.0	5.0	5.0
Displ'mt (in.^{3})	284	300	318
Rated rpm	1,200	1,200	1,200
Compression ratios-mark:			
Distillate	INA*	INA*	4.2-K
Gasoline	---	---	4.7-G
Dist. (high alt.)	---	---	5.2-H
Gas (high alt.)	---	---	5.5-B
Applications:			
Tractors	United	U 7405-	U 12001>
	U prior	12000	---
	7405	UC 1269-	UC 2282>
	UC prior	2281	IU
	1269	M prior	M 2942>
	---	2942	42 S.Patrol
Power units	U	U40	U318
	---	---	UM Cooper

* Information not available.

Specifications for the Model B15 Blower*

Bore and stroke (in.)	3.375 × 3.50
Displacement (in.^{3})	125.3
Rated rpm	1,400
Operating range (rpm)	1,200 to 1,600
Max. brake horsepower	19
Clutch	8.5-in. dry disk
Fuel supply	Own 6 gallon or remote supply
Electrical	6 V with battery, starter, and generator
Fan blades	8
Fan diameter (in.)	36
Fan capacity at 1,500 rpm	20,000 ft^3/min

* The specifications for the B15 blower were dated May 1945. It was designed for drying hay and other crops. It was based on the B15 power unit and had similar specifications, however, there is no record of any production. The B15 blower project was under development at the same time as the Models D, F, and H tractors.

West Allis Gasoline Engines

		3.50-in. Stroke			3.875-in. Stroke		
	Waukesha	**B**	**R**	**C/CA**	**138**	**149**	**160**
Cylinders	4	4	4	4	4	4	4
Bore (in.)	3.0	3.25	3.375	3.375	3.375	3.0	3.625
Stroke (in.)	4.0	3.50	3.50	3.50	3.875	3.875	3.875
Displ'mt (in.^{3})	113	116	125	125	138	149	160
Rated rpm	1,400	1,400	1,500	1,500/1,650	1,650	1,650	1,800-2,000
Compression ratios-mark:							
Distillate	INA*	4.6-K	4.7-K	4.7-K	----	5.2-S	---
Gasoline	---	4.9-G	5.0-G	5.7-G	7.8-R	7.8-R	8.0-R
Dist. (high alt.)	---	5.7-B	5.7-B	5.7-B	---	---	---
Gas (high alt.)	---	6.0-B	6.0-B	6.2-G	----	8.5-T	9.0-S
Applications:							
Tractors	B-	B	RC	B	D10	D14,D15	D15II
	first	IB	---	C	D12	D10,D12	I60
	96	---	---	IB	---	H3,I40	I600
	---	---	---	CA	---	I400	H3
Power units		B15	B125	B125	G138	G149	G160
Aux. engines		---	B reduction	B reduction	---	D14 aux.	---
		---	R reduction	R reduction	---	---	---

* Information not available.

West Allis Gasoline Engines with 4.0-inch Bore

	Waukesha	**W**	**D/F***	**WD**	**45 Gas**	**17 Gas**	**170/175 Gas**	**180 Gas†**
Cylinders	4	4	4	4	4	4	4	4
Bore (in.)	3.625	4.0	4.0	4.0	4.0	4.0	4.0	4.0
Stroke (in.)	4.50	4.0	4.0/5.0	4.0	4.5	4.5	4.5	4.5
Displ'mt (in.^{3})	186	201	201/251	201	226	226	22	226
Rated rpm	1,300	1,300	1,300/1,450	1,400	1,400	1,650	1,800	2,000
Compression ratios-mark:								
Distillate	INA‡	4.6-KA	4.2-K	4.6-K	4.7-KA	---	---	---
Gasoline	---	5.0-G	5.0-G	5.7-PA	6.5-G	7.2-M	---	---
Dist. (high alt.)	---	5.0-HA	5.5-B	5.0-HA	5.8-KA	5.8-T	---	---
Gas (high alt.)	---	---	---	6.6-BA	7.2-GH	8.0-Z	8.0-Z/8.2-V	8.0-Z
LPG, Nat. gas	---	---	---	---	---	8.2-V	---	---
Applications:								
Tractors	W,WC-	WC,WF	D/F*	WD	WD45G	D17G	170G/175G	180G
	First	Maintainer	---	Cotton Picker	Cotton Picker	Cotton Picker	---	---
	25	S.patrol	---	---	---	707XTB	---	---
Combines	---	---	---	SP100	Super 100	---	---	---
	---	---	---	---	---	E	---	---
Power units	---	W25	---	W201	W226	G226	---	---

* Models D and F tractors were never placed in production.
† Replaced by six-cylinder 265-in.^{3} gas in production.
‡ Information not available.

Industrial Tractors and Equipment

The Speed Patrols

From 1932 through 1940 the LaPorte Works assembled a line of motor graders called the Speed Patrols. They were powered by West Allis tractor engines and power trains. The two smallest units, the WC Speed Maintainer and the W Speed Patrol, were based on the WC tractor (for more information see that section).

After the initial offering of the Model H in 1932 and 1933, six models were produced. The 42S (single drive axle) and 42T (tandem drive axle) were built around the Model UM engine and drive train, and rated at 42-brake horsepower. The 54S and 54T had the E 510-cubic inch engine and the KO54S and KO54T had the KO 563-cubic inch oil engine, both rated at 54-brake horsepower. The larger engines ran 1,050 rpm and the 42 engine was rated at 1,200 rpm. There were four forward speeds ranging from 2.3 to 10.0 mph. Twelve-foot blades were standard with 14- and 16-foot sizes available. The 42, 54, and KO54 Speed Patrols were phased out prior to World War II, and after the war were replaced by more modern designs built at the Springfield plant.

Oil Field Equipment

Fred E. Cooper represented the Rumely Company in Oklahoma. About 1918, convinced that the outdated methods of servicing oil wells could be improved, he designed and built the first OilPull-mounted winch for the cleaning, repairing, and swabbing operations of oil wells. From that small start, Fred E. Cooper, Inc., of Tulsa, Oklahoma, became a major supplier of oil field equipment. Though acceptance was slow at first, the OilPull winch tractors and stationary winches became a significant Advance-Rumely business. A total of 27 oil companies were on the user list of the OilPull winch tractors and stationary winches.

After Allis-Chalmers acquired Rumely and the OilPull was discontinued, A-C tractors took over the job. The new tractors were welcomed into the oil well servicing industry because rubber tires permitted quick transit between wells. Tractor-based skid units for stationary or truck mounted winches also were built in the West Allis tractor plant.

For a period of years from about 1930 to 1960, Fred E. Cooper, Inc., was perhaps the largest OEM account for the tractor plant. Besides the tractor-based components, Cooper was also a big dealer for power units starting with the E, L, and U engines up through the G149 and G226 engines. Many engines burned natural gas, since the fuel was ever-present in the oil fields.

From 1936 through 1952, Cooper accounted for well over 1,100 tractor winches and skid winches. Tractors that Cooper converted to winches included the Model E, A, U, WC, WD, D14, and D17. Some of

The Advance-Rumely winch tractor was rated at 35 hp. The concept was developed by Fred E. Cooper and it grew into a substantial business for both Cooper and Advance-Rumely. A stationary winch which consisted of the engine and winch components without the tractor power train and chassis was also offered.

The Speed Patrols maintained the country roads. Originally all single-axle machines, tandem axles soon made up the majority of production. All controls were mechanical with gear-driven screws for adjustments.

This 1939 Model E winch tractor was one of many models supplied to Cooper. The Models U, E, and A were supplied with both 15- and 30-mph road gears. A few were built with the KO engine in 1937 and 1938.

The W winch was equipped with WD45 gas or diesel engines, both rated at 60-brake hp. It used the tractor transmission and differential but not the offset final drives. The transmission gave a choice of winching speeds depending on length of tubing to be hoisted.

The UE skid winch had a Model U tractor power train with an E engine. It powered the Cooper double-drum oil well service and clean-out rig in the late 1930s.

Cooper modified the Model WC as a tractor winch. This 1950 model is equipped with the "mud front end". These were sold in the oil fields of Illinois, West Virginia, Pennsylvania, and New York.

This version of the W winch had a standard front end. From the cornfields to the oil fields, the WC was a workhorse that handled all kinds of jobs.

The D14 was one of the last tractors to be modified into a winch tractor. It appears that truck-mounted winches took over that market segment.

the early tractors, the U, E, and A, were built with road gears of either 15 or 30 mph. Skid winches were made from the same tractors for mounting on trucks or stationary units. These were engine, transmission, differential, and axle units with the tractor's rear axles driving the winches. Later truck-mounted winches were driven by G149 or G226 closed power units driving D14 or D17 transmission and differential assemblies which in turn drove single- or double-drum winches.

As oil wells went deeper, Cooper outgrew tractor components. The company went on to build portable derricks and servicing winches for wells 14,000 feet deep, powered by 250-horsepower diesel engines.

The Yellow Line
West Allis Industrial Tractors

After World War II, the construction of new homes, schools, and shopping centers created an exploding demand for light industrial equipment and a new industry was born. Loaders, backhoes, bulldozers, scarifiers, and other equipment required tractor power matched to their capacity. Small manufacturers of these specialized machines grew at an astonishing rate, and soon attracted the attention of the major farm equipment makers. Many acquired the successful upstarts rather than develop their own lines.

Allis-Chalmers, like most other manufacturers, came into the industrial tractor business through the back door by modifying farm tractors. Beginning in 1954 there were a few dozen WD and WD45 tractors "less hydraulics and PTO for Henry backhoe" listed in the production records, and these continued until they were replaced by the D17.

About 1959, an industrial equipment sales department was formed, separate from farm tractors, to promote the sale of the D Series as utility tractors. The D14 and D17 tractors were offered as base tractors without Power-Director, power take-off, hydraulics, or Snap-Coupler, for industrial equipment such as loaders or backhoes.

At this time A-C pursued a policy of "Cleared Companion Equipment" wherein tractor engineering approved of certain loaders, backhoes, and other

The D17 farm tractor with an industrial loader was an early attempt to move into the industrial equipment market. This approach was typical of the industry at that time.

The "Cleared Companion Equipment" policy allowed tractor engineering to approve, but not test, other manufacturers' equipment for use on A-C tractors. This was how the company and the Henry Mfg. Co. got together.

The new yellow line was shown to dealers at French Lick, Ind., in the fall of 1960. It was the real beginning of the A-C industrial line and was a high point of the show.

industrial attachments for use on Allis tractors. Some of the brand names included Henry, Wagner, and Pippin. This left the dealer with the chore of buying from another supplier, mounting the equipment on the tractor, and all the attendant problems. Since other light industrial manufacturers supplied tractor-backhoe-loader units fully assembled, priced, and floor-planned as a unit, A-C dealers were at a disadvantage and the sales records confirmed it.

An excellent working relationship with Henry Manufacturing Company, of Topeka, Kansas, developed and their equipment became part of the new A-C industrial line in 1960, though the company did not acquire Henry until 1968. The 1960 yellow industrial equipment line convinced dealers that A-C was serious about the industrial business. About two dozen units were added and the company became a bona fide light industrial supplier overnight.

The D15 gas and diesel tractors were new. Besides the well-known Power-Director, both the new D15 and D17 could be ordered with the Shuttle-clutch, that provided quick and easy reversing without using the foot clutch. This provided four forward and four reverse speeds and changed a farm tractor into an industrial tractor. The Shuttle-clutch was a simple design, accomplished by adding an idler in the gear train of the low range side of the Power-Director. Pulling the shuttle lever back engaged a reverse equal to the forward speed. A power take-off was available, but it ran a non-standard speed of 875 rpm. A new draft-sensitive category 1-2 three-point hitch made the D15 a lot more attractive to landscapers and others who used rear-mounted equipment.

Other than the new paint job and the availability of the Shuttle-clutch, there were few differences between the industrial models and their agricultural counterparts. Unfortunately many customers still perceived the line as yellow farm tractors, not light industrial machines.

Even the D10 and D12 found industrial employment with a Henry loader and backhoe. They were the right size for cemetery operations.

The yellow D15 industrial tractor had a shuttle clutch, fixed front axle, and industrial tires but the rest was farm tractor.

The I60 and I600 tractors had the look and features of a true industrial tractor. I60 engines were the 160-in.^{3} gas and 175-in.^{3} diesel, but the I600 diesel received the 200-in.^{3} diesel. Shuttle clutch gave 15 percent faster reverse.

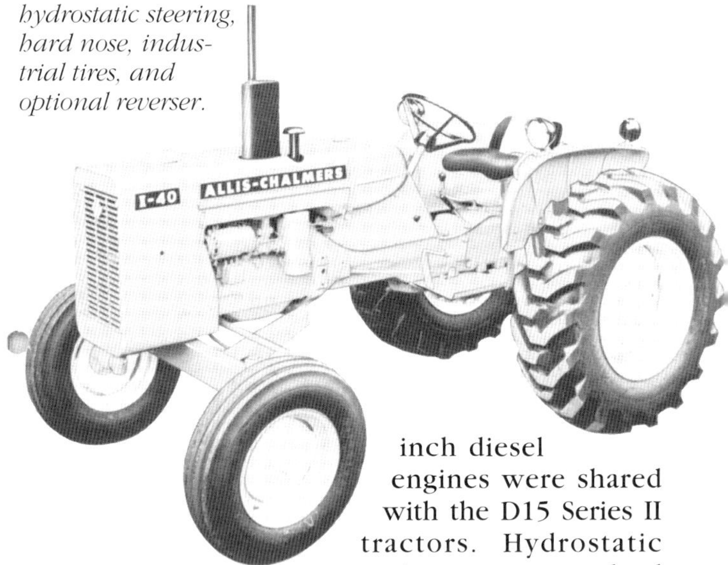

The I40 and I400 were D10-based but with heavy-duty front axle, hydrostatic steering, hard nose, industrial tires, and optional reverser.

Real Industrial Tractors

With the introduction of the I40 and I60 during the winter of 1964-65, the painted farm tractor image was left behind. The rugged punched-steel plate grill, husky one-piece non-adjustable front axle, larger front tires, and large R4 rear tires gave these new industrial tractors a businesslike image.

The I40 shared nearly every major mechanical component with the Series III D10, plus some of its own. For instance, the Series III D10 had an optional low range, providing eight forward speeds. The I40 could have this feature, too, or could be equipped with a reverser. This was a gear shift, not a Shuttle-clutch. Clutching was required but reversing was easy, with reverse about 23 percent faster than the forward gear. The I40 had hydrostatic power steering, mandatory for a loader-tractor. In April 1966, the I400 replaced the I40 but mechanical specifications were similar.

The I60 tractors had three transmissions available: standard four-forward one-reverse, Power-Director, or Shuttle-clutch. The 160-cubic inch gas and 175-cubic inch diesel engines were shared with the D15 Series II tractors. Hydrostatic steering was standard equipment. The I600 replaced the I60 in April 1966, the same day that the I400 went into production. The I600 diesel had the 200-cubic inch diesel engine.

The Model 500 and 600 rough-terrain forklifts were based on the D15 power train, including the 160-cubic inch gas and 175-cubic inch diesel engines. As the 175-cubic inch diesel was phased out of production the 500 forklift was changed to the more modern four-cylinder 200-cubic inch direct-injection diesel. The 500 forklifts were available with 10-, 12-, 14-, or 21-foot masts. The 600 forklift, a heavier machine, was offered with same sizes plus a 30-foot mast.

The 615 loader-backhoes (LBH) had a D15 tractor under the loading and digging works. They started

The D15 chassis was the basis for both the 500 and 600 forklifts. The 500 offered masts up to 21 ft and the 600 could handle a 30-ft mast.

The 615 loader-backhoe had the D15 power train with shuttle clutch and the 160-in.^{3} gas or 175-in.^{3} diesel engines. At Serial No. 5001 the diesel was changed to the 200-in.^{3} engine. The 615 could dig 15 ft deep and had a 1-yd loader bucket.

with the 160-cubic inch gas and 175-cubic inch diesel engines, but the diesel was changed to the 200-cubic inch engine at Serial No. 5001. The 615 LBH had a 15-foot backhoe and 1-yard loader bucket. After 1969, West Allis built the basic stripped chassis or the skid unit for forklifts and backhoes which were completed at the Topeka plant.

The 918 loader-backhoe was a big rig with an 18-ft backhoe and 1 1/2-yd loader bucket. A torque-converter smoothed reversing and eased operation.

The Model 918 Loader-Backhoe

The big Model 918 LBH was based on the West Allis-built One-Ninety tractor engine and power train but little else was common with the farm tractor. The Power-Director was replaced by a torque-convertor and hydraulic shuttle that had four forward and four reverse speeds. The experimental units did not have the torque-convertor but testing proved the it was needed to protect the power train from shock loads encountered in the loading cycle. The frame, front axle, grill, and hood were of industrial design and strength. An optional cab alleviated bad-weather operation. After industrial equipment engineering and assembly was moved in 1969, West Allis continued to build the basic 918 tractor and it was completed in the Topeka plant.

A prototype of the beach sanitizer pulled by a D19 Beachmaster tractor working on a California beach. It could make a day at the beach a much more pleasant event.

The Beach Sanitizer

Anyone who has had a day at the beach marred by all the debris that can accumulate on the sand would appreciate the A-C beach sanitizer. The sanitizer would elevate the sand and trash over a chain mesh raddle which allowed the sand to drop back to the ground, then delivered the separated debris to a bin on the back of the machine. The result — a clean beach.

The beach sanitizer dumped the collected debris into a waiting truck. The One-Ninety Beachmaster tractor had a mid-mounted scarifier to loosen sand for the sanitizer.

The beach sanitizer was developed at the Oxnard plant which was close to the California beaches. Only one D19 diesel beach tractor was built in 1963 to power the prototype sanitizer, followed by an undetermined number in 1964. When the One-Ninety came out that year, the beach tractor was based on it. A 265-cubic inch gas and naturally aspirated 301-cubic inch diesel, both rated at 70-PTO horsepower, were supplied. While many sanitizer rigs were sold on both coasts, it was a small market with limited replacement. This resulted in a short production life.

The H3 and HD3 Crawlers

When the H3 and HD3 crawlers were introduced in the fall of 1960, they stole the show. They were just what was needed to encourage industrial dealers to meet their competitors head on. With their full line of loaders and dozers, along with many other attachments, these tough little crawlers were ready to take on the world of construction.

The H3-HD3 crawlers were offered in both industrial yellow for that market and Persian orange No. 2 for the farm market. Two grill styles were available; a rugged perforated steel plate for the industrial applications and a cream mesh grill which matched the farm tractors. The tougher steel plate grill quickly became the more popular, and except for a few in the beginning, most tractors were painted yellow.

The H3 had the 149-cubic inch gas engine and the HD3 used the 175-cubic inch diesel, both rated at 1,650 rpm. The H3 tested 32.11-PTO horsepower and the HD3 developed 32.53-PTO horsepower. During 1963 at Serial No. 6001, the H3 received the 160-cubic inch engine with 8.0:1 compression ratio, still rated at 1,650 rpm. Both tractors were equipped with a dry air cleaner at that time. In 1965 (at Serial No. 8698 for the gas and Serial No. 8694 for the diesel) the rated engine speed was increased to 1,800 rpm.

Since industrial tractors were not tested at Nebraska, the tractors did not return to the torture track in Lincoln. Industrial tractors were rated by engine horsepower. The H3 had 47.9 engine horsepower and the HD3 developed 43.0-engine horsepower. Crawlers were also rated by maximum draw-

At Nebraska the HD3 pulled 8,207 lb, 107.5 percent of its 7,645 lb. The H3 pulled 7,893 lb, 106.8 percent of its 7395-lb weight. They set new records for crawlers.

The H3-HD3 farm crawlers had D15 sheet metal and Persian orange No. 2 paint. Early gas tractors had 149 in.^{3}, later raised to 160 in.^{3} The diesels had 175 in.^{3}

The T100 prototype became the H3 and HD3 in production. The HD designation was chosen for consistency with the Springfield line and to benefit from their excellent reputation.

bar pull, and the H3-HD3 had set new records at Nebraska by pulling more than their own weight:

Tractor	Weight (lb)	Max. Pull (lb)	Percent
H3	7,395	7,893	106.8
HD3	7,645	8,207	107.5

Based on their maximum pull at rated speed, this is the equivalent of a phenomenal 54- to 55-percent torque rise.

The power train, including the Power-Director, Shuttle-clutch, power take-off, and engine clutch, was similar to the D15 and over 50 percent of the parts were common between the tractors. The hydraulic pump was driven from the belt pulley drive. Steering clutches were operated by levers, while the brakes were applied with pedals, therefore each could be controlled separately as required by conditions or safety. This was an advantage over tractors with combined steering clutch and brake controls.

The track assemblies were built in the Springfield plant, home of the big Allis crawlers, and shared one of their features, the "hunting tooth" drive sprocket. With an odd number of teeth, each sprocket tooth contacted a track pin only once every other revolution, improving sprocket life. Four-roller truck frames were standard, with five-roller frames optional, both with 34 track shoes on each side, laying 63.75 inches of track on the ground. Tread (or gauge) was 48 inches.

A full selection of full- and semi-grousers on 7-, 10-, 12-, and 14-inch track shoes was available. The HD3 Muckland Special with 18-inch track shoes provided high flotation on muck and peat soils. A cane version of the HD3 was built for Thomson Machinery of Thibodaux, Louisiana. Three ski-slope tractors were also produced. The cane and ski-slope tractors had modified tracks. In about 1963, H3 track assemblies were used in grouser traction research studies by the National Tillage Laboratory.

Both 15- and 25-gpm hydraulic systems were optional on early models, but in later versions an 11.5-gpm system with a one-spool valve was standard equipment. The smaller systems used an internal

A non-draft sensitive three-point hitch allowed the H3-HD3 twins to handle mounted implements, but the farm market for this size crawler was extremely limited.

Doing what it did best, the HD3 worked with a 1066 loader and 1003 scarifier on the construction site. The loader and scarifier were produced in the LaCrosse plant which also made the bulldozers for the H3 and HD3.

The Henry-built Mark V backhoe could dig to 13 1/2 ft and load to over 10 ft. With backhoe and loader the 3 Series tracklayers were strong competitors in the industrial market.

reservoir, but the higher output pump had an external tank.

A category 2 lift-lower three-point hitch was intended to adapt the tractor to farming but the farm market for the H3 and HD3 never materialized. The use of three-point hitch was limited in the industrial market. The Nos. 1066 loaders, 900 straight dozers, 1000 angle dozers, and 1003 scarifier were built at LaCrosse. The Henry-built Mark V backhoe, No. 400 winch, fireline plow, and mobile drop hammer rounded out the line of H3-HD3 equipment.

The H4 an HD4 Crawlers

Industrial operators are just like farmers and growers — they can always use a little more power. Shortly after H3-HD3 production began, Allis-Chalmers needed a larger crawler and the more powerful H4-HD4 line was developed. The 2200 Series four-cylinder 200-cubic inch engines were chosen for power. The gas engine delivered 59 horsepower and the direct-injection open-chamber diesel developed 52 horsepower at 2,100 rpm.

Other necessary features, besides more power, included improved operator visibility, greater ease of operation, power reversing, more cooling capacity, more track on the ground, and increased hydraulic output. The new models were not enlarged H3-HD3's although they had many parts in common.

The increased cooling capacity required a larger radiator, which if located at the front of the tractor, would have reduced operator visibility. Another major consideration was to put the radiator in a less vulnerable position. H3-HD3 tractors were subject to radiator damage from debris encountered in logging and dozing. A forward sloping hood was considered important and after much testing, the radiator was placed between the operator and the engine. A suction fan pulled air away from the driver. The radiator was completely protected and the sloped hood vastly improved the operator's view of the loader bucket or

The H4-HD4 tractors with their industrial loaders were at home in big feedlots and earthmoving jobs on large farms. The 200-in.³ gas and diesel engines powered the pair, but the gas model was dropped as users converted completely to diesel power.

The H4-HD4 mid-mounted radiator provided high-cooling capacity for the engine, torque converter, and hydraulic system yet allowed the sloped hood which improved operator visibility.

Torque-converter drive and high operator's station combined with an all hydraulically controlled dozer provided easy operation and high output.

dozer blade. Visibility was also enhanced by the high operator's station.

Power-reversing was provided by the power-shift Shuttle-clutch, which was a hydraulically-actuated clutch similar to that in the 918 LBH. A single-stage torque converter between the engine and the Shuttle-clutch cushioned the shock loads of direction-changing.

An 8.5-gpm hydraulic pump provided charge oil to the torque converter and actuated the Shuttle-clutch. The shuttle control was a short, easily operated lever located next to the throttle in a console at the driver's left, leaving his right hand free for the hydraulic valves. The hydraulic system delivered 15 or 25 gpm at 1,625 psi. Like the Springfield tractors, the H4-HD4 had separate steering clutches and brakes for precise control and safety. Steering clutches, brakes, differential, and final drives were upgraded versions of the H3-HD3.

Track assemblies had five-roller truck frames with 64 or 70 inches of track on the ground. A variety of 10-, 12-, and 14-inch shoes and grousers were available. Tractor weight was approximately 8,500 pounds bare, up to 12,400 pounds when equipped with a loader and rear mounted ripper. The loader was a rugged 1-yard unit with a breakout force 10,900 pounds and a lifting capacity of 5,000 pounds. A variety of hydraulically controlled inside and outside push beam dozers matched the increased power. A backhoe, winch, and other attachments rounded out a line of matched equipment for the H4 and HD4 crawlers.

Production began at Serial No. 1001 in September 1965, and ended at West Allis at Serial No. 4332 in June 1969, when production was transferred to the Industrial Tractor Division, the result of Allis-Chalmers' purchase of the Henry Manufacturing Company, Topeka, Kansas, in late 1968. Following this shift, West Allis supplied just the power trains. The tractors were continued in the industrial line as the 650 series crawlers with 59-horsepower diesel engines only.

This HD4 Series II prototype at the Racine County proving ground was transferred to Topeka when the Industrial Tractor Division was formed and became the 650 Series crawler. The 200-in.^{3} diesel had 59-engine horsepower.

Nebraska Test Results and Specifications for the H3, HD3, and HD4 Crawlers

	H3	HD3	H4	HD4
Test No.	793	794	None	None
Year	1961	1961	---	---
Fuel	Gas	Diesel	Gas	Diesel
Max. pto hp	32.11	32.53	---	---
Max. dbr hp	25.48	26.17	---	---
Weight (lb)	7,395	7,645	---	---
Max. pull (lb)	7,893	8,207	---	---
Percent pull	106.8	107.5	---	---
Engine	A-C	A-C	A-C	A-C
Model	G149	D175	G2200	2200
Engine hp	47.9	43.0	59.0	52.0
Cylinders	4	4	4	4
Rated rpm	1,650*	1,650*	2,100	2,100
Bore (in.)	3.50†	3.62	3.875	3.875
Stroke (in.)	3.875	3.87	4.25	4.25
Displ'mt (in.^{3})	149†	175	200	200
Comp. ratio	7.5:1†	15.5:1	8.15:1	16.25:1
Carburetor (in.)	0.875	DNA‡	INA§	DNA‡
Avg. working weight (lb)	7,000	7,100	12,200	12,250

* 1,800 rpm, effective Serial No. 8698 (gas) and Serial No. 8694 (diesel).

† Bore, 3.625 in.; 160 in.^{3}; 8.1:1 comp. ratio effective Serial No. 6001.

‡ Does not apply.

§ Information not available.

Serial Numbers and Production Data for West Allis Industrial Wheel Tractors

Serial Numbers:

	I40	I60	I400	I600	600 FL	918 LBH
1964	1055-INA*	---	---	---	---	---
1965	INA*	1005-INA*	---	---	---	---
1966	INA*-1608	INA*-2340	1003-1280	1002-1746	---	---
1967	---	---	1281-1494	1747-2708	1001-1114	---
1968	---	---	1495-1750	2709-3056	1115-1714	1001-1031
1969	---	---	---	---	---	INA*

Production Dates:

Tractor	First Serial No.	Date	Last Serial No.	Date
I40	1005	10-64	1608	4-66
I60	1005	2-65	2340	5-66
I400	1003	4-66	INA*	
I600	1002	4-66	INA*	
600 FL	1001	9-67	INA*	
918	1001	9-68	†	

* Information not available.

† Production moved to the Topeka plant in 1969.

Specifications for West Allis Industrial Wheel Tractors

	I40 Gas I400	I60 Gas I600 500 FL 600 FL 615 LBH	I60 Dsl 615 LBH 500 FL	I600 Dsl 615 LBH* 500 FL*	One-Ninety Beach Gas	One-Ninety Beach Diesel	918 Diesel
Engine model	149	160	D175	2200	G2500	2800	2800
Engine hp	40.6	56.8	44.0	57.0	70†	70†	84
Cylinders	4	4	4	4	6	6	6
Displ'mt (in.^{3})	149	160	175	200	265	301	301
Rated rpm	2,000	2,000	2,000	2,000	2,200	2,200	2,200
Comp. ratio	7.5:1	8.0:1	15.5:1	16.25:1	8.0:1	16.25:1	16.25:1

* 200-in.^{3} diesel effective Serial No. 5001 for 615 LBH, serial number not available for 500 FL.

† PTO horsepower.

Serial Numbers and Production Records for H3, HD3, H4, and HD4 Crawlers

H3-HD3 Serial Numbers and Production Data:

	Serial No. to Serial No.	Production Gas	Production Diesel	Total	HD3* Muckland	HD3* Thomson	H3/HD3* Ski-slope
1960	1001-1195	116	80	196			
1961	1250-	1,041	926	1,967	18	4	---
1962	3199-4294	529	1,683	2,212	23	8	---
1963	6001-6944	382	561	943	---	5	3
1964	6945-7889	407	538	945			
1965	7890-8855	---	---	965			
1966	8856-9482	---	---	627			
1967	9483-9699	---	---	217			
1968	9700-9949	---	---	250			
Total				8,322			

H4-HD4 Serial Numbers and Production Data:

	Serial No. to Serial No.	Production
1965	1001-1243	243
1966	1244-2689	1,446
1967	2690-2937	248
1968	3001-3905	905
1969	3906-4332†	427
Total		3,269

H3-HD3 Running Changes:

Tractor	Change	First Serial No.	Date	Last Serial No.	Date
H3	Production	1001	10-60	9699	7-67
HD3	Production	1001	10-60	9949	9-68
H3	160-in.^{3} engine	6001	1-63		
H3-HD3	Dry air cleaner	6001	1-63		
H3	1,800-rpm engine	8698	11-65		
HD3	1,800-rpm engine	8694	11-65		

* Included in grand total.

† Transferred to Topeka plant.

Speed Patrol Serial Numbers and Production Data

Serial Numbers:*

	H	42S	42T	54S	54T	KO54S	KO54T
1932	2001-2049	---	---	---	---	---	---
1933	2050 only	2501-2643	---	---	---	---	---
1934	---	2643-2775	---	6501-6502	---	---	---
1935	---	2776-2941	---	6503-6597	---	---	---
1936	---	3031-3160	3161T-3199T	6628-6763	6639-6777	---	---
1937	---	3276-3325	3200T-3361T	6821-7002	6825T-7108T	6817-O--7127-O	6778-OT--6977-OT
1938	---	3455-3516	3362T-3488T	---	7021T-7186T	---	7053-OT--7102-OT
1939	---	3517-3567	3489T-3625T	---	7187T-7227T	---	7228-OT--7260-OT
1940	---	3568-3575	3626T-3675T	---	7261T-7277T	---	---

Production Data:

	H	42S	42T	54S	54T	KO54S	KO54T
1931-2	50	---	---	---	---	---	---
1933	---	42	---	---	---	---	---
1934	---	133	---	2	---	---	---
1935	---	166	---	95	---	---	---
1936	---	130	39	136	141	---	---
1937	---	50	111	79	74	23	90
1938	---	42	106	---	91	---	50
1939	---	8	70	---	41	---	33
1940	---	---	8	---	17	---	---

* 42S and 42T serial numbers are mixed. 54S, 54T, KO54S, and KO54T serial numbers are mixed.

Tillage Equipment

The Lacrosse Plow Company

In 1860, John M. Barclay and I. S. Bantan founded the Barclay and Bantan Implement Works to make plows and other farm implements for the farmers of the LaCrosse, Wisconsin, community. Only five years later, on May 17, 1865, young Albert Hirshheimer bought the operation for $1,530. Albert Hirshheimer left Blairsville, Pennsylvania, when he was 16-years-old and worked his way toward the Mississippi River. He saved enough money to buy an old steamboat, the "Eclipse," which he operated between LaCrosse and Hokah, Minnesota. Some time later, having enough of the river life, he sold the boat and bought the plow shop from Barclay and Bantan.

An early Hirshheimer walking plow was displayed at a special event at LaPorte. It is on permanent display at a museum in Prairie du Chien, Wis.

The factory he bought measured 20 by 50 feet and had two forges, anvils, a drill press, and an emery wheel — tools for making a walking plow and a few other implements. He operated the firm under the name of A. Hirshheimer and Company for some time. The company prospered, acquired more land, and built a new factory on that land. In 1893 the concern was incorporated as the LaCrosse Plow Company with Albert Hirshheimer as president. Sales grew from $60,000 in 1887, to $200,000 in 1894, and then to $1,000,000 in 1909.

Hirshheimer's two sons, Harry and Louis, were involved with the company. Harry became president in 1925 when the elder Hirshheimer died, and ran the company until it was purchased by Allis-Chalmers on October 1, 1929. The business had declined and there were only 60 employees at that time. When A-C LaCrosse celebrated its silver anniversary in 1954, employment had grown to over 1,000, five buildings had been built, and a foundry was purchased.

The Model 10-18 was teamed with a three-14-in. LaCrosse tractor plow on a North Dakota farm. The LaCrosse name appears on the plow lever.

The LaCrosse line included moldboard and disk plows, disk harrows, cultivators, and grain drills. LaCrosse pioneered mechanical power lifts for plows

A LaCrosse-built No. 43 three-14-in. plow is being pulled by an air-tired Model U. The Nos. 54, 44, 45, 64, and 65 plows were similar in appearance to the No. 43.

Orchard plows were built in two-, three-, four-, and five-bottom sizes. Beams and axles were common with standard plows, only wheels and controls were special for orchards.

The Model C with a one-16- or one-18-in. pull-type plow was a big improvement over horse-drawn plowing.

and drills. Although it had good products, business had declined to the point where the company could no longer survive without outside capital. General Falk acquired the LaCrosse Plow Company for a modest $275,000.

In the implement business there is an old cliche, "Without an implement, a tractor is good only for going after the cows." That observation is credible. General Falk and Harry Merritt must have agreed with it when they acquired the LaCrosse Plow Company for a line of implements to complement the expanding tractor line. Fewer than one dozen major tractor manufacturers survived the big shakeout of the 1920s and 30s, and each had a line of implements to sell alongside their tractors. Not that an implement line was the sole key to success — far from it — but a good line of implements was just one of the factors that identified a successful farm equipment company.

Moldboard Plows

LaCrosse Plow was appropriately named. A catalog from the 1930s showed tractor-drawn moldboard plows in 12 models and 27 sizes. The No. 16 was a one 16-inch two-wheel plow. The LTP came as two and three 14-inch bottom three-wheel plows. The No. 43 and No. 54 were close-coupled, medium weight plows. The 43 was a three-bottom plow, and the 54 was a four-bottom size, each reducible by one bottom. The No. 64 and 65 plows had regular fore-and-aft spacing, while the No. 44 and 45 had wide spacing. Both were available with 12- or 14-inch bottoms and all were reducible by one bottom. These were real heavyweight plows weighing from 1,800 to over 2,100 pounds. The five-furrow plows were matched to Model K crawler power.

Today no one would think of using a moldboard plow in an orchard or vineyard, but at one time the practice was quite common. The A-C 142, 143, 164, and 165 orchard plows fit right into that culture. Although they had normal vertical clearance, everything above the beams was as low and smooth as possible to prevent snagging low branches and vines. Levers were replaced with screw cranks for depth and leveling adjustment. Solid wheels also protected trees and vines. They were supplied in two-, three-, four-, and five-bottom sizes. When growers finally realized what deep tillage was doing to their fine groves and orchards, the orchard moldboard plow was "not long for this earth."

The No. 1 Series were lighter weight one- and two-bottom plows matched to the Model B, RC, and C tractors. The No. 1 plow was replaced by the No. 4, a two-wheel plow of similar weight. The No. 4 appeared in the price list as late as 1960.

A No. 2 two-bottom plow with plow-down fertilizer attachment handled two jobs at once. The tractor is a flat-radiator WC.

The No. 53 three-bottom square-beam plow with pin hitch was an improvement over the WD pickup plow, but hitching was like threading a needle in the dark with the hitch-point hidden under the tractor. The Snap-Coupler fixed that problem.

The No. 2 was the high volume unit of this period, and had a long production life. This three-wheel plow was available in two-, three-, four-, and five-bottom sizes with 14-, 16, and 18-inch bottoms. The 18-inch plow bottoms were not widely used; most farmers preferred the nice even furrows of a 14-inch bottom while others liked the trash capacity of a 16-inch. The 18-inch bottom needed speed with all the soil it had to lift and turn, and true high-speed moldboards with the power to pull them were well into the future. The weight of No. 2 plows ranged from 921 pounds for a two-bottom to 1,879 pounds for a five-bottom model.

Mounted Plows

Prior to the WD, A-C mounted plows were limited to the pickup plows for the Models B and C. Even these early models were led from a single hitch point, a principle that A-C plows would employ until the Snap-Coupler was phased out in the late 1960s.

The Model WD tractor required a mounted moldboard plow to utilize the Traction-Booster system. The WD two-bottom pickup plow, known to company people as the "round beam" plow, was essentially a mounted version of the No. 2 pull-type plow. It was a significant improvement over a pull-type plow because the Traction-Booster system reduced slippage and the whole rig was more maneuverable. A large flat coil spring at the hitch point provided some protection but the plow suffered many bent beams in rocky soils.

About 1950, LaCrosse produced the "square beam" plows of bolted construction with the vertical beam forming a right angle with the frame. If the plow hit an obstruction a shear bolt would break, which allowed the bottom to swing back, and prevented further damage. The good thing about shear bolts was that they were cheap. Initially they were identified as the CA or WD plow, but were soon given the 50 Series designations of 52, 53, and 54 spinner. All had an improved pin hitch, called Quick-Hitch. It was a bit tricky to align the hole in the plow tongue with the hitch point under the tractor since the operator could see neither. The Snap-Coupler eventually solved that problem.

The 50 Series plows were short-coupled and had limited trash clearance. The Bullet Blade bottoms appeared with the 50 Series. The Bullet Blade was Allis-Chalmers' version of the throw-away plow share which became popular in the 1950s. The 52 and 53 prices ranged from $193 to $321 and the 54 spinner listed from $428 to $487.

This WD tractor has a WD pickup plow. It could be set up as a 14- or 16-in. frame, depending on bottoms. A large flat-coil spring at the hitch point provided limited protection from damage when obstructions were struck.

As tractor power increased, larger plows and greater trash clearance were needed. This led to the 60 and 70 Series plows. When the WD45 hit the market in 1953, it was followed shortly by the Big Capacity implements, one of which was the No. 64 four-bottom 14-inch semi-mounted plow. Big Capacity implements utilized the remote-ram principle for semi-mounted and trailing implements. The transport valve on the tractor permitted the Traction-Booster system to transfer weight hydraulically, but withheld pressure from the remote cylinder. When the plow was lifted at the furrow end, the transport valve delayed pressure to the cylinder, allowing the bottoms to leave the ground progressively, providing even headlands.

By 1955 the 62 and 63 fully-mounted plows had been added. The 65 five 14-inch plow came out in 1957, matched to the D17 tractor. Three-point hitch was offered as it became available on A-C tractors. The 60 Series were good values for the farmer who had only moderate crop residue. However, the market was demanding more clearance to handle trash and the 70 and 80 Series plows were developed to meet that requirement.

When the No. 73 semi-mounted plow joined the line, the sales folks could start to talk about trash clearance. Until then the subject of trash clearance was to be avoided because the 50 and 60 Series plows had limited clearance and were noted for their trash-plugging. With 24-inch fore-and-aft spacing and about 2 feet of vertical clearance, the 70 Series throat was easily 25 percent larger than the 60 Series.

The 80 Series came out in 1960, about four years after the first 70 Series plows, but did not replace the 70. The 70 Series had shear-bolt protection while the 80 Series had spring-trip beams. When an 80 Series plow hit an obstruction, the Roller-Lock spring trip released and allowed the bottom to swing back to

The Big Capacity four-bottom No. 64 for the WD45 was A-C's first semi-mounted plow. Transport valve delayed lift of the tail wheel for even headlands.

The One-Ninety easily handled a 70 Series two-way plow. The 70 and 80 Series two- and three-bottom spinners had adjustable frames and hydraulic rotation.

A D19 hitched to a No. 85 pull-type plow with the Traction-Booster drawbar for weight transfer had greatly-improved field performance. The 70 and 80 Series pull-type plows came in three-, four-, and five-bottom sizes.

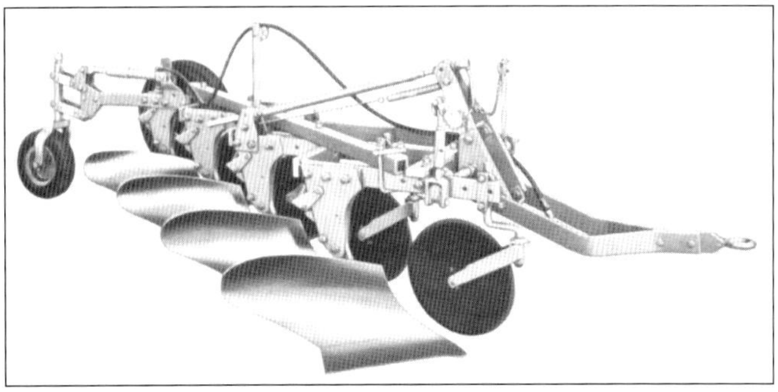

Semi-mounted Snap-Coupler plows like this No. 84 came in three-, four-, and five-bottom sizes. The 70 Series were the same plows with shear-bolt beams.

The D21 handled a seven-bottom 9000 pull-type plow very nicely. The plow was a base five-furrow expandable to six or seven with in-furrow hitching on smaller units and on-land hitching for the larger ones.

avoid damage. To resume plowing the operator just backed the tractor until the Roller-Lock latched into position, then took off again — much better than replacing shear bolts all the time. Several models and sizes up to five 16-inch bottoms, including pull-type, were added to the 70 and 80 Series as tractor horsepower increased.

A hydraulic actuator rotated the 70 and 80 Series two- and three-furrow spinners. The actuator was a vane-type hydraulic cylinder which turned the spinner shaft through 180 degrees to bring the opposite bottoms into working position in just 6 seconds. A double-acting remote valve operated the actuator. The entire line of 70 and 80 Series plows was manufactured until about 1964 when all but the fully-mounted plows were replaced by the new 9000 Series plows.

The 9000 Series Plows

The D19 diesel had 67-PTO horsepower and when the D21 came out in 1963, the top horsepower leaped to 103 — a 50 percent increase. Larger implements were needed to utilize the big power.

When steel-wheeled tractors prevailed the normal plowing speed was comparable to horse plowing at about 3 mph so 8 to 10 horsepower (belt or PTO) per bottom was adequate. Then farmers began to desert the 14-inch bottom for the 16- and 18-inch bottoms so they could plow deeper to cover the increased crop residue from higher yields. The old rule of thumb was that a plow bottom worked best at about half or slightly more of its width. That pegged a 16-inch bottom at 8 to 9 inches. Furthermore, they wanted to plow faster, and speeds were up to around 5 mph when the D21 was introduced. Now at least 15 horsepower per bottom was required and that eventually increased to around 20. The 70-80 Series design had reached its optimum with five bottoms. That was all right for the D19 but the 103-horsepower D21 needed six or seven, depending on soils.

The 9000 Series semi-mounted plows came in four- or five-bottom in-furrow plows or a larger six-expandable to seven-bottom in-furrow or on-land hitched model. They were matched to One-Ninety or D21 power.

To meet all these requirements LaCrosse produced the 9000 Series, a line of four- or five-bottom semi-mounted plows, a larger six- expandable to seven-bottom semi-mounted plow, and a large pull-type unit of five- expandable to six- or seven-bottoms. Fourteen- and 16-inch bottoms were offered.

With 28 inches, the 9000's had the needed trash clearance. Spring-trip beams with an improved latch replaced the Roller Lock. It was adjustable for tripping load and could be released by hand for changing shares. The 9000 line was built from 1963 until 1970.

The Monoframe Plows

Large-dimension structural steel tubing was rigid enough to withstand high stresses without bracing. That simplified frame design and reduced the number of parts, as well as costs. When a plow was built of these materials, it was clean with no extraneous members and was wide open for easy access to all bottoms. It needed a special name and the name that best described it was Monoframe.

Designed at LaCrosse, production of the Monoframe was transferred to LaPorte. The Monoframes went into production in early 1970 and replaced the 9000 Series. The Monoframe was offered only with 16-inch bottoms and two new bottoms with 45 degree share angles were introduced. The more abrupt share and moldboard angle provided improved scouring and turning of the furrow slice. With 28-inch fore-and-aft spacing and 28 1/2-inch vertical clearance, the new plows had plenty of clearance to handle heavy crop residue.

The Monoframe plows had two important new features. The rubber-mounted coulter had a rubber disk bonded between two steel plates, one of which was bolted to the shank pivot, the other attached to the coulter arm. When an obstruction was struck, the rubber disk acted as a torsion spring allowing the coulter blade to ride over the obstruction without damage. This was a proprietary design which A-C engineers also applied to the No-Til coulter for planters. The rubber-mounted coulter contributed to non-stop plowing by saving stops to replace shear bolts.

An all-new hydraulic-reset beam allowed true non-stop plowing. When the share hit a rock, the bottom could move upward and backward to clear it, then a hydraulic cylinder pulled the bottom back into normal working position. All the individual cylinders on the bottom linkages were connected to an accumulator which was charged by the tractor hydraulic system. The Allis design was unique in that the share point did not travel below the furrow bottom; it

The Two-Twenty FWA and 3000 Series eight-bottom plow turning cornstalks under at the rate of 50 acres per day.

"Roll out" of the new Monoframe 3000 on-land plow. (From left to right) Les Wedeking, Lee Granum, Bill Tanke, Dick Moe, and Frank Mnuk. Taken at LaCrosse with the Mississippi in the background.

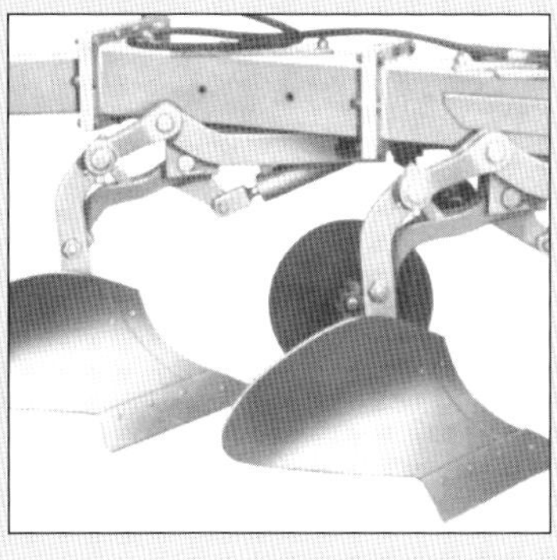

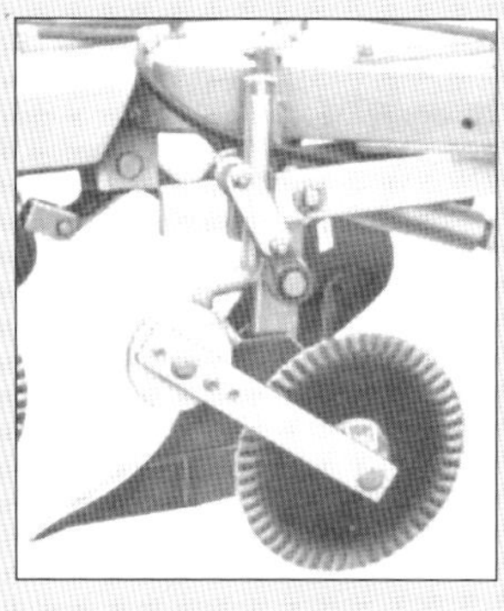

Important new features of the Monoframe plow were hydraulic reset beams and rubber mounted coulters for non-stop plowing.

The 2018 four 18-in. bottom plow was an ideal match for a Model 185 tractor in most conditions. It was a base three expandable to four bottoms.

	In Furrow			On Land	
	5	6	7	7	8
A	X	X	X	X	X
B	X	X	X		
C				X	X
D		X		X	
E			X		X

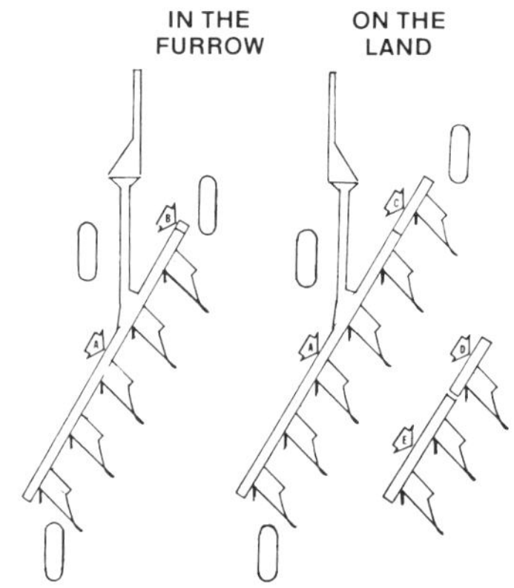

The base five-bottom 2500 pull-type plow could be expanded to six- or seven-bottom in-furrow and seven- or eight-bottom on-land hitched plows.

always moved up and back at about a 40 degree angle to the furrow bottom, then returned by the same route. Other designs allowed the share point to go below the furrow bottom which could aggravate contact with the obstruction and result in more damage.

The Monoframe plow was supplied in two series, the 2000 and 3000. The 2000 was a semi-mounted plow in three-, four-, or five-bottom sizes, each expandable by one bottom. With category 2 or 3 hitch, it was an in-furrow type. The 3000 was a semi-mounted six-bottom plow expandable to seven or eight bottoms, and was designed for on-land hitching with either category 2 or 3 hitch. Plow depth was controlled by a pivoting front furrow wheel and a depth gauge wheel running on the land.

The second phase (in the late 1970s) of the Monoframe plow line saw the 3000 dropped and the 2000 expanded with an in-furrow six- expandable to seven-bottom model. The spring-trip, shear bolt, and hydraulic-reset beams had been redesigned to provide 32 inches of vertical clearance. Three 18-inch plows known as the 2018 Series were then added. These included a four-bottom and a five-bottom, each expandable by one bottom. The third plow was a six-expandable to seven- bottom plow. These 18-inch plows had a fore-and-aft spacing of 31 inches with 32-inches vertical clearance.

The 2500 pull-type plow was similar to the discontinued 3000 semi-mounted plow. This was a base five-bottom plow expandable to six or seven bottoms for in-furrow operation or to seven or eight bottoms for on-land hitching. Offered only with 18-inch bottoms the 2500 had 31- by 32-inch throats that could handle lots of trash. The larger versions of the 2500 were built for the Allis-Chalmers four-wheel-drive tractors.

The 7580 four-wheel drive tractor put 220-engine hp in front of this Monoframe 2500 eight-18-in. bottom plow to turn 70 acres per day.

This is one of a new line of moldboard plows engineered and tested but never placed in production. There were three basic plows: 5- to 7-bottom in-furrow, 6- to 8-bottom on-land, and 9- to 13-bottom hinged-frame on-land plows.

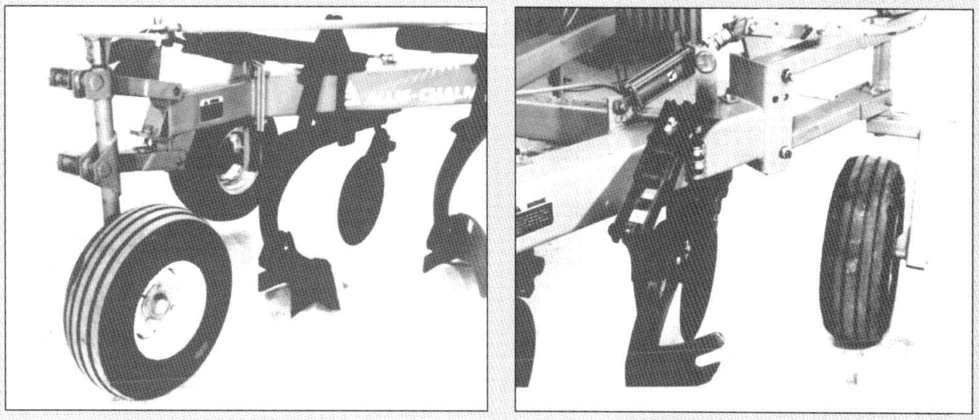

The new line of plows had a choice of beam protection — a single-pivot spring-reset beam or a toggle-trip manual reset beam.

The moldboard plow line was discontinued when the LaPorte plant was closed. Although much of the implement line was continued with production by Vermeer Manufacturing Company, Pella, Iowa, the moldboard plows had been replaced by chisel plows and conservation-tillage tools.

Plows of the Future

In the early 1980s LaPorte engineering designed a new line of moldboard plows. There were three models: a five- to seven-bottom in-furrow plow, a six- to eight-bottom on-land plow, and a huge nine- to thirteen-bottom unit with a hinged frame. All of these plows were of the Monoframe design. A new single-pivot spring reset beam and new toggle-trip manual reset were important advancements. All the prototypes were built and tested, and they were ready for production. The new plow had problems, however, none mechanical. Between insurmountable tooling costs and a shrinking moldboard plow market, these good-looking plows never saw production. This was indeed the end of an era.

An 8550 tractor with the experimental 13-bottom hinged plow working near Marshall, Mich., in the spring of 1980. With 16-in. bottoms it cut over 17 ft.

The S7 and S9 wheatland disks were made in LaCrosse during the 1930s. Each lever raised a disk gang and the width-of-cut was controlled by the rear wheel.

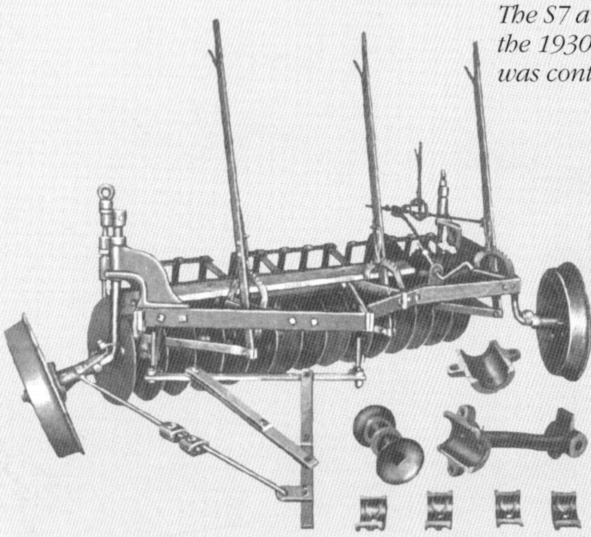

The LaCrosse-built No. 260 Texas "Beater" was a heavy-duty disk plow for work in mesquite brush and gumbo soils. The five-disk size weighed nearly 3 tons.

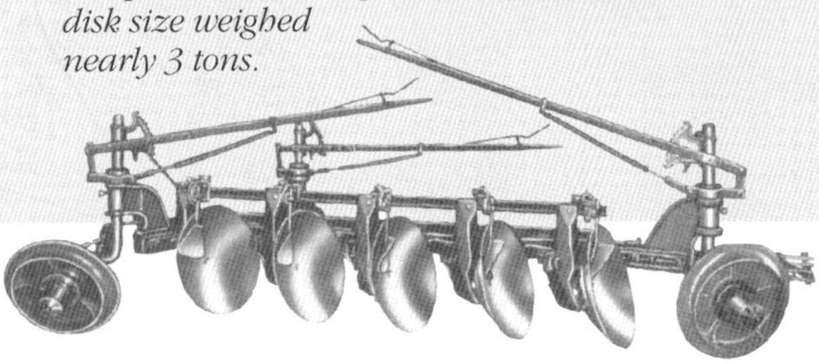

A flat-radiator WC pulling an Athens two-disk plow. Still called an Athens product in 1939, it was branded an Allis-Chalmers unit shortly after.

Disk Plows, Disk Tillers, and One-Way Disks

Disk plows and disk tillers were commonly used in the Southeast, mostly because they were better adapted to the soils and conditions than moldboard plows. They would roll over rocks and roots in newly cleared land with little or no damage. The disk implements also worked well in sticky soils where moldboards wouldn't scour.

A disk plow had large 26- to 32-inch disks, each having its own support and bearing assembly, and was capable of working to normal plowing depths. A disk tiller or a one-way had one or more gangs of 24- or 26-inch disk blades mounted on a common shaft with two bearings, similar to a disk harrow, and worked at shallow to medium depths.

Allis-Chalmers offered the No. 260 Texas "Beater" and the 420 Series disk plows during the mid-1930s. The Texas "Beater" was available in two- to five-disk sizes with weights ranging from 2,900 to 5,930 pounds — truly a heavy weight for plowing in mesquite and gumbo conditions. The 420 Series was offered for less demanding conditions in two- to six-disk models. They weighed 2,200 to 3,350 pounds, considerably less than the "Beater."

The S7 and S9 disk tillers were medium weight machines intended for use in lighter soils and shallow disking. The S7 had 12 disks and the S9 had 15, 22-inch disks spaced 8 inches apart. They weighed 2,073 and 2,173 pounds, respectively. LaCrosse-built disk plows and tillers apparently were discontinued by the late 1930s, only to be replaced by the Athens line.

By 1939 the WC was teamed with Athens disk plows and tillers. By 1941 these same implements were being sold under the Allis-Chalmers brand. This buy-sell relationship was the most durable of all that A-C established over the years. Contrary to some accounts, Allis-Chalmers did not own any part of Athens Plow. There was no formal contract between the two companies and business was done on a handshake.

Disk implements depend on their weight for soil penetration, indicated by their heavy cast-iron wheels and frame weights. There is no "suck" as in a moldboard. Suck is an old farm machinery term that describes the downward force on a plow bottom caused by soil weight and dynamic forces as it moves through the ground. A tractor would normally pull the same number of disks as plow bottoms. A Model WC handled a two-disk plow, each disk cutting about 10 inches, substantially less than a 14-inch plow bottom.

Single-disk and two-disk plows were furnished for the Models B and C tractors. At extra cost, both could be equipped with hydraulic lift, which was probably

Two- and three-disk plows with Snap-Coupler hitch were used with CA, WD, WD45, D14, and D17 tractors. Disk blades had a 26- in. diameter and each cut 10 to 11 in.

A D17 with a No. 355 five-disk plow preparing a deep seedbed. The 350 was the last of the A-C disk plows.

A WD45 tractor plowing with a Big Capacity No. 340 four-disk plow. The land wheel gauged the depth and carried the plow in transport. The transport valve allowed the Traction-Booster system to transfer plow weight to the tractor while withholding pressure from the wheel-lift cylinder.

easy to sell. Two- and three-disk plows were available for CA and WD tractors. All of these had the pin hitch, used before the Snap-Coupler hitch was offered. The CA plows had 24-inch blades and the WD plows had 26-inch blades. Prices for CA and WD disk plows ranged from $250 to $370.

Snap-Coupler hitch disk plows were furnished for the CA and WD, and later for the D14, D15, and D17 tractors. They were supplied in two- and three-disk sizes and had 26-inch disk blades. All were equipped with a heavily weighted stabilizer-coulter to keep the plow running straight and maintain full width of cut.

One- and two-disk plows were matched to the Model B and C tractors. Manual lift was standard with hydraulic lift optional.

The Big Capacity implement line for the WD45 tractor included the 340 four-disk plow. The semi-mounted 340 had a transport-gauge wheel for good tractor stability. With the transport valve the Traction-Booster system functioned normally, but the wheel lift cylinder didn't get pressure until the hydraulics were activated for lift, therefore the wheel did not affect plow depth except as a gauge wheel. Twenty-six-inch blades were standard, but 28-inch blades were optional for deeper plowing.

The 340 was replaced by the 350 Series disk plows in 1959. The 354 four-disk and 355 five-disk plows were more modern in design and appearance. The frame, a single three-inch square solid steel beam, provided strength and weight for penetration. The transport wheel ran on the land for positive depth control. Disk blades were 26 or 28 inches with up to 5/16-inch thickness. Width of cut was 44 or 55 inches. The 354 weighed 1,655 pounds and sold for $712 while the 355 tipped the scales at 1,865 pounds and had a list price of $801.

The big disk plows met the same fate as the big moldboard plows as minimum tillage became standard practice in agriculture. As tractor power grew the wide-cutting plowing disk harrows or chisel plows took over and disk plows faded away.

Disk Tillers

Athens disk tillers showed up at the same time as their disk plows, about 1939. Early pull-type units had heavy cast furrow and tail wheels and heavy frames for soil penetration. When rubber tires came into general use, the wheels still had heavy cast centers or wheel weights.

The Athens-built A-C tillers came in three- to six-disk sizes with 24- or 26-inch blades, spaced about 10 inches apart. The three-disk size was recommended for the B tractor, the four-disk model for the Model C, and five or six for the WC. Manual lift encouraged tilling around the field which required less lifting. Early models had the main beam along the left side of the disk gang. As designs evolved, the beam was relocated above the gang for improved trash clearance.

Improved disk tillers had the Snap-Coupler hitch for CA, WD, WD45, and D-Series tractors. Sealed Bal-Pak bearings helped reduce draft and required no servicing. Four-, five-, and six-disk sizes had 26-inch disks and scrapers were extra cost. Prices ranged from $342 for a four-disk to $400 for a six-disk tiller. Weights varied from 832 to 1,040 pounds.

The 320 disk tiller was part of the Big-Capacity implement line and it utilized the transport valve-remote ram principle for weight transfer. Originally a 10-disk size, 8- and 12-disk models were added for D14 and D17 tractors. The 26-inch blades were spaced 10 5/8 inches apart in two separate gangs, each carried on a pair of self-aligning Bal-Pak bearings. A front furrow wheel and the on-land rear transport wheel controlled depth while the stabilizer coulter controlled width-of-cut and working angle. An eight-disk unit sold for $667 and the 12-disk size listed for $798. They weighed 1,799 and 2,116 pounds, respectively.

A WD45 with a Snap-Coupler-hitched five-disk tiller working in stubble. These tillers were equipped with the A-C sealed Bal-Pak bearings which did not require service.

A styled WC pulling an Athens five-disk tiller. The rear wheel had a coulter with a heavy cast weight to counteract the implement's tendency to go toward the left.

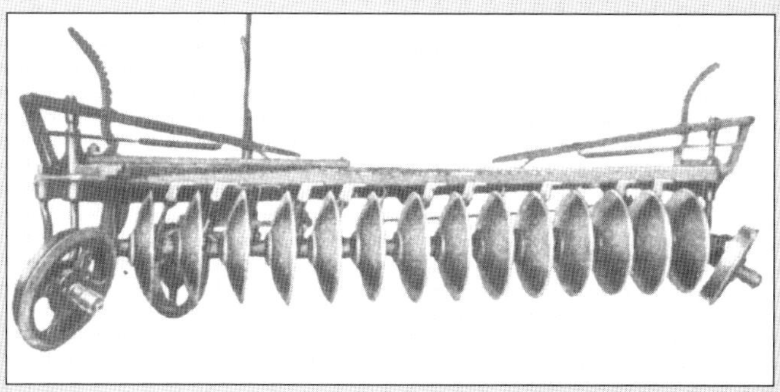

The 1932 WD 26-14 wheatland disk cut a 10-ft swath.

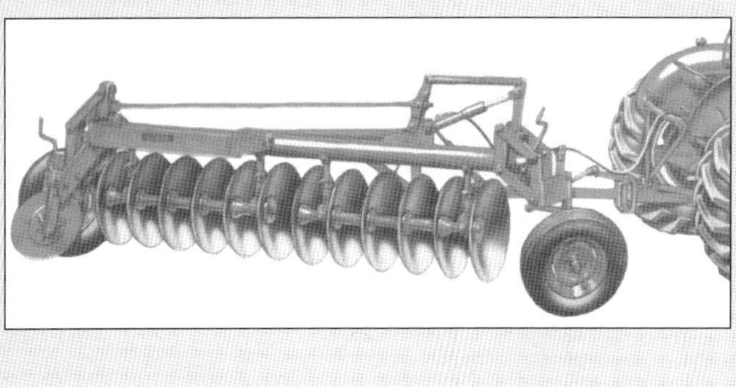

The 320 disk tiller was built in 8-, 10-, and 12-disk sizes. The 26-in. blades were spaced 10 5/8 in. apart.

The 310 one-way looked like the 320 but had sixteen 24-in. blades spaced 9 3/8 in. apart and normally worked at shallower depths. One-ways were mostly used for small grain seedbed preparation.

A WC tractor working with a No. 114 wide-type single-action disk harrow. The seat revealed its horse-drawn past.

A WC and a 7-ft No. 8 disk harrow disking Illinois cornstalks. The No. 8 also came in 6-, 8-, and 10-ft sizes.

The No. 10 heavy-duty disk harrow was built in 5- to 10-ft sizes with 16-, 18-, or 20-in. blades.

Although a one-way disk had much in common with the disk tiller, the two implements had different applications in distinct markets. The one-way was at home in the wheat country of the central plains. It was widely used for shallow tillage for wheat and summer fallowing. As wind erosion was recognized as a serious problem and implements leaving stubble on the surface were commonly accepted, the one-way lost favor except when equipped with seed boxes for planting wheat. Only the A-C disker-drill was equipped for planting.

The 310 one-way was similar to the 320 disk tiller and had many interchangeable parts. It had 16, 24-inch blades spaced 9 3/8 inches apart — 1 1/4 inches less than a disk tiller. It was a Big Capacity implement utilizing the remote-ram principle and transport valve for Traction-Booster weight transfer. Transport width could be reduced by adjusting the angle of the rear transport wheel. WD45 or D17 power was recommended. A 310 one-way sold for $950 with Snap-Coupler hitch, trash bars, and scrapers. It weighed 2,371 pounds.

Lacrosse Disk Harrows

During the 1930s many tractor-drawn disk harrows were horse-drawn single-action disks with a tandem added. The heavier models designed for tractor operation were just arriving. That would describe the A-C disk harrows, too.

From a catalog description of the No. 114: *"Like all A-C disk harrows the Wide Type Disk Harrow is equipped with electrically heat-treated disks, carefully ground and polished, and made from special steel of unusual toughness and wear-resisting quality. They retain their sharpness and will not chip or break under the most severe usage. These disks work in any ground and will clean in the stickiest soils. No better disks are built, and farmers will appreciate their long life and efficient performance in this fine disk harrow."*

They just don't make'em like they used to . . .

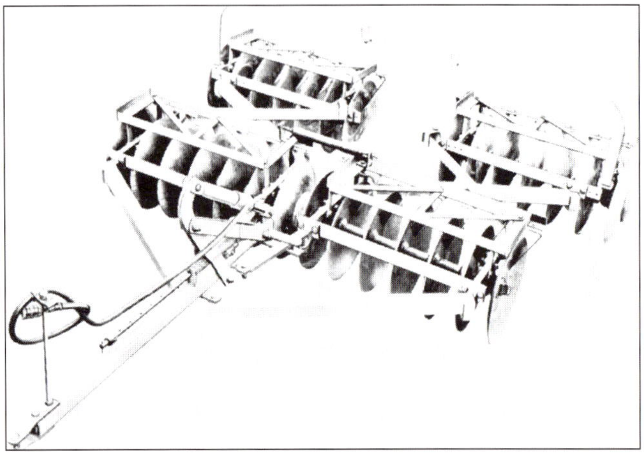

The 11B was the first A-C disk harrow with hydraulic control of the gang angle. The cross linkages de-angled the rear gangs in turns to eliminate ridging.

Allis-Chalmers was one of the first disk harrow manufacturers to use the inside-bevel grind which not only cut better than outside grind, but retained its sharpness longer. The Bal-Pak was replaced by a new bearing with triple-lip seals on the 2100 disk harrow. It could be lubricated to flush out dirt.

Mounted Disk Harrows

The first A-C tractor-mounted disk harrows made their appearance in 1953 for use with the CA and WD tractors. Both were single-hitch point implements with pin hitch. Later they were called the CA20 and WD20, and gained the Snap-Coupler hitch around mid-1955. These harrows had Bal-Pak bearings, which had a sealed ball bearing inside a sleeve with double-lipped seals on each side. No lubrication was required.

The No. 36 Heavy Duty Disk Harrow

The Athens-built heavy-duty disk harrows were single-action 6- or 7 1/2-foot harrows. Sometimes called bush and bog harrows, they were effective in brush, hard soils, tough stubble, corn or cotton stalks, and pasture renovation. With the gangs reversed, they could build beds and levees. With gangs adjusted to a V, they could make ditches.

The early 7 1/2-foot No. 30 was offered for the WD tractor. Later the No. 36 was built in both 6- and 7 1/2 foot sizes. The 6-foot No. 36 had eight 24-inch heavy gauge blades

The heavy-duty No. 40 was matched to WD power in 6- and 7-ft sizes. Blades had 18-, 20-, or 22-in. various spacings. It had adjustable gang angle and Bal-Pak bearings.

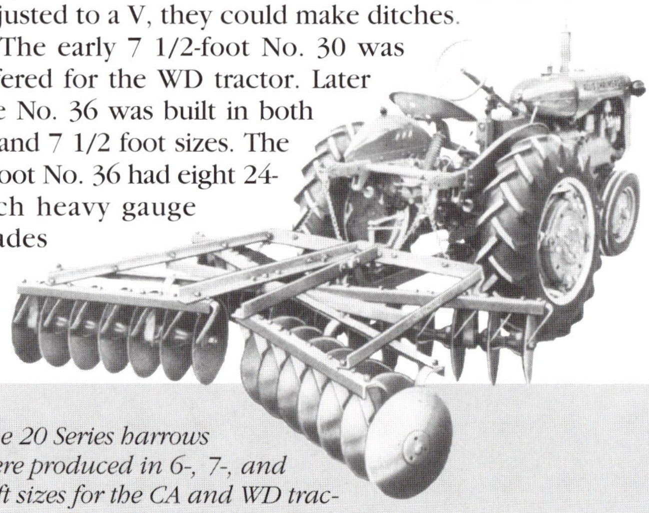

The 20 Series harrows were produced in 6-, 7-, and 8-ft sizes for the CA and WD tractors. Early models had pin hitch, later changed to Snap-Coupler.

The 100 Series harrows came in four sizes from 5 to 10 ft with Snap-Coupler or three-point hitch. A beaming crank was added to make fore-and-aft leveling more convenient.

The 2100 Series was the last of the A-C mounted disk harrows. Single-tube gang frame provided good trash clearance. There were four sizes from 5 to 10 ft.

spaced 10 inches apart and 139 pounds per blade for penetration. The 7-foot size had 10 blades. They were equipped with Bal-Pak bearings and Snap-Coupler hitch. A 6-foot No. 36 harrow with scrapers weighed 1,024 pounds and sold for $430. The No. 38 was a larger and heavier harrow similar in design and matched to the D19 tractor.

The Athens-built No. 36 heavy-duty disk harrow was made for working in brush, hard soils, and heavy stubble. It could also build levees and dig ditches by rearranging the fully-adjustable gangs.

Wheel-Type Disk Harrows

Allis-Chalmers' first wheel-type disk harrow was the No. 213, a 12 1/2-foot unit which was part of the Big Capacity implement line for the WD45. This unique implement broke new ground, and not just the surface of the earth.

It was the first wheel-type disk harrow that could utilize hydraulic weight transfer and was an A-C exclusive. The harrow was close-coupled to the tractor with Snap-Coupler and a chain yoke that connected the tractor lift arms to the harrow hitch. The transport valve allowed the 1,200 psi Traction-Booster system to exert tension on the chain yoke to transfer weight to the tractor's rear wheels while disking but

The WD45 tractor could easily handle the No. 213 12 1/2-ft disk harrow because it utilized the Traction-Booster weight transfer system. It was the first wheel-type implement to accomplish this through the unique A-C transport valve.

A D19 with a deep-working No. 214HD heavy-duty harrow. The 32 20-in. blades were spaced 10-in. and the harrow cut 13 ft. It weighed 2,408 lb with 75 lb per blade for good penetration. It was an early A-C approach to a deep-cutting disk harrow for the Middle-West.

kept pressure from entering the lift cylinder. Full system pressure of 3,500 psi opened the transport valve to lift the harrow. Since there was no load on the Snap-Coupler spring to call for weight transfer, the chains went slack to permit turns at the headland. This relatively simple system put the WD45, a 4,500-pound tractor, solidly into the four-plow class and allowed it to outwork much larger and heavier tractors. It was the beginning of vastly improved tractor performance.

The 200 Series disk harrows shared the Bal-Pak bearings and general gang design with the 100 Series mounted disk harrows. Gang angles were fixed at the optimum cutting angle for its 16- or 18-inch blades. With 40, 18-inch blades, the 213 weighed 1,960 pounds or 49 pounds per blade. Throughout its production life the 200 Series was built in six sizes from 8 to 15 feet. The two largest models were available as folding harrows with a transport width of less than 10 feet.

When the D21 came out, wider disk harrows to match its power were needed. LaCrosse made a squadron hitch which combined two-13 1/2- or 15-foot 200 Series disk harrows for a total width of 27 or 30 feet and capacity of 150 to 180 acres per day. This concept was fine for a farmer whose farm consisted of one extremely large field, but was not quite so good if transporting was required. Squadron hitching was more acceptable in the wide open plains.

The 2000 Series disk harrows were based on the 200 Series gangs but it built in folding and non folding sizes up to 20 feet. Both the 200 and 2000 Series disk harrows were replaced by the 2200 Series in sizes ranging from 9 to 20 feet. The 2200 line had single-tube gang frames and adjustable gang angle. They retained the Bal-Pak bearing.

The 300 offset tandem grove harrow had bissel blades for positive depth control in sandy groves. It was built in 10- and 14-ft sizes and is shown here with the Traction-Booster drawbar.

In 1972 the 2300 Series replaced the 2200 line. The gang frame was a single structural steel tube and bearing hangers were flat steel plates for more trash clearance. The bearings were self-aligning and flange-mounted with triple-lip seals and could be lubricated. The 2300 was built in 9- to 21-foot sizes with manual or hydraulic folding for the larger units. Blade sizes of 18, 20, and 22 inches were offered. Regular blade spacing was 8 1/2 inches and the weight per blade varied from 65 to 93 pounds. Total weight ranged from 1,990 to 3,888 pounds. Prices (1984) were from $3,466 to $8,140. The 2300 Series remained in the line until the LaPorte plant closed in 1985.

The 3000 Series disk harrows had up to 140 pounds per blade for good penetration. They were available in 10- to 31-foot widths, in single- or double-offset configurations, and folding or non-folding models. The 3000 Series harrows were built for Allis-Chalmers by American Products (Crustbuster) of Spearville, Kansas. They were offered from 1973 until the 2600 Series became available.

The heavy 2600 Series tandem disk harrows were designed to work with 100- to 250-horsepower tractors. There were 11 sizes from 16 to 32 feet in two frame types, center-fold and double-fold. The two-sec-

The D21 with two squadroned No. 215 harrows could disk up to 180 acres per day. Difficult to move from field to field, squadrons worked best in wide open country.

A bar-grill One-Ninety XT pulling a 20-ft No. 2000 disk harrow. Similar in design to the smaller 200 Series harrows, it was built in 13 1/2-, 18 1/2-, and 20-ft sizes and was the first of A-C's large wheel-type tandem disk harrows.

A Model Two-Twenty tractor pulling a 20-ft 2200 Series disk with coil-tine harrow attachment. The 2200's had single-tube gang frames for improved trash clearance and Bal-Pak bearings.

The Model 200 tractor and 2300 disk harrow working near San Antonio, Tex., in 1972. Both had just been introduced to the dealers.

The rugged 2600 center-fold disk was heavy enough for primary tillage. It had six transport wheels and folded to 14 ft for safe transport.

The outer gangs folded forward and the wings folded upward so the 2500 Series 33 1/2-ft disk was less than 14 ft wide and 14 ft wide in transport. At least 200 hp was recommended.

A Model 7060 and 2600 double-fold disk could cover up to 150 acres per day. Eight wheels provided flotation is soft soils. The harrow could fold to 13 1/2 ft for transport.

A 7030 and a 3000 Series heavy-duty disk working in cornstalks. With up to 140 lb per blade, the 3000's could cut deep and eliminate plowing.

tion center-fold hinged in the center for flexibility and transport, while the three-section double-fold had a wing on each side which provided flexibility and narrow transport width. These big disks were intended for primary tillage, cornstalk disking, and general seedbed preparation. The 24-inch blades were spaced 9 or 11 inches. Blade loading ranged from 118 to 173 pounds. The smallest 17-foot 2600 center-fold weighed 6,040 pounds and had a list price of $10,632. The largest 32-foot 2600 double-fold with cushion gangs weighed 12,685 pounds and had a list price of $21,573. The 2600 Series disk harrows remained in production until LaPorte closed.

In late 1978 the 2500 Series disk harrows were added to the line. Built for secondary tillage and seedbed preparation and somewhat lighter, the 2500 Series looked very much like the 2600. Rigid, center-fold, and double-fold models were built in 10- to 33 1/2-foot sizes. There was a choice of 22-inch blades spaced 9 inches or 20-inch blades spaced 7 1/2 inches apart. With working weights from 3,000 to 10,000 pounds, blade loading ranged from 95 to 116 pounds per blade. The smallest 10-foot rigid 2500 disk harrow sold for $5,763 in 1984. The largest 33 1/2-foot double-fold with cushion gangs sold for $19,858. The 2500 Series disk harrows remained in production after the LaPorte plant closed, being produced for Deutz-Allis by the Vermeer Manufacturing Company of Pella, Iowa.

Deep Tillage Equipment — The Oxnard Line

Allis-Chalmers bought the Monarch Corporation in 1928, redesigned the crawler line, and by the mid-1930s was selling the Springfield tractors through distribution outlets acquired with Advance-Rumely in 1931. However the dealers still did not have a full line for their customers. If a large offset disk harrow or subsoiler was needed the dealer had to purchase it from another supplier. The company needed a line of implements to participate in the big package deals common in that market.

The Brenneis Manufacturing Company

The Brenneis Manufacturing Company, Inc., was founded by Louis Brenneis in 1898. It was located in Oxnard, California, which is about 100 miles northwest of Los Angeles, and was well-located to serve its primary markets in California and Arizona.

The original factory was located in downtown Oxnard. By 1930 the line included disk harrows, ridgers, field cultivators, chisel cultivators, subsoilers, road rippers, spring tooth harrows, pulverizer-pack-

An early Brenneis offset grove harrow.

A Brenneis four-section regular disk harrow with 36 blades pulled by a Best tractor.

A United tractor pulling a Brenneis harrow with offset hitch (dated 1930).

ers, and beet and bean planters and cultivators. In 1938 Allis-Chalmers acquired the Brenneis Manufacturing Co. and it became the Oxnard Works. In 1946 the company purchased 15 acres on the eastern outskirts of Oxnard, 1 acre of which was transferred to a railroad. The remaining 14 acres became the site of a new factory which was occupied in 1950.

The Oxnard Disk Harrow Line

When A-C bought the line it was complete and ready to go. Brenneis offered disk harrows in about 10 types in several dozen sizes, both tandem and offset. There were straight, as well as squadroned, harrows. The Heavy Double Action disk harrow came in 8-, 9 1/2- and 11-foot sizes that weighed from 1,753 to 2,431 pounds — about 90 pounds per blade. Power requirements ranged from 25 to 50 horsepower. The Model G offset disk harrow was furnished in three sizes, the largest was 5 feet 3 inches and weighed 1,049 pounds. They required 15 to 30 horsepower. The full line of disk harrows filled 10 pages of the catalog.

The J and JM offset disk harrows were built for open field and orchard work where level disking and elimination of furrows were primary requirements. The JM came in three sizes from 6 to 7 1/2 feet and the larger J was offered in nine sizes from 6 to 18 feet with squadrons to 24 feet. The 22-, 24-, or 26-inch blades were spaced 9 inches with up to 100 pounds per blade for good penetration. There was a choice of hydraulic, mechanical turn bar, or one-way trip-rope closing.

Late Model L squadrons were in-line front-gang-coupled in 31- and 36-ft sizes.

A propane-powered Model L crawler and two 12-ft squadroned Model J harrows cutting 24 ft preparing a seedbed in California.

The Model L offset disk harrow was built in sizes up to 25 ft with weight per blade up to 128 lb.

The Model L offset harrow replaced the J harrow around 1959. The Model L had a single gang member above and ahead of the disk gang, which reduced trash plugging. It came in eight sizes from 9 to 21 feet. As tractor power increased the size range became 15 to 24 feet. Squadrons of 15 1/2- and 18-foot sizes produced 31 and 36 foot harrows with in-line front gangs and staggered rear gangs. Earlier squadrons had staggered front gangs and in-line rear gangs. In-line front gangs simplified hitch design and operation. Mechanical control was available but hydraulic control was more popular for quick gang closing and right turns. Gang construction consisted of 22-, 24-, or 26-inch blades spaced 9 inches apart by heavy cast spools on 1.5-inch gang bolts, and carried on sealed anti-friction bearings. This gang construction was used on all disk harrows except the largest GW and GWO models.

The Model L harrow was big, rugged, and heavy. A 21-foot model tipped the scales at over 3 tons and the largest 36-foot squadron weighed about 7 tons. Weight per blade from 120 to 150 pounds assured good trash cutting and soil penetration. The 21-foot harrow sold for $2,626 in 1960 but this was up to $10,959 by 1981. The 36-foot squadron sold for $21,737 that year. The L was built for heavy soils and crawler power, and it handled them well.

The Model K wheeled offset disk harrow was a popular Oxnard product because it came in several sizes for tractors of 40 to over 100 horsepower. Early sizes were 6 to 10 1/2 feet, but by 1981 the range was 9 to 13 1/2 feet as tractor power grew. The K was a low profile harrow for working in the groves. Wheels gave the positive depth control, easy transport, and quick right turns. With about 100 pounds per blade it had good penetration. A 9-foot Model K weighed 2,448 pounds and sold for $911 in 1960. By 1980 the list price had risen to $4,555.

The Model K was about the only Oxnard product to be used east of the Rockies due to the high cost of shipping. However, the Memphis A-C branch folks found that this disk could perform well in the Delta buckshot soils and dealers in the Mid-South began selling it. E. M. "Coach" McCance (All-American, Tulane University, 1929), sales manager at Memphis, was sold on the Model K and preached its virtues in every customer meeting. Now the K, like any offset harrow, required very specific adjustments if it were to follow the tractor, and some users had difficulty in making those adjustments. When the farmer complained about his field performance, Coach always asked him, "Do you have the scrapers on?" If the customer said yes, he was told to "take'em off;" if no, he was told to "put'em on," and the meeting resumed. Coach sold quite a few K harrows.

The Models KT and WKT were tandem disk harrows. The KT was built in 12-, 13 1/2-, and 15-foot sizes and the WKT in sizes up to 18 feet. A 12-foot Model KT with 32 blades weighed 3,700 pounds and sold for $6,091 in 1981.

More powerful tractors required larger offset harrows, plowing disk harrows, and two-way plows. Tractors with up to 300 or more engine horsepower were capable of operating at field speeds easily twice that of crawlers. Oxnard engineering had designed several plows, harrows, and subsoilers matched to the higher power levels.

The KT, a tandem version of the Model K, was built in 6- to 13 1/2-ft sizes. A 12-ft model had 115 lb per blade.

The Model K wheeled-offset harrow was one of the few Oxnard units used east of the Rockies. With about 100 lb per blade it was effective in tough soils and heavy trash.

The WK wheel-type offset disk harrow was offered in 13 1/2-, 15-, and 16 1/2-foot sizes. Heavier main and gang frames provided weights per blade from 124 to 137 pounds and gang angle was adjustable from 36 to 46 degrees. Each dual transport wheel assembly had its own hydraulic lift cylinder, and a hitch linkage kept the harrow level when lifted. A 16-foot Model WK had 44 blades, weighed 5,460 pounds, and had a 1981 list price of $9,279.

The Model XK offset disk harrow was a larger version of the WK with an even heavier main frame in three sizes of 16 1/2, 18, and 21 feet. Like the WK, dual transport wheels had separate lift cylinders and self-leveling in transport. A 21-foot XK had 56 blades, weighed 7,563 pounds and sold for $12,463 in 1981.

Many West Coast operators were turning to very heavy or plowing disk harrows for their primary tillage because these big disks could work deep at higher speeds than two-way moldboard plows. They were also less susceptible to damage when obstructions were encountered. The GO pull-type offset and GWO wheel-type offset disk harrows were tough enough to handle these conditions. These huge harrows had an enormous disk gang. The 32-inch notched disk blades were 3/8-inch thick and spaced 13 inches apart by 9-inch diameter cast iron spools, all held together by a 2 1/4-inch square gang bolt. The 8- × 8-inch gang frames could be filled with bal-

The 21-foot XK harrow and the 22-ft mulcher-packer made an excellent seedbed for any small seed. With 135 lb per blade the XK could also work deep in tough soils.

A Model Two-Twenty and an 18-ft Model WKT working in heavy trash near Hondo, Tex. For narrow transport the outer gangs folded forward and backward.

The WK was a larger version of the Model K and the gang angle was adjustable from 36 to 46 degrees. A linkage leveled the harrow for transport.

The GWO was a plowing disk harrow with up to 480 lb per 32-in. blade. It could replace a two-way moldboard plow.

last for greater penetration. The GO pull type had a rugged hinge with a 3-inch hinge pin to keep gangs aligned. Hydraulic control allowed quick closing for turns. The GWO wheel type had 4 1/2- × 8-inch main frame members for great strength and gang alignment. Gang angle was adjustable. Transport wheels were 11-16 for good flotation and positive depth control. Hydraulic leveling kept the harrow level in both working and transport positions. Both the GO and GWO were built in 13- to 21-foot sizes weighing from 9,238 to 16,203 pounds with weight per blade of 350 to 480 pounds. List prices ranged from $21,012 to $31,731.

Oxnard Two-Way Plows

Two-way moldboard plows were added to the Oxnard line about 1965 when three series, the 700, 800, and 900 were placed in production. Up until that time, Allis-Chalmers offerings of two-way plows were limited to a few models built at LaCrosse, none of which were large or rugged enough to meet West Coast requirements. A two-way plow is used to throw all the soil the same direction and eliminate dead furrows. This was absolutely required in areas where flood irrigation was used, and most of those were located west of the Rockies. West Coast A-C dealers needed two-way plows to meet their customers' requirements, and had been selling plows built by competitive manufacturers.

The patented gauge wheel automatically repositioned itself during rotation.

Plow bottom and beam production facilities already existed in LaCrosse, so Oxnard's task was mainly that of designing frames and a hydraulic rotating mechanism. LaCrosse had a hydraulic rotator, but it did not have enough torque capacity to spin the big West Coast units.

The Oxnard plows had two proprietary design features. A double-acting hydraulic cylinder with a patented valve rotated the plow. After the plow was lifted clear of the ground, actuating a remote valve would contract the cylinder to start plow rotation. As the plow went over top dead center the unique valve would reroute the oil pressure to extend the cylinder, completing the spin and putting the plow into the opposite working mode. It was necessary to start the cycle quickly so that the plow would have enough momentum to pass top dead center. Otherwise the plow could stop on dead center, an unhappy circumstance which an operator allowed to occur only once.

The other proprietary device was the gauge wheel mounting. The wheel was attached to a swinging beam which would reposition itself by gravity as the plow rotated, then lock in that working position until the plow was spun the opposite direction. It was one of those utterly simple designs that really worked well.

The 700 two-way plow was a three- expandable to four-bottom plow. Later a two-three bottom was added. Beams were spaced for 16-inch bottoms, but 18-inch deep tillage bottoms could be used for deeper plowing. Either shear-bolt or spring-trip beams could be used. The 700 plow was designed for in-furrow plowing with tractors from 80 to 110 horsepower.

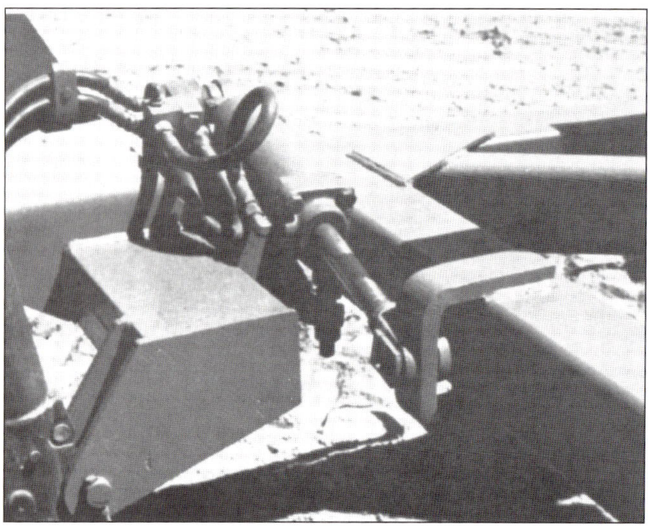

An exclusive valve-and-cylinder device rotated the large spinner plows. The piston rod was retracted then extended as the valve rerouted pressure at the mid-point of the rotation.

The 700 Series two-way plows were produced in two-three and three-four bottom in-furrow types for small and medium-sized tractors.

The huge 950 six-bottom spinner was matched to four-wheel drive tractors like the 7580. The plow weighed nearly 3,500 lb.

The 800 also used the LaCrosse shear-bolt or spring-trip beams. It was an on-land four- or five-bottom plow. Later this plow was redesigned with a heavier main frame and became known as the 950. The four-bottom base was expandable to five or six bottoms. This plow had 18-inch high-lift plow bottoms.

The 900 plow could be mounted on a 400 tool carrier to work with a crawler tractor. A BeGe hydraulic system lifted the carrier and rotated the plow.

The 950 rigid-beam two-way plow was an on-land type in three-, four-, and five-furrow sizes. The rigid beams were 1 1/2×4-in. double-ended members.

The 900 was a larger unit with rigid beams. The 900 came as a three-bottom in-furrow and three-, four- or five-bottom on-land plow. Later as the 950 Rigid Beam it was built as a four- or five-bottom on-land type. The 900 could be mounted on an HD6 mounted toolbar or 400 pull-type tool carrier. The 950 was equipped only with category 2 or 3 three-point hitch. Prices in 1981 ranged from $4,142 for a 700 two-bottom, to $10,070 for a 950 six-bottom plow.

An HD11 Series B farm crawler with five subsoiler shanks on a 600 tool carrier ripping hardpan in California.

This HD3 (1942) was pulling an early version of the 400 tool carrier with a two-disk border-ridger.

Oxnard Tool Carriers

Tool carriers were a big part of the deep-tillage equipment business. The versatility of a tool bar and its attachments made it a natural for West Coast agriculture. Irrigation, particularly flood, tends to pack the soil. Without the freeze-and-thaw cycles of hard winters, the central valleys of California and the southwestern United States are faced with compaction problems unknown in the rest of the country. This situation brought about the use of long subsoiler shanks to shatter the soil compaction. Where flood irrigation was used it was necessary to build borders and levees to control the water. Crops were planted on beds so roots wouldn't become waterlogged. Add cultivating and all these operations led to the widespread use of tool bars. As crawlers were the primary source of big power, the pull-type carriers were first to appear.

The HD6 toolbar was reversible and handled an 8-ft dozer on the front or a 4 1/2 x 8-in. toolbar on the rear for peak versatility.

Oxnard built two pull-type tool carriers, the medium weight 400 and the heavyweight 600. The 600 was a massive carrier built for the largest tracklayer power. It carried 6- × 12-inch toolbars in either 10- or 15-foot lengths. Hydraulic cylinders of 5 to 6 inches controlled the big machine, but cable control was available for tractors with winches. Attachments included 28- or 36-inch subsoiler shanks, chisels, and coil shanks with a variety of ground working tools. A 600 carrier with hydraulic control, 10-foot toolbar, and five subsoiler shanks weighed 2 tons and had a list price of $2,020 in 1960 and $8,150 in 1981.

The 400 carrier hauled 4 1/2- × 8-inch tool bars of 79 to 180 inches. It had hydraulic control only, and the wheels could be adjusted to 40- or 60-inch treads. Some of the many tools available included: 28-inch subsoilers, lister bottoms, chisels, coil shanks, border ridgers, and bedding disks. With subsoilers it weighed 1,200 pounds and sold for $850. As larger four-wheel-drive tractors with three-point hitch took over, the 400 carrier was discontinued.

Mounted toolbars came in even more variety than pull-types. The 200 mounted toolbar was a 6- × 12-inch bar with category 3 hitch in either 10- or 15-foot lengths for use with four-wheel-drive tractors up to 200 horsepower. It carried 24- to 36-inch subsoilers plus ground working tools like those for the 600. With five shanks and gauge wheels it weighed 1,050 pounds and sold for $3,000.

The HD6 toolbar carrier was an extremely versatile implement. When the outside hitch members were swung forward, they could be equipped with an 8-foot straight dozer blade. With them swung rearward, they could carry

The D19 Snap-Coupler toolbar frame could be adapted to 2 1/4-in. or 4 1/2 x 8-in. toolbars with their many attachments.

either 2 1/4- or 4 1/2- × 8-inch toolbars with their wide variety of ground working tools. The tractor had to be equipped with a BeGe hydraulic control unit. Two 5- × 8-inch double-acting hydraulic cylinders provided lift-lower control, and hydraulic tilt was available for the dozer blade. The HD6 tool bar became a popular implement wherever the HD6 was sold. It even helped sell tractors.

The D19 toolbar frame put the D19 tractor to work in deep tillage as the HD6 declined. Equipped with Snap-Coupler hitch, it was an easy-on easy-off implement. It used the same toolbars and attachments as the HD6 toolbar. Oxnard made several Snap-Coupler carriers for mounting 2 1/4-inch tool bars on the Models CA, WD, WD45, D14, and D17.

As three-point hitch tractors dominated the Allis line, three-point hitch tool carriers replaced the Snap-Coupler versions. The 45 toolbar carrier for 2 1/4-inch bars was a low cost unit recommended for wheel tractors up to 60 horsepower. The 100 toolbar carrier used 4 1/2- × 8-inch toolbars and was for tractors up to 110 horsepower.

V-Rippers and Subsoilers

The Oxnard-built chisels and subsoilers were V-rippers in that the toolbar formed a large V with the point toward the front. The benefit of this design is that the front shank shatters the soil first, then each following shank encounters less resistance and pulls easier, reducing power requirements. With straight tool bars this was done by having one front, two middle, and two rear shanks to form a V, but five shanks were the limit. With the welded V-bar, units with 7, 9, or 11 shanks maintain this advantage.

The Model 600 Chisel and Subsoiler Line

Model	620	630	640
Type	chisel	subsoiler	subsoiler
Toolbar (in.)	4 1/2 × 8	6 × 12	6 × 12
Shank (in.)	30	36	36
Depth (in.)	20	26	26
Hp per shank	15-20	25-30	25-30
Tractor (hp)	to 160	to 200	to 300
Typical:			
Shanks	7	5	9
Weight (lb)	2,030	2,800	5,233
Price	$3,515	$4,420	$7,970

Toolbar carriers for Snap-Coupler tractors used 2 1/4 or 2 1/2-in. toolbars. This WD45 and two-bar spring-tooth harrow are preparing a seedbed in Arizona.

A seven-shank 630 subsoiler could rip 26 in. deep when pulled by an 8550 tractor.

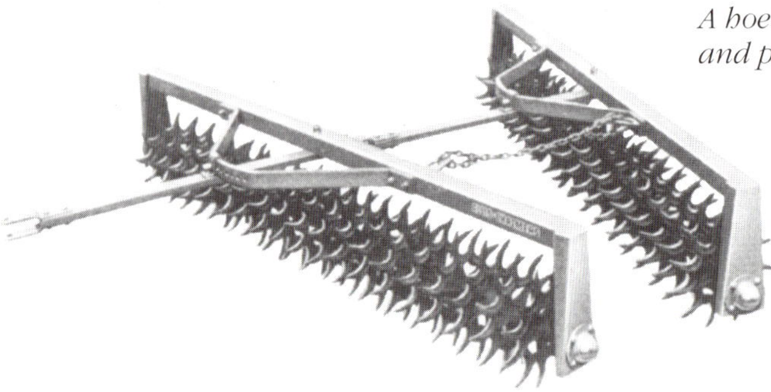

A hoe-treader worked by pushing the stubble into the soil to anchor it and prevent erosion. Pulled backward it would pack seedbeds.

The Mulcher-Packer and Hoe Treader

The mulcher-packer made pulverized seedbeds and helped to assure good stands of high value crops. With 180 to 210 pounds per foot the packer had plenty of weight to crush clods and pack the soil to conserve moisture. The unique Allis-Chalmers packer wheel had 23 fins on each side so even small clods were crushed. Each wheel floated independently. There were nine sizes from 6 to 25 feet, later reduced to six sizes of 14 to 25 feet. These sizes had weights of 2,590 to 4,630 pounds with prices ranging from $2,763 to $4,823.

The hoe treader was used in stubble fields and summer fallow land of northwestern wheat country for moisture retention, reduced soil erosion from wind and rain, and weed control. If reversed, it would firm seedbeds for better stands.

The 18-inch 11-tooth wheels weighed 34 pounds each and were spaced 5 inches apart. Each gang had 19 wheels, 38 total. Increasing the working angle produced a more aggressive treading action. Working width at maximum angle was 93 inches and was 96 inches at minimum angle. The hoe treader weighed 1,285 pounds. The usual arrangement was to have two or three units in a squadron.

The mulcher-packer with its exclusive packer wheel design crushed even the smallest clods to prepare fine seedbeds for vegetables and other high value crops.

Subsoilers

The Athens Plow Company made two types of subsoilers for Allis-Chalmers. The first was a group of pin-hitch single-shank units for the CA and WD. These were simple, curved-shank implements with no beaming adjustment. These were followed by Snap-Coupler hitched models for the same tractors. The No. 3 was for the WD, and the No. 4 was a slightly heavier version for the larger tractors.

The second type was the Snap-Coupler hitched No. 44 twin-shank subsoiler for the D17. A third shank could be added. The Nos. 4 and 44 had hand-wheel-operated beaming screws for leveling the point at working depths as deep as 18 inches. Coulter and mole attachments were available for both. Prices ranged from $55 to about $255. Oxnard made a Snap-Coupler single-shank subsoiler that was capable of working 18 inches deep. It had a curved shank

This WD subsoiler shattered plow pan to help control run-off. The later No. 4 had a handwheel for a beaming adjustment. It was made by Athens Plow Co.

This single-shank subsoiler was made in Oxnard and sold on the West Coast. It was distinguished by its beaming screw and curved shank.

The No. 44 was a two-shank subsoiler with an optional center beam. Built by Athens, it was matched to D17 and D19 Snap-Coupler tractors.

and no handwheel on the beaming screw which distinguished it from the Athens-made No. 4. The Oxnard unit was sold in the western states.

Chisel Plows

The chisel plow had been accepted for primary tillage in the high plains and other areas with soils subject to wind erosion. If the company wanted to sell implements in that market, a chisel plow was required. Allis-Chalmers elected to purchase a line of chisel plows rather than developing it at the LaCrosse plant.

Chisel plows from Jeoffroy of Amarillo, Texas, became part of the A-C implement line in 1959. Jeoffroy was well established in the chisel plow business and had grown enough to need a greater market outlet. The line included mounted, pull-type, and wing-type plows from 7- to 38-foot widths. Rigid clamps were used in rock-free areas and were available at a lower cost. Spring clamps with two springs protected the shanks and ground-working tools from obstructions. Shanks of 22- and 32-inches were offered. The longer 32-inch shanks provided more clearance for heavy crop residue. Ground working tools were 2- × 16-inch chisels or 16-inch sweeps. With shanks spaced 12 inches, the sweeps had good overlap for complete tillage.

The 600 mounted chisel plow had two tool bars and came in 7-, 9-, and 11-foot sizes with three-point hitch. With extensions mounted, plows up to 15 feet were available. Gauge wheels provided accurate depth control.

The 400 pull-type plows were built in 7-, 10-, and 13-foot base units, expandable to 17 feet. Seamless steel tubing frames had three bars for staggered shanks spacing and greater trash-handling capacity. Rod weeder attachments were also available. A 13-foot pull-type plow with Traction-Booster hitch, spring clamps, 32-inch shanks,and chisels weighed 2,388 pounds and sold for $959 in 1960.

The wheatland D17 with Traction-Booster hitch and an 11-ft pull-type chisel plow could till up to 50 acres per day.

The D19 and an 11-ft mounted chisel plow made a fast-stepping easily-handled outfit. Earliest mounted chisel plows had two-bar frames and were furnished in 7-, 9-, and 11-ft sizes. All could be expanded by 2 or 4 ft.

The Model 220 tractor was teamed with a 20-ft No. 610, A-C's largest mounted chisel plow. The 610 had a three-bar frame which provided good trash clearance.

The 800 wing plow came in 22- and 28-ft base sizes and the 850 was built in 28- and 34-ft base units. All could be expanded with 2-, 4-, 6-, 8-, and 10-ft extensions to provide any size from 22 to 44 ft. These were the last of the Jeoffroy chisel plows.

The D21 with a 28-ft wing plow could chisel 110 to 140 acres per day, depending on working depth. A single hydraulic cylinder activated all four transport wheels through a cable arrangement.

Wing Plows

The big wing plows were designed for larger farms and tractors. Various combinations of 7-, 10-, 136-, and 16-foot base units with 6- or 9-foot wings and extensions would provide 19- to 38-foot wing plows. With a power requirement of 8- to 10-drawbar horsepower per shank, only big tractors need apply. A single, large hydraulic cylinder and cable arrangement lifted the plow. Another cable lift folded the wings for transport. The smallest, least-equipped 19-foot wing plow weighed about 3,900 pounds and sold for $1,513, while the largest, best-equipped 28-foot model weighed nearly 7,600 pounds, and had a list price of $2,983.

The 610 mounted, 410 pull-type, and 710 wing-type chisel plows were second-generation Jeoffroy-built implements which replaced the prior models in the late 1960s. The larger 610 three-bar mounted plows came in widths up to 20 feet. The 410 pull-type plows had new frames and hitches. The 710 wing plow had an improved lift system which provided hydraulic lift for the wings.

The 800 and 850 wing plows were added in 1971. The 800 had an 11-shank center section and with wings and extensions provided 22- to 38-foot sizes. The 850 had a 17-foot center section. With various wings and extensions, sizes from 28 to 44 feet were available. The Jeoffroy chisel plows were replaced by the LaPorte-built 1600 Series in 1972.

The 800 and 850 wing plows were transported by raising the entire frame to vertical, then folding wings forward along the hitch.

1600 Series Chisel Plows

The 1600 Series chisel plows were introduced in 1972, and were still in production when the LaPorte plant closed. There were four types in the series: mounted with three-point hitch in 8- to 20-foot sizes, pull-type in 8- to 16-foot sizes, short wing pull-type in 18- to 24-foot sizes, and long wing pull-type in 26- to 30-foot sizes. Frames were 4- × 4-inch tubular steel toolbars of all-welded construction. The mounted and rigid pull-types had three-bar frames, while the wing types had four-toolbar frames for wider shank spacing and trash-shedding ability. Rigid, single-spring, or dual-spring clamps, 26- or 32-inch shanks, and sweeps, reversible double point, or twisted shovels were available across the line. Shanks were spaced 12 inches apart.

Larger, long-wing pull-type plows had dual transport wheels on the center section and singles or duals on the wings. On wing-type plows a 4- × 24-inch double-acting cylinder folded the wings upward, reducing transport width to 20 feet. Each transport wheel had an 8-inch cylinder for lift and depth control.

Some typical 1984 prices:

Model	Weight (lb)	List Price ($)
Mounted 8 ft	1,143	2,144
Rigid pull-type 12 ft	2,457	4,628
Short wing pull-type 24 ft	5,360	11,075
Long wing pull-type 30 ft	7,994	16,611

The 1600 long-wing plow had a four-bar frame for widely spaced shanks and maximum trash clearance. This plow could be equipped with four, six, or eight wheels. Walking-beam duals provided even depth in rough land.

A 7060 tractor easily handled this 1600 22-ft short-wing folding plow. It was equipped with rigid clamps for working in soils without rocks or roots.

A 7060 and a 1600 16-ft rigid pull-type chisel plow going over a field the second time for a fine seedbed. This plow had single-spring clamps for moderately rocky conditions and 32-in. shanks for good trash clearance.

The 1500 Min-Til conservation tillage tool left crop residue on the surface for erosion control and permitted primary tillage at a rate three times that of a moldboard plow.

The 1500 Min-Til Conservation Tool

As progressive farmers faced their soil erosion problems and looked for ways to reduce the time and cost of tillage, the chisel plow gradually replaced the moldboard plow. However, in heavy cornstalks the chisel plow became an extremely effective rake, defeating any anticipated reduction in time. Front-mounted coulters solved this problem and the chisel plow evolved into the coulter-chisel tool.

The Min-Til Conservation Tool was A-C's version of the coulter-chisel. The coulter gangs had 20-inch blades spaced 7 1/2 inches apart to double-cut the crop residue ahead of the shanks which were spaced 15 inches apart. The 4- × 4-inch tubular steel frame was similar to the chisel plow, except that the front toolbar directly over the coulters was solid steel for extra weight to keep the coulters cutting. The coulter gangs were cushion-mounted to protect blades from rocks. Rigid, single-spring, or double-spring clamps and 32-inch shanks with a choice of ground working tools equipped the Min-Til for a wide variety of soil and crop residue conditions and to provide the degree of trash coverage needed. The 4-inch twisted shovels were a popular choice, leaving the trash well anchored and providing a wide trench to catch surface water.

The Min-Til was built in five sizes from approximately 6 to 16 feet, with working weights from 3,300 to 7,450 pounds. Prices ranged from $4,700 to $9,800. The Min-Til was still in production when LaPorte closed and continued to be built for Deutz-Allis by Vermeer.

Bedders, Listers, and Bedder-Planters

For the benefit of "flat-landers," some definitions may be necessary. A bedder was a toolbar implement with a set of bottoms which shaped the soil into ridges or beds. It was also known as a middlebuster or row splitter. Add a planter to each bedder bottom and it became a lister or bedder-planter. Over the years this same implement, with modifications, was also known as a plateau-planter, mulch-planter, minimum-tillage planter or till-planter.

There are several situations which called for the use of these machines. In semi-arid areas listing or planting in the furrow bottom put the corn, maize, or cotton seed in moist soil for quicker germination. After the crop had emerged, cultivation brought soil from the ridges to cover small weeds and help retain moisture in the furrow bottom. Ridges helped to reduce wind erosion and protect the young crop from the wind. This practice was common on the high plains of Kansas, Oklahoma, and Texas.

In high moisture soils, ridges or beds were made with the bedder (no planting attachment) and the soil was allowed to dry until it was right for planting. The seed was planted on top of the beds with a conventional planter equipped with "bed knockers" or skids which flattened the beds. The furrows remained to catch and hold excess water. This practice was used for cotton in the Mississippi Delta and other high soil moisture areas.

To meet the wide variety of conditions in these important crop-growing areas, most companies developed an extremely broad line of listing and bedding equipment and Allis-Chalmers was no exception. Early listers were horse-drawn pull-type implements, while tractor-drawn types were mostly unchanged. As with other equipment, row-crop tractors, tractor hitches, and hydraulics rapidly changed lister-bedders to mounted designs.

A One-Ninety XT with a four-row till-planter equipped with fertilizer attachment doing a once-over planting job in stubble. Two- and six-row sizes were also available

The 1935 L-50 independent-beam wheatland lister came in three-, four-, and five-row sizes. Planter attachments were quickly mounted on lister beams. Five seed plates for corn and maize were furnished.

Plateau, Mulch Planting, and Till-Planting

Mulch tillage was the practice of planting the new crop directly in stubble from the previous crop. It was a once-over operation with a minimum of seedbed preparation. The plateau planter was an adaptation of a rotary-moldboard lister unit which formed a 10-inch plateau on which the crop was planted. A furrow on each side of the plateau drained excess water away from the crop yet it was protected from wind erosion by the ridges between rows.

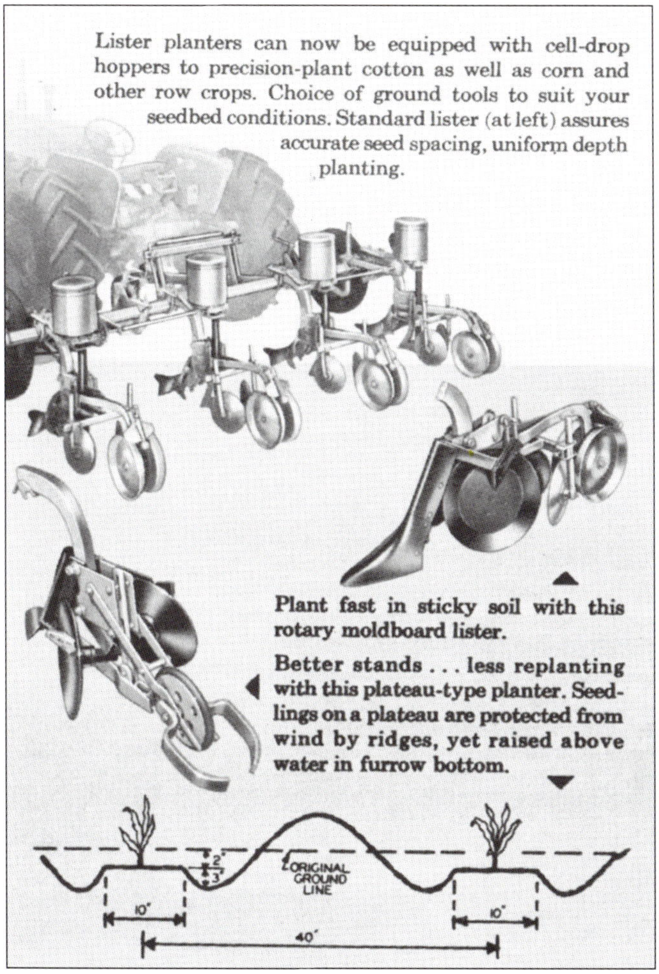

Till-planting was a minimum-tillage method for corn and grain sorghum that conserved soil and water as well as reducing costs up to 50 percent. The first step was to shred old stalks thoroughly to reduce residue problems in cultivation. Then the new crop was planted in the old crop ridge where the soil was warm and well-drained. Cultivation threw soil and residue into the row for weed control. The crop was grown with only three field operations. A-C listers were easily adapted to till-planting.

Two-row bedders and listers were available for both the Models UC and WC. One-row rigs matched the Model B, two-row for the RC and C tractors. Starting with the WD, two-row front-mounted bedders made their appearance. A rear center attachment gave three row capacity. The WD45 and D17 brought four-row rear-mounted listers, bedders, and bedder planters. The D19 upped capacity to six rows, and the D21 raised that to eight rows.

A WD45 with four-row bedder-planter could plant up to 60 acres per day. Large hoppers were supplied for corn-only or corn-and-cotton types.

This No. 257 bedder-planter for the WC tractor was equipped with flatland planting attachments. Depth skids leveled the surface ahead of the openers and controlled depth of seed.

In the early 1960s a new kind of bedder, the disk bedder (called "hippin' ridger" in the mid-South) permitted higher operating speeds. A-C built four- and six-row sizes.

The heavy-duty front carrier from LaCrosse could be fitted with four bottoms in front and one in the rear for five-row bedding capacity. Four-row planting attachments and four-row cultivator rigs made this unit a highly versatile implement.

Two types of planter hoppers were available: the regular plate-type singulation for corn and the combination type for linted cotton or corn. As acid-delinted cotton seed replaced the use of linted seed, the combination hoppers became obsolete. With a catalog full of bedder bottoms, lister shares, special sweeps, rotary moldboards, and covering equipment, the variety of lister-bedder equipment was endless.

Field Cultivators

The early A-C field cultivator (quack digger in Wisconsin) was built in two types, spring-tooth and stiff-tooth. The spring-tooth was preferred for quack grass killing, if there was such a thing. The spring teeth could bring the quack rhizomes to the surface, but the killing depended on several days of sunny, dry weather to dry them thoroughly. If a nice spring rain came before the drying was complete, the whole operation turned into mere transplanting. A more certain procedure was to rake and burn the roots, but on most farms that would have been a Herculean undertaking. Not limited to quack digging, these cultivators were used for alfalfa renovation, secondary tillage, and summer fallowing.

The One-Ninety with a four-row disk bedder (hippin' ridger) could throw up beds at a much faster rate than conventional middlebuster bottoms. A six-row sizes was also available.

The One-Ninety front-mounted implement carrier could be equipped as a five-row bedder, four-row planter, or four-row cultivator.

The D21 tractor with a 320-in. 5× 7 toolbar equipped with eight bedder bottoms and planting attachments was good for 120 acres per day.

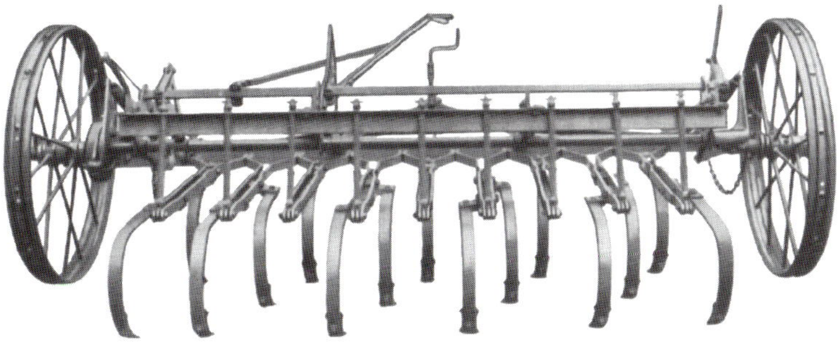

This 1935 LX1216 7 1/2-ft spring-tooth field cultivator was one of five sizes from 6- to 12-ft. Spring tooth version had 2-in. double-pointed shovels. The stiff-tooth version was built in comparable sizes and had 12-in. sweeps.

Both types were built in 6-, 7 1/2-, and 9-foot sizes with a single power lift, and 10 1/2- and 12-foot sizes with double power lifts. The spring-tooth version had double-pointed shovels spaced 6 inches apart. The stiff tooth, in reality a spring-loaded shank, model had sweeps spaced 9 inches apart. Both had staggered teeth which allowed trash to pass through.

Coil-Shank and Spring-Tooth Field Cultivators

The pull-type quackdiggers of the 1930s and '40s surrendered to mounted field cultivators as the drawbar capitulated to hitch and hydraulic systems. Among the first for Allis-Chalmers, the Model B rear-mounted field cultivator set the pace for change. It had ten double-pointed shovels on spring teeth and took a 6-foot swath. In the absence of hydraulics a single gauge wheel controlled the depth.

Coil-Shank Field Cultivators

The coil shanks were made of heat-treated spring steel and were usually spaced 12 inches apart. Twelve-inch double-pointed shovels were normal equipment but 10- to 15-inch sweeps could be substituted. The CA model had five coil-shanks, while the WD unit had seven. When the WD45 entered the line, 9- and 11-shank models became available. In 1960 the coil-shank line consisted of three sizes: 9, 11, and 15 shanks. They weighed from 595 to 1,110 pounds and sold for between $237 to $357 with Snap-Coupler hitch.

The Model B with its field cultivator was only 6-ft wide and could work in narrow rows of tall staked crops such as this vineyard.

The Model CA with a five-tooth coil-shank field cultivator was good for pasture renovation. With rigid lift arms, the cultivator had little lateral movement, unlike later true single-hitch-point implements.

Nine-, 11-, and 15-shank coil-shank cultivators were supplied for the WD45, D14, and D17 tractors. With Snap-Coupler single hitch-point these field cultivators could move laterally to follow the tractor on contours.

A D17 tractor with the largest 15-ft mounted spring-tooth field cultivator. Ten- and 12-ft sizes were also offered for the D14 and WD45 tractors.

Spring-Tooth Field Cultivators

On spring-tooth cultivators, the shovels were spaced much closer at 7 inches than the coil-shank cultivator. Spring-tooth cultivators for the smaller tractors had lighter teeth with about 100 degrees of curvature. Larger cultivators, for more powerful tractors, had much heavier spring teeth with about 280 degrees of curvature for extreme flexibility.

The CA spring-tooth field cultivator had 12 teeth spaced about 7 inches apart and cut 7 feet. The WD spring-tooth unit had 15 teeth and cut 8 1/2 feet. For the WD45, a larger 12-foot cultivator with 18 teeth was available. For the D14 and D17 tractors, three sizes were available: 10-foot, 15 teeth; 12-foot, 18 teeth; and 15 foot, 21 teeth. Weights ranged from 684 to 1,014 pounds and prices from $270 to $368 with Snap-Coupler hitch.

The Silent Weed Killers

The LaCrosse implement engineering folks seemed to have a penchant for utilizing rubber in unusual ways. First was the invention of zero-pressure rubber tires which would shed moist dirt and became standard in the industry for gauge wheels of all kinds. Bill Tanke, chief engineer, gets credit for that one. Another was the torsion rubber mountings for coulters which eliminated shear bolts and springs in that application. Perhaps the most important was the exclusive rubber-cushioned clamps for spring teeth on field cultivators. Here was a clamp which would permit a heavy, heat-treated spring shank to move in any direction necessary to dodge field obstructions, then immediately return to work without damage — and do it quietly.

When the new 1000 Series field cultivators with rubber-cushion clamps hit the market in 1968, many dealers refused to stock it because they had no faith in the rubber clamp. Those that did stock it anticipated a huge replacement parts business and really stocked up on the rubber cushions. After all, there were four cushions per clamp — bound to be good parts money in that. Strangely most of those cushions were still laying in the parts bin 20 years later. They just did not wear out. If a dealer finally sold a few, it was because the bolts eventually came loose and the cushions were lost. In any event the rubber-cushioned clamp proved itself, the non-believers started believing, and for a few years the company had about 20 percent of the field cultivator market. Another highly successful Allis-Chalmers innovation.

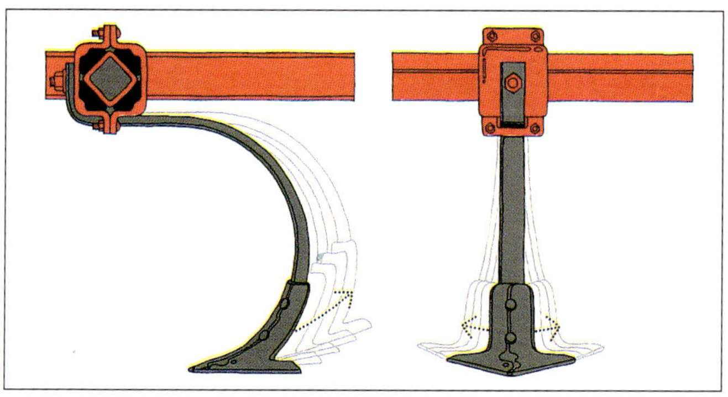

The patented rubber-cushioned clamp eliminated metal-to-metal contact so it earned the name, "The silent weed-killer." The clamp allowed the shank to move in any direction to dodge obstructions, furthermore, it didn't wear out.

The 1979 Silent Weed Killer line consisted of three series with cutting widths of 10 1/2 to 39 1/2 ft. Their good work, quiet operation, and long life made them market leaders.

The 1000 Series field cultivators were first offered in mounted three-point hitch and pull-type cultivators in a 10-foot base unit with various wings and extensions to provide sizes from 10 to 21 1/2 feet. Shanks were spaced 18 inches apart on each of three tool bars for an effective spacing of 6 inches. Double-pointed shovels and 4- or 6-inch sweeps were supplied. Later this series would be redesignated as the 1200 Series when the larger 1300 Series came out.

The One-Ninety XT easily handled a 16 1/2-ft 1000 Series field cultivator. Sizes up to 21 1/2 ft were furnished.

The 1300, which came out in about 1978, was a pull-type base unit of 11 feet with wings and extensions to provide sizes from 23 1/2 to 32 1/2 feet, all with four wheels. Transport width was 16 feet. The 1350 was an upgraded up version of the 1300 with two more wings in sizes from 33 1/2 to 39 1/2 feet with four or six wheels. The 1350 Series would eventually be redesignated as part of the 1300 Series.

Extremely successful, the 1200 and 1300 Series remained in production until the LaPorte plant closed and the 1300 was produced by Vermeer for Deutz-Allis. A 10-foot mounted 1200 field cultivator sold for $2,062, while a 21-foot pull-type sold for $4,802. The larger 23-foot 1300 had a price tag of $6,905 and the 40-foot size had a list price of $10,326.

The 1400 Chiselvator

The 1400 Chiselvator was an extension of the field cultivator line utilizing a much heavier version of the rubber-cushion clamp. With 3- × 3-inch tubular steel toolbars and 0.75- × 1.75-inch heat-treated shanks, it could work as deep as a light chisel plow, hence the name "Chiselvator." It had four toolbars for plenty of clearance around the shanks which could be spaced either 7 1/2 or 9 inches apart. The base unit was 12 feet wide and when combined with inner wings, outer wings, and extensions, would provide working widths of 21 to 61 feet. Ground working tools included double-pointed spikes or 6- to 12-inch sweeps.

Dual transport wheels had walking-beam suspension for more precise depth control. The number of

The 1400 Chiselvator was a heavy-duty field cultivator built in 21 to 61-ft sizes. Though not as heavy as a chisel plow, it could perform the same function in shallow stubble work and could replace the lighter field cultivator in secondary tillage.

The Chiselvator rubber-cushioned clamp was a larger and heavier version of the field cultivator but provided the same quiet operation, flexibility, and long life.

A three-point hitch mounted four-section coil-tine harrow. Coil-tines could vibrate in all directions for unusually effective action to remove small weeds. Sections came in 5 1/2-, 6-, and 7 1/2-ft widths with 10-, 12-, and 14-in. teeth.

wheels varied from 6 on the smallest 21-foot size, to 14 on the largest 61-foot model. Hydraulic folding reduced transport width to about 13 feet and to 16 1/2 feet on sizes over 49 feet. Coil tine harrow attachments were available for all sizes of the Chiselvator. The 21-foot sized model weighed 4,345 pounds and sold for $7,720, while the 61-foot size tipped the scales at over 11,200 pounds, and had a list price of $21,325. Chiselvator production was continued by Vermeer for Deutz-Allis after the LaPorte plant closed.

Tractor hydraulics could fold a 61-ft Chiselvator to 16 ft for transport in just a few seconds.

Spring-Tooth and Spike-Tooth Harrows

Spike and spring-tooth harrows fall into that broad category of, "If you've seen one, you've seen'em all." Ever since manufacturers learned how to make them out of steel, changes have been minimal if not nonexistent. Harrows, especially spike-tooth, are basic. Immediately after the first agrarians invented plows, they discovered a need for an implement which would smooth the soil enough to plant seeds. Turning again to trees as they had for their plows, they fashioned wooden triangles through which they somehow attached wooden spikes. Then with man-woman-slave-oxen power they prepared a seedbed. The more affluent who owned an ox could pull a bigger harrow and get his seeds planted sooner. Thus began farmers' age old demand for more power and larger implements.

Although designs have changed little, better materials have contributed to longer life and more satisfactory performance. Ground-working tools last a lot longer. The biggest changes involve transporting the harrows. The first field cultivators were essentially spring-tooth harrows on wheels. Tractor-mounted harrows brought easier operation and faster transport. Hydraulically controlled wheel carriers made wider harrows practical. Until minimum tillage became the accepted practice, nearly every farm had and used their spike and spring-tooth harrows.

The LaCrosse Plow Company was building both types when A-C acquired it in 1929, and they were continued doing so for some years afterward. In a 1951 price list a three-section spring-tooth harrow sold for $69 and a five-section for $119. Sometime after that, the LaCrosse plant stopped producing harrows.

This four-section spring-tooth harrow is mounted on a hydraulic cart that permitted easy trash-clearing and fast transport. Sections were supplied in 4- or 5-ft widths.

Spike-tooth sections came in 5 1/2- and 6-ft widths. This six-section rig cut 33 ft. There were other sizes up to nine sections and 54 ft.

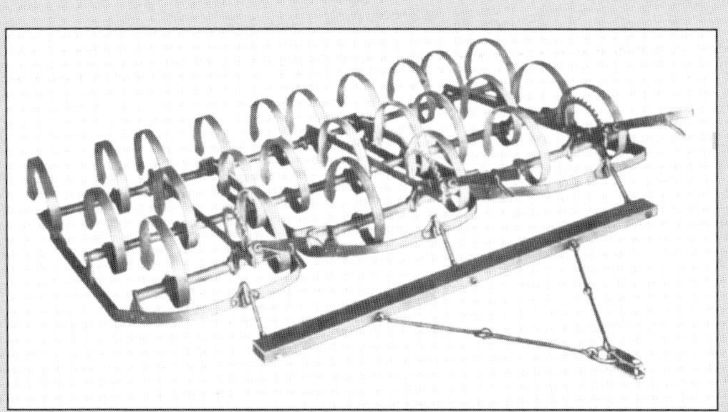

This three-section spring-tooth harrow was typical of the 1935 LaCrosse line. Five different types of teeth were offered and the largest size was an eight-section 65-tooth unit which cut about 22 ft.

The Oxnard "Controlled Tillage" spring-tooth was rope-controlled to rotate teeth to transport position, dump trash, and return to work. Five sizes were available from 6- to 10-ft.

The 1100 soil conditioner had a combination of spring-teeth and coil-tines to prepare a fine seedbed. A standard hydraulic cylinder lifted the entire rig to vertical then outer wings could be folded forward for narrow transport.

Brenneis was manufacturing spring-tooth harrows when A-C acquired that firm and the Oxnard Works continued to build two types. An automatic revolving harrow could clear itself of trash when the triprope was pulled, allowing the teeth to make a complete revolution than return to work. It was made in five sizes from 6 to 10 feet. A regular lever-controlled spring-tooth was built in 5- to 10-foot sizes. Both appeared in catalogs as late as 1955.

In the early 1960s, Allis-Chalmers reentered the harrow market by acquiring a line from the John R. Kovar Company of Anoka, Minnesota. The Kovar Company was well-established in manufacturing harrows and their products enjoyed a good reputation. It proved to be a fine working relationship which lasted for many years. Kovar furnished an extensive line of spike-tooth, spring-tooth, and later, coil-tine harrows, along with hitches, drawbars, and carriers. Only a few examples of the dozens of combinations are shown here.

The 1100 Soil Conditioner

The 1100 Series soil conditioners joined the line in early 1973. They were used for seedbed preparation and weed eradication, much the same as a field cultivator. The "S"-type spring teeth provided good trash clearance. For transport the standard 8-inch cylinder rotated the frame to vertical, then the outer wings were folded forward along the hitch to reduce overall width. Available sizes ranged from 12 to 36 feet. The 1100 Series was built by the Kovar Company.

Planting and Cultivation Equipment

Where the Harvest Begins Row-Crop Planters

George Schreiber, long time company associate and friend, once produced a film on A-C planters entitled, "Where the Harvest Begins." It was a well-told presentation of the planter line and how it effectively performed the jobs of seed placement, soil firming, and fertilizer and pesticide application. The title did, however, say it all. The planter can make or break a crop. Of all the machines on his farm, a farmer will tolerate no compromises in his planter and justifiably so. If the planter does not perform to the farmer's expectations there will be a new one, probably a different color, in his barnyard come next spring. The infinite variation of crops and soil conditions make it nigh impossible for one planter to be all things to all farmers.

The most unusual method of planting corn that the writer ever observed was on a Caribbean island. The practice was to bury a handful of corn, take five or six steps, then bury another, and so on. When the corn came up there were all these "corn bushes" spaced about like a dwarf orchard. Needless to say, the native farmers wondered why their corn did not produce much. For some reason, probably dating all the way back to hand planting, dropping seeds in hills was an established practice. This planting in hills was carried into modern mechanized farming.

Allis-Chalmers' first planter appeared in mid-to-late 1930s catalogs. It was a horse-drawn check planter with the seat removed and a short tractor hitch installed. These planters used a check wire with buttons every 40 to 42 inches, matching the row width, stretched across the field to trip the drop mechanism in the planter to plant a hill of corn. If the farmer was skilled in stretching the wire, the hills were aligned in rows crosswise of the field enabling him to cultivate the crop both ways with a much better chance of controlling weeds. The writer grew up on a farm where only absolutely straight rows with near perfect cross-checking was acceptable. (In contrast the writer farmed in a time when the standards became, "so what if the rows are a little crooked; only gonna use'em one year anyway").

A Model WC planting corn with a two-row check planter in 1938. Checking corn was limited to a horse's pace, about 3 mph.

Cross-checking lost favor for several reasons. First, it was slow. With horse-drawn planters, the horses got a little rest while the farmer moved the check wire stake. As tractors took over the planting operation, stopping to reset the stake was a time-consuming aggravation that only served to delay the urgent job at hand — getting the seed in the ground. Different forms of cultivation, including the high-speed rotary hoe, improved cultivator sweep design, and chemical weed control, all combined to eliminate the need to cross-cultivate corn. Closely-spaced single stalks were more easily mechanically picked than hilled corn, too. Since all check planters would also drill, it was just a matter of resetting the planter and the farmer's thinking, then leaving the heavy roll of checkwire in the toolshed.

Hill-dropping cotton was at one time as widespread as checking corn. The practice of hill-dropping cotton survived longer than checking or hill-dropping corn, but soon disappeared as mechanical cotton pickers took over the harvest. Cotton pickers worked better when they didn't have to handle the crop in large bunches.

A six-row pull-type Row-Flex planter with fertilizer attachment. Seed hoppers, fertilizer attachment, and herbicide attachment were all driven by the unit's press wheel. Unit design allowed many different row widths.

A Model WD with a mounted four-row drill planter with fertilizer attachments. The four-row was two, two-row planters with a four-row hitch.

The high-speed rotary hill-drop valve grouped the seeds and was opened by the cam. Seed was ejected backward at a speed approximately equal to the planter's forward speed which made tightly-spaced hills.

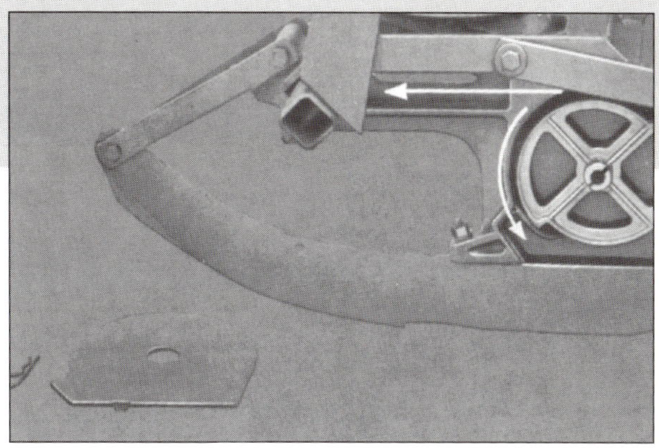

The Drill Planter

The first phase of post-World War II A-C planter design was the two-row mounted drill planter for the WD and CA, which soon expanded to a four-row for the WD and WD45. Initially these were equipped with runner openers and later with double-disk openers. Edge-drop or combination hoppers adapted the planter to a wide variety of seeds and crops. The seed plates of the A-C and the Deere 999 planter were interchangeable. How this occurred is not known, but it might be the last time the two companies ever agreed on anything. Non-corrosive fiberglass fertilizer hoppers replaced the metal hoppers about 1959. The two- and four-row drill planters were kept in the line for many years to provide a low-cost planter.

Row-Flex Planters

The Row-Flex unit planters came out in 1959. Units were supplied with either runner or disk openers, drill or hill-drop boots, and edge-drop or combination hoppers. Here the term "unit" indicates that the seed and fertilizer mechanisms were driven by the unit's own wheel, not by an external power source. The Row-Flex planters were built in two- and four-row mounted and four- and six-row drawn planters.

The high-speed hill-drop was an A-C exclusive in the Row-Flex units. A spring-loaded nylon valve collected the seed in a pre-set quantity determined by the seed plate. Cam action of the impeller wheel opened the valve and ejected the seed toward the rear at a speed approximately equal to the planter's forward speed. This reduced the seed bounce and kept the seed tightly grouped for accurate placement in the hill. The high-speed hill-drop was used for

The 400 cotton planter had either the high-speed hill-drop or drill openers. A second man rode the platform when linted cotton seed was planted with picker-wheel seed hoppers.

The All-Crop planter could be adapted to nearly any row crop. It was the first A-C planter which placed the fertilizer below and to the side of the seed by mounting the fertilizer attachment and its openers ahead of the seed openers.

planting corn and delinted cotton seed. It was a big improvement but it came out just as hill-dropping was phasing out.

The 500 Series Planters

The 500 Series planters brought increased fertilizer capacity with a 500-pound capacity hopper for each pair of rows. A six-row planter had two 500-pound and two 250-pound hoppers. Fertilizer was metered by the "Force-Flo" mechanism similar to that used in the All-Crop drill. Fertilizer openers were located ahead of the seed openers to place nutrients deeper than the seed. The frames and front-mounted fertilizer attachments made a new unit design possible. These were shorter and narrower than the Row-Flex units, which were still offered for cotton planting and to those who preferred them. Four types of units were available for the 500 — runner or double-disk, with or without the high-speed hill-drop.

The Wheel-Track Planter

During the early 1960s, agronomists began developing planting methods which finally led to minimum tillage systems. They reasoned that a fine, firmed seedbed was needed only in a small area immediately next to the seed and that the rest of the soil required tilling only to help control weeds. Out of this concept came several methods of limited seedbed preparation, one of which was wheel-track planting. The ground was plowed with a moldboard

A One-Ninety XT with a mounted coil-shank field cultivator and a 500 four-row planter. An early approach to minimum-tillage, it saved a trip over the field. The fertilizer hoppers held 500 lb and were driven by a traction wheel running against a transport wheel, permitting the use of a lighter weight, more compact planting unit.

The wheel-track planter used tractor wheels and two heavy packer wheels to firm narrow seedbeds in freshly-plowed ground, eliminating secondary tillage operations. It was another important step toward minimum tillage.

A second-generation wheel-track planter was built on the 600 frame with transport wheels arranged to pack the outside rows. A third wheel on the planter packed the third row, and the tractor wheels packed rows two and four. This planter is equipped with dry fertilizer and liquid herbicide attachments.

The adaptable 600 planter frame could be set up to wheel-track plant six-30-in. rows. The tractor wheels packed rows two and five, special packer wheels firmed rows three and four, and the transport wheels packed rows one and six.

plow, then the crop was planted immediately with the wheel-track planter. Wheels firmed the soil where it was needed for the seed to germinate, and the rough unworked soil between the tracks caught and held rainfall for reduced run-off. Soil compaction was reduced by eliminating the disking and other tillage operations.

LaCrosse built a wheel-track planter to perform this operation. The planter had an offset hitch so that rows one and three were planted in the tractor wheel tracks, while rows two and four were planted in tracks made by two heavy rubber-tired wheels mounted on the planter frame. The standard A-C Row-Flex units were heavy enough to place the seed at the proper depth. Chemical weed control reduced the need for cultivating the crop.

Second generation wheel-track planters were based on the 600 frame with a four-wide row and a six 30-inch row planter. On the six-row planter, the tractor wheels were set as close as possible to a 90-inch tread to pack rows two and five. Two heavy frame-mounted wheels packed rows three and four, and the transport wheels were set at the extreme outside to pack rows one and six. Wheel-track planting was successful in many soils and conditions, but was supplanted by minimum tillage methods which replaced the moldboard plow altogether.

Twenty- and Thirty-Inch Rows

During the latter part of the 1960s, narrow rows took over in the corn and soybean country. At first, 36 inches was considered a narrow row spacing, and compared to the horse-width 40- and 42-inch rows common for a century, it was. The horse was long gone — so who needed horse-wide rows?

Progressive agronomists and farmers alike realized that population was the key to greater corn yields and that new varieties of hybrid corn could tolerate high populations. It was better if the higher population were gained by narrowing the row spacing rather than crowding more seeds into wide rows. There was also considerable evidence that soybeans, long planted in 7- or 8-inch rows with a grain drill,

This Michigan farmer is planting eight 20-in. rows, applying fertilizer, and spraying herbicides in bands. This rig carried a ton of fertilizer while herbicide tanks were mounted on the tractor.

would bush out more extensively and set more seed if they were planted in wider rows. This brought about a search for a row spacing that would serve both crops and eliminate the need for an extra set of planting and cultivating equipment and its substantial investment.

Allis-Chalmers planter and cornhead engineers designed equipment for both 20- and 30-inch rows. Planter units were jammed together on toolbars for 20-inch rows. Single-gang cultivators rear-mounted on tractors with skinny tires, along with chemicals, helped control weeds. Cornhead engineers designed cornheads for the A, C, and E combines to harvest four-, six-, or eight 20-inch rows.

Although there were many farmers who were able to benefit by using 20-inch rows, particularly for soybeans, 20-inch rows did not gain general acceptance in corn. However, 30-inch rows were becoming more popular. Corn yields were not significantly less than for the 20-inch rows but other factors were simplified. Larger tires would fit middles, which benefited both tractors and combines. Cultivation was easier with 30-inch rows, and it wasn't necessary to have a second set of special wheels and narrow tires for the tractor. Thirty-inch rows soon dominated in corn-soybean country, and are still the general rule for corn. Soybean growers who preferred 15- and 20-inch rows found that toolbar planters or split 30-inch planting met their narrow-row planting requirements. Wide rows continued to be broadly used in high-yielding cotton and to a limited degree in corn. Allis-Chalmers built planters that could plant rows from 15- to 40-inch spacings, and more if necessary. Special units for sugar beets and edible beans were also available.

This versatile 600 planter frame has 11 units and No-Til coulters set up for narrow rows for soybean planting.

A One-Seventy tractor with toolbar-mounted beet and bean planter units. Toolbar units could be spaced as close as 18 in.

No-Til Planters

Wheel-track planting had confirmed that only a small area of fine, firmed soil was needed for germinating the seed. Planter specialists were asking the next logical question, "Why do we need to upset 500 to 1,000 tons of soil per acre with a moldboard plow when in reality we will use only about 8 percent of it for a seedbed?" The answer was found in the No-Til coulter.

The No-Til coulter was developed by Maynard Walberg of implement engineering. It was a deep-fluted 17-inch coulter that would cut through crop residue or sod to prepare a 2 1/2-inch wide seedbed

The No. 700 planter was A-C's first common-drive conventional planter. In common drive all planter units were driven from a single shaft driven from the transport wheels. Planting rate could be changed by selecting one pair of sprockets which saved time.

There were three types of No-Til coulter blades for different soils and conditions. The full 2 1/2-in. flute (left) was the most commonly used.

ahead of each planter unit. It incorporated a heavy rubber torsion-spring (the same as plow coulters) to give good penetration in tough soils, but provided protection for the blade in rocks or other obstructions. Extensive testing proved that the system would produce high-yielding crops in all but perhaps a few difficult soils, and would eliminate all plowing and tillage operations. Machine costs were dramatically reduced by using the No-Til coulter and the time saved meant more timely planting in wet seasons.

In milder climates, the No-Til planter was used to plant soybeans after small grain had been harvested. This double-cropping system could greatly enhance a farmer's income and profit opportunities. This practice has been extremely successful and continues to be used as farmers seek ways to beat the perennial cost-price squeeze.

To accommodate the No-Til coulters, A-C designed planter frames with three tool bars: the first was equipped with the No-Til coulters, the second carried the fertilizer boxes or tanks and openers, and the planter units were attached to the third. Although most were set up for 30-inch rows, the frames could accommodate a wide range of row widths from

No-Til planters ushered in a new age of minimum tillage. The No-Til coulter prepared a narrow seedbed in stubble or sod, eliminating plowing and disking completely.

around 20 inches to the old standard 40 inches. Frames for 2 to 12 rows were available.

The No-Til planter was recognized by the Equipment Manufacturers Institute as one of the 100 most important developments in agriculture in 1993. It changed planting row-crops completely for thousands of farmers and helped put money in their bank accounts. First an A-C exclusive, the concept was soon copied by most planter builders. It sure was tough on the moldboard plow and disk harrow business.

The 300 planter frame came in three widths for two-wide up to eight-narrow rows. It could use No-Til coulters and Air-Champ or plate-type units.

The No-Til planter made double-cropping soybeans after small grain a time, money, and equipment-saving operation.

The 600 folding frame planter could plant 8-wide or 12-narrow rows. Liquid fertilizer attachment with a 500-gal tank was often used.

A 7045 with a six 30-in. row planter equipped with Air-Champ units and liquid fertilizer planting in a conventional seedbed.

The Air-Champ Planter

Seed singulation had been the bane of planter people since day one. Hundreds of methods had been tried but only a few survived. Edge-drop plates, cell-drop plates, and inclined plates were the most successful, and had planted most field crops for perhaps a century. The edge-drop hopper, the most popular for corn and soybeans, required dozens of different plates to accommodate a broad range of seed sizes, shapes, and populations. Furthermore, this method lost accuracy as ground speed increased because peripheral speed of the plate didn't allow sufficient time for the cells to fill. Every planter manufacturer had this problem and all were searching for a method of singulation which would provide the needed seed versatility while permitting faster planting speeds. Out of this search came vacuum, air-pressure, and mechanical-finger singulation. Allis-Chalmers chose to develop an air-pressure design later named the Air-Champ.

The Air-Champ planter had a small electric blower, powered by the tractor's 12-volt system, which built up a relatively low pressure in the seed chamber under the planter box. A plastic seed disk, which had a row of seed-sized cups around its circumference, rotated through the seed chamber. Each seed cup had a tiny hole through which air could escape. The flow of air held a single seed in the cup until it passed out of the seed chamber. Then held in each seed cup by a soft pad, the seed traveled around and down to within 5 1/2 inches of the seed furrow bottom where it was released and planted.

This fully-equipped six-row Model 330 common-drive conventional planter had Air-Champ units and granular fertilizer, herbicide, and insecticide attachments for a once-over operation. The 330 also came in a four-row size.

Although about one dozen different seed disks were finally needed for all crops, this was far fewer than the one hundred or so required by the plate-type hoppers. The Air-Champ also met the design criteria of higher speed planting, permitting in some cases planting speeds up to 7 mph with the required accuracy.

The No-Til planters with Air-Champ or improved plate-type units were the leaders of the A-C planter

The 770 Air-Champ common-drive planter came in four-wide or six-narrow row sizes. Plate-type units were available for customers who preferred them.

The Air-Champ singulation system had a small blower to supply low-pressure air in the seed chamber. Air flow held a single seed in each seed-sized cup in the seed disk. The seed disk rotated around to the furrow and released the seed for a short 5-in. drop.

The 390 with 12, 16, or 24 units was big-time planting. This enormous rig folded to only 13 1/2 ft wide for transport. At least 160 hp was required for a 24-row model.

The 5x7-in. hydraulically-folded mounted toolbar carried units for 6- or 8-wide rows or 8- to 12-narrow rows.

The implement carriers were most commonly used to transport planters. Two sizes, 21x7 ft and 28x9 ft, with 10,000 lb capacity were offered. Electric brakes, ball hitch, and turn and taillights were available. They were built be an outside supplier.

line throughout the late 1970s and early '80s. During the latter part of that period LaPorte engineering developed a new planter line. It continued the proven No-Til and Air-Champ principles and incorporated new offset double-disk openers, more seed hopper capacity, improved chemical application equipment, and new frames for 4 to 12 rows. The new line was ready for production when the company was sold and it was manufactured by Vermeer for Deutz-Allis.

Grain Drills

The 1935 Allis-Chalmers grain drill was offered in 6-, 7-, and 8-inch spacings from approximately 8- to 11-foot sizes. Several sizes of horse-drawn drills were also listed. There was a choice of single-disk, double-disk, and shoe openers, all of which were staggered for trash clearance. The A-C drill featured fluted-feed seed distribution, a design which has survived due to its accuracy and ability to handle delicate seed with less damage. These were plain drills (no fertilizer) and grass seed attachments were offered. Power-lift and power-down pressure were standard, although lever lift was available. Earlier models had wood spoke wheels which were replaced by steel wheels in mid-1930s. Wood seed boxes were discontinued in favor of steel boxes about 1938. These drills disappeared from the implement line during World War II.

The 20x6 single-disk drill with power lift and power down-pressure. Steel wheels were standard by 1935 and a steel seedbox followed about 1938.

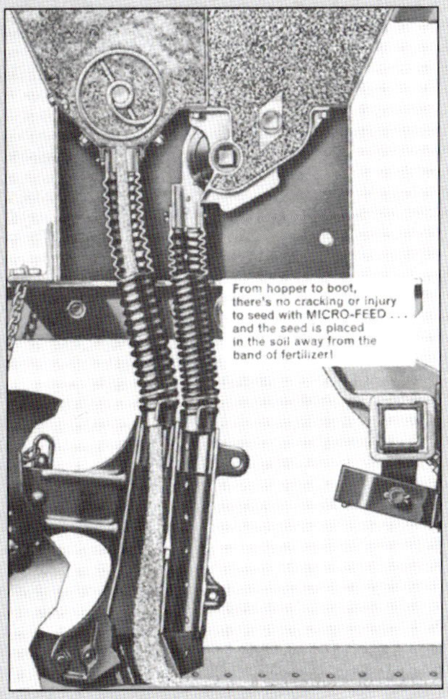

The Micro-Feed adjustable-rim seed-metering mechanism was an Allis-Chalmers exclusive. Force-Flo fertilizer feed had tempered-steel spirals to deliver fertilizer at a constant rate. Twin-Boot design placed seed and fertilizer in separate bands to protect seed from fertilizer burn.

A D17 tractor with a 16x10 plain grain drill covered 65 acres per day. The drill is equipped with Traction-Booster drawbar and 1 3/4-in. press wheels.

This D17 and 15x7 fertilizer-grain drill could seed up to 45 acres per day. This one has the close-coupled Traction-Booster hitch and is equipped with 2 3/4-in. wide zero-pressure press wheels and grass seed attachment.

The 11x7 mounted All-Crop fertilizer-grain drill on a WD45 tractor had a daily capacity of 25 to 35 acres per day. A highly-maneuverable outfit, it was ideal for small fields and irregular plots.

The All-Crop Drills

Allis-Chalmers reentered the grain drill market in the early 1950s with the All-Crop mounted fertilizer-grain drills for the Models CA and WD. These were 9- and 11-run units, respectively, with 7-inch disk spacing. A-C had designed an exclusive adjustable-rim feed mechanism called Micro-Feed which was capable of precision seed-metering. The Force-Flo fertilizer feed had tempered-steel spirals to auger fertilizer to the adjustable outlets. Single- or double-disk openers featured the unique Twin Boot which placed the seed and fertilizer in two separate bands to protect the seed from fertilizer burning. Torsion springs supplied consistent down pressure on the openers for seed depth control. The mounted drills had Snap-Coupler hitch, and were chain driven from the tractor's rear wheel.

About 1957, pull-type fertilizer-grain drills in 15, 18, and 24 × 7 sizes were added to the line. These shared the design features of the mounted models and were offered with Snap-Coupler or drawbar hitch. Grass seed attachments and both narrow and wide zero-pressure press wheels were available for all models. In 1958 some sizes of plain (no fertilizer) grain drills were added. In 1959 the drill line consisted of the following models:

Fertilizer-grain: 9 × 7*, 11 × 7*, 13 × 7, 15 × 7, 18 × 7, 24 × 7
Plain: 13 × 7, 15 × 7, 16 × 10, 18 × 7, 24 × 7
* both mounted and pull types.

Price range (1960):
Mounted 9 × 7 fertilizer-grain drill: $386
Pull 24 × 7 fertilizer-grain drill: $1,200

The 12-ft 24-blade disker-drill powered by a D17 tractor seeding spring wheat in Manitoba, Canada.

The Disker-Drill

The Allis-Chalmers All-Crop drills were not adapted to the small grain growing practices of western Canada. There the accepted machine for planting spring wheat was the disker-drill, which essentially is a one-way disk tiller equipped with a plain drill seed-box driven from the land wheel.

In 1959, Allis-Chalmers entered into an agreement with Kilberry Industries, Winnipeg, Manitoba, Canada, for them to modify their disker-drill with A-C Micro-Feed seed metering and Traction-Booster draw-bar hitch. It was offered in 12-foot 24-disk and 15-foot 30-disk sizes. The 18-inch blades were spaced 7 inches on the gang for approximately 6-inch row spacing. Seed hoppers held 1 1/2 bushels per foot. In addition to the disker-drill, Kilberry also supplied a pair of pull-type swathers in 12- and 16-foot sizes which the company sold in the western Canadian provinces. Like many arrangements with outside suppliers, the Kilberry agreement did not last long.

During the late 1960s, experimental work was done with the No-Til drill which had the same coulters as the No-Til planters. Initial results were satisfactory but no more work was done when drills were discontinued and implements moved to LaPorte.

A four-section rotary hoe mounted on a One-Ninety tractor. Hydraulics folded outer sections up to reduce transport width.

A One-Ninety pulling a four-section rotary hoe mounted on a Kovar-built hydraulic cart. The hydraulic cylinder lifted the hoe to 90 degrees, then outer sections could be folded forward to minimize width.

The Rotary Hoe

The A-C rotary hoe joined the implement line in 1960. It was built at LaCrosse, originally as a pull-type unit with two, three, or four sections. Later, a three-point hitch version was offered and it was also adapted to the Kovar wheeled carrier.

Each hoe section was 42 inches wide and had fourteen 12-tooth wheels spaced 6-inches apart so that teeth worked the soil every 3 inches to uproot little weeds and grass. Ideal for the first cultivation, rotary hoes were most effective at relatively high speeds of 6 or more mph. The rotary hoe was dropped when LaCrosse was closed in 1969.

A WC with a two-row cultivator working in cross-checked corn. The cultivator was the drive-in type for quick easy mounting on the tractor. Both hand and mechanical power lift were available.

The D10, like the earlier Model B, was a one-row tractor especially well-suited for tobacco farming. With a 111 cultivator and fertilizer attachment it destroyed weeds and side-dressed with fertilizer in one operation.

The UTC-24 standard four-row cultivator for the Model UC had ball bearings in the caster wheels for easier steering. It was available for 38-, 40-, and 42-in. rows.

The Model B was a handy cultivating tractor and many versions of the one-row cultivator were offered for it. This unit had hand lift but hydraulic lift was optional.

Row-Crop Cultivators

Until rear-mounted cultivators became the general rule, any new row-crop tractor had to have a front-mounted cultivator available very early in its sales life. The cultivator was so important that its design and features had a significant impact on the tractor-buying decision of many farmers. Invariably the cultivator was part of the original tractor sale, although delivery might not have been made until cultivating season. Its importance to tractor sales necessitated a close working relationship between tractor and implement engineers. Cultivating visibility greatly influenced tractor sheet-metal size and shape. Row-crop tractors tended to have narrow or tapered hoods, while their standard tread counterparts had more square hoods with larger fuel tank capacity. The UC and WC didn't have tapered hoods, but the seat was offset to the right to enhance cultivating visibility.

Another important factor in front-mounted cultivator design was the ease of mounting and dismounting it. A-C met that challenge head-on with the Drive-In design of the All-Crop tractor two-row cultivator. So easy "a boy can attach or detach the cultivator unit in five minutes time." A four-row unit for the UC had a pair of large caster wheels to carry the frame and appeared to be well beyond that boy's capability.

The WD brought four-row equipment back to the A-C line. After the UC was discontinued in 1941 until the WD arrived in 1948, the company had no four-row tractor in the line.

The 400 Series four-row cultivator on a D14 could clean weeds from up to 60 acres per day. This basic cultivator was available in two-, four-, and six-row models for the D14, D15, D17, and D19 tractors.

The 500 Series two-row rear-mounted cultivator with either Snap-Coupler or three-point hitch on a D12 was an ideal rig in the field. Large coulter-gauge wheels carried the toolbar and provided lateral stability to keep the cultivator on the rows.

The 500 Series four-row rear-mounted cultivator with Snap-Coupler hitch on a D17 Series III tractor was quick to the field and easy to handle. In two-, four-, and six-row sizes, it was also offered with three-point hitch. Easy-hitching rear-mounted cultivators rapidly displaced front-mounted cultivators for all makes.

The 8000 Series 12-narrow-row cultivator came in two basic configurations: 18 to 22-in. rows and 24 to 30-in. rows, both in 4-, 6-, 8-, and 12-row sizes. Hydraulic folding was optional on the 8- and 12-row units. The 6000 was the wide-row version of this design with two gangs per row-middle and was supplied in two-, four-, six-, and eight-row models.

The 90 Series cultivators were equipped with the exclusive A-C elastomer clamps already proven in the field cultivator line. This Model 200 tractor is working a four-row Model 95 with five shanks per middle for wide rows. It was offered in two-, four-, six-, and eight-row sizes. The Model 93 had three shanks per middle in 4-, 6-, 8-, and 12-narrow-row sizes. The rubber-mounted clamp brought quiet operation and long life to row-crop cultivation.

Cultivating speed rose to seven or eight mph when high-speed rotary cultivators entered the picture. Slicer-tine spider-wheels cut, turned, and pulverized the soil and weeds.

The 571 cultivator for six narrow rows featured a new parallel-linkage gang with wider bearing areas for better gang stability. Narrow-row cultivators had only one gang with three shanks in each row middle.

The 7020 tractor with a 12-row S-tine (Danish) cultivator could cover acres at a 5- to 7-mph clip, good for 150 to 200 acres per day. The S-tine shank had an aggressive vibrating action to break up crusted soil, shake off trash, and dodge rocks while destroying weeds. This unit is mounted on a 7x7-in. hydraulically-folded toolbar. The S-tine cultivators came in 4-, 6-, 8-, and 12-row sizes.

The WC cultivator continued the drive-in mounting arrangement. Both the WC and UC had mechanical power lift available, so the back-wrenching job of lifting cultivator gangs at row ends was avoidable. When the Model B came out in 1937 it had an optional hydraulic lift, ushering in a new age of operator convenience and control. It was 11 years before the larger tractors could match that feature. Early B cultivators were mounted by backing the tractor astraddle the rig, then attaching the lift links — pretty handy.

The WD and WD45 brought four-row cultivating back after an absence of several years due to the discontinuation of the UC. The D Series tractors had a completely new line of 200, 400, and 600 Series row-crop cultivators with two- and four-row models for the D14 and up to six-row sizes for the D17 and D19. The Model C and CA tractors had front-mounted cultivators and A-C promoted this feature heavily because two major competitors, Ford and Ferguson, had only rear-mounted types. A-C's small tractors had front-mounted cultivators until the D10, D12, and D15 tractors were phased out in 1968.

In 1959, A-C moved the cultivator to the other end of the tractor with the 500 Series two- and four-row, later six- and eight-row rear-mounted Snap-Coupler and three-point hitch cultivators. That foretold the eventual demise of all but a single special-purpose front-mounted cultivator in the A-C stable. Rear-mounted cultivators, ridiculed when Harry Ferguson and Henry Ford first produced them, ultimately took over in all but the vegetable crops.

Narrow rows required some major redesign of cultivators, particularly those used in 20-inch rows, with one gang-per-row middle equipped with only three shanks. By the early 1970s, the conventional spring-trip and spring-tooth shanks had been retired to the parts bins and a new breed of cultivators took over the fields. Spring teeth with elastomer clamps, S-tine (Danish) shanks, and high-speed rotary cultivators devastated the weeds and ruled the row-crops. All emphasized high speed, more effective work, greater ease of adjustment, and all were rear-mounted.

Vegetable Cultivators

The vegetable cultivator was built by the Oxnard plant, which was close to its major markets in California and Arizona. Models were offered to fit the CA, WD, WD45, D14, and D17 tractors. Heavy frames featured parallel lift gangs, both front and rear, which

This high-clearance D17 with a special-crop cultivator from Oxnard is working in a California cut-flower operation. Who says you can't smell the roses while working in the field?

The Oxnard-built FM carrier with cultivator gangs on a high-clearance single-front wheel D17 working in the market-basket country of southern California.

assured precise control of working depth of tools. Front tool bars ranged from 68 to 136 inches and rear bars from 96 to 136 inches in length to match a wide variety of row and bed widths. A generous selection of ground-working tools equipped the vegetable cultivator for most any crop requirement.

FM Carrier

The FM carrier, new in March 1960, was a heavy-duty four-row cultivator designed for the D14 and D17 tractors and later adapted to the D19 and Hundred Series tractors. Although adaptable to many crops, its primary use was cotton cultivation. With six rugged parallel-linkage front gangs, 18 or 21 shanks, ground working tools, and other attachments, it was well-matched to this application.

Lacrosse Implement Carrier

The last LaCrosse-built front-mounted carrier was a versatile tool. It was a ruggedly built implement that could be set up as a four-row cultivator, four-row front-mounted planter, or with a rear center attachment, a five-row bedder. A hydraulic lift cylinder for each side allowed individual control of front sections. It fit the One-Ninety and 200 tractors.

Sprayers

Allis-Chalmers had two separate field sprayer programs in addition to chemical application equipment for the planter line. The first was a line of sprayers designed and built at the LaCrosse plant beginning in 1960. There were two boom-type trailer sprayers, a boom- type tractor-mounted sprayer, and a tractor-mounted wand or broadcast sprayer. The top of the line was the trailer sprayer with a 200-gallon fiberglass tank. This unit had a jet-type agitator which kept chemicals in suspension. The same trailer with three 55-gallon plastic-lined drums was less expensive, and a tractor-mounted boom sprayer with two 55-gallon plastic-lined drums was an economy model. The wand or broadcast sprayer was ideal for open field spraying. The well-known Hy-Pro pumps were used, and nozzles and drops came from specialized chemical equipment manufacturers. After a few seasons the LaCrosse-built sprayer was discontinued, primarily because specialized sprayer builders had more complete lines and the A-C line was not competitive.

A-C Sprayers from Hahn

The second sprayer program was a purchased-finished-product line from Hahn, Inc., Evansville, Indiana, who was well-known for a complete line of quality products. A-C started marketing Hahn-built sprayers in 1966.

The heavy-duty front carrier with cultivator gangs on a One-Ninety working in Arizona cotton. It could also be equipped as a four-row planter or a five-row bedder.

The LaCrosse-built 200-gal boom-type trailer sprayer with nozzles arranged for overall coverage working in a Wisconsin field.

A-C needed a self-propelled high-clearance sprayer that could handle a wide variety of chemical applications, and the 2300 filled that need. With 72 inches of crop clearance and rear tread adjustable from 76 to 120 inches, it could fit into most any row crop. It had a 30-horsepower Wisconsin V4 engine and a four-speed transmission with speeds of 2.5, 5.0, 10.0 and 20.0 mph so it could cover the acres in short order. With hydrostatic steering and hydraulic boom control the operator's job was easy. It had an aluminized steel tank and a huge variety of booms, pumps, drops, nozzles, liquid nitrogen knife-in equipment, etc., which adapted it to any application requirement.

The 400T trailer sprayer came in two versions: a liquid fertilizer applicator with a 400-gallon corrosion-resistant tank and stainless steel boom, or a less expensive model with an aluminized steel tank for regular work. Both had adjustable frames for 56- to 84-inch treads. The 200T trailer type had a 200-gallon tank. Tractor-mounted sprayers came with 100- or 200-gallon tanks or with racks for two 55-gallon barrels. Side-mounted tractor racks could carry two 100-gallon tanks. A catalog full of pumps, valves, booms, drops, nozzles, and other accessories adapted these sprayers to nearly any kind of field-crop spraying.

The 402F bulk fertilizer spreader rounded out the line. With 4,000-pound capacity, ground drive, large tires, and adjustable spreading pattern, the 402F was an ideal substitute for heavier truck-mounted spreaders when the ground was soft. Automatic hydraulic brakes permitted relatively fast road speeds with greater safety.

This No. 2300 self-propelled sprayer is equipped with full fender equipment and a dry boom. Front outrigger safety-wheels were available for operation in hilly terrain.

400T Trailer Sprayer

Side Tanks and Rear Mounted Boom

200-Gal. Rear Mounted Sprayer

402F Bulk Fertilizer Spreader

Farmstead Equipment

A 1950 prototype front spreader was equipped with sides and a top to become a forage trailer. Lack of capacity may have ruined its chance for production.

Manure Spreaders

Manure spreaders, unlike new tractors or combines, seem to lose all their appeal after the first use. A manure spreader was a necessary abomination to the livestock feeder. Anyone who had to deal with a frozen or broken conveyer chain, always when the machine was fully loaded, no doubt wondered why he ever wanted to feed livestock in the first place.

Allis-Chalmers' first approach to the manure spreader was like their undertakings with several other machines — slightly unorthodox. The spreading works were on the front end. The A-C and the Galloway of many, many years ago were the only spreaders that salesmen could really stand behind. That little joke was badly overworked within a day or two after the Allis spreader hit the market. Front-spreader prototypes dated back to 1950 and some interesting concepts were investigated during that time. One version was fully enclosed for use as a silage wagon behind a forage harvester. Another model had a rear-mounted self-loader to load the spreader as it was backed in to the manure pile. A rear-mounted loader on a WD was tested in 1951.

Two prototypes in action in 1951. The WD rear-mounted loader and the newly-designed front-unloading spreader were two innovative concepts that didn't catch on. Progress moved the loader to the front of the tractor and the works to the rear of the spreader.

This experimental front spreader was equipped to self-load by backing into the manure pile. No doubt there were many conditions where it couldn't get a grip on the material.

The 110 Spreader

First just called the Allis-Chalmers manure spreader, it later gained the designation No. 110 for its 110-bushel capacity. The conveyor chain ran forward to the unconventional spreading mechanism which consisted of two beaters (three effective Serial No. 601) of normal design shredding the material and feeding it to the "three high speed turbine-type spreaders" which threw it out each side in a 12-foot swath. It had power take-off drive which was not yet widely used at that time. This spreader could be loaded from the rear, as well as from both sides. As the partial load moved forward, weight was transferred to the tractor for traction. With quality features like all roller-chain drives and penta-treated box, it should have been a winner. The only problem was that it wasn't. The conventional spreaders of specialized builders were well established and many A-C dealers held their contracts. The 110 spreader was introduced in 1956, and its last price (1960) was $938.10. A total of 1,500 units were built.

When spreading with a No. 110 the WD45 gained more traction as the load moved forward.

The No. 110 spreading mechanism consisted of three beaters (after Serial No. 601) and three turbines to spread a 12-ft width.

A D17 Series III and No. 180 spreader could move manure to the field fast and spread it quick. This 180 was the largest of the line which consisted of the 95G, 95M, 140, 140S, 180, and 180S spreaders.

Conventional Spreaders

By 1960 the A-C spreader line included the 95G, a 95-bushel ground drive unit, and shortly after, the 95A, a power take-off-driven version of the 95G. In 1962 the 140 three-beater and 140S single beater, both power take-off driven, expanded the line. By 1963 the line was complete when the 180 three-beater and single-beater spreaders became available. These spreaders were of conventional design and their model numbers indicated capacity in bushels. The 1960 price of a 95G spreader was about $630, or $300 less than the 110. The 95, 140, and 180 spreaders were built at LaPorte, and some models were sold to other companies on an OEM basis. Some 1964 prices included: the Model 140S, $916; 140, $992; 180S, $1,162; and 180, $1,257. In early 1970 the 95-140-180 line was replaced by two new models, the 281 (81 ft^3, 140 bushels) and 299 (99 ft^3, 180 bushels). Both were offered as single beater machines and an upper beater was optional.

A Model One-Sixty tractor with a 460 loader pile it on the No. 299 (99 ft, 180 bushels) single-beater spreader. The 281 and 299 were the last of the A-C spreaders.

Although the company offered good products, A-C was unable to capture a significant share of the spreader market. Many dealers had competitive lines which already had high customer acceptance and they were reluctant to stock and sell the A-C machines. The spreader line went the same route as so many other good products — discontinued because of low sales volume.

Farm Loaders

One fact is quickly learned in the tractor business: Unless a farm tractor is much too large for a farm loader, it is important to introduce a suitable loader the same day as the tractor. Otherwise one or more short-line specialized loader manufacturers will have one for it the next day, or so it seems. Dealers had three questions about a new tractor under 100 horsepower: "What's the horsepower? How much does it cost? Do you have a loader for it?" Sometimes the last question was the first asked.

Although a prototype rear-mounted loader for the Model WD was built in 1951, there was no production. The company first went into the farm loader business in about 1955, when they added the Models 8 and 9 Quick-Change loaders for the CA, WD, and WD45. These were tubular frame and boom construction with single-action lift cylinders and mechanical bucket dump. The Model 8 had an 8-cubic foot and the Model 9 had a 9-cubic foot bucket. Lift heights were 8 and 10 feet, respectively. When the D14 and D17 tractors came out in 1957, the Models 14 and 17 were also introduced, and shared the specifications of the Model 9. Since

The No. 90 loader put the WD45 tractor to work in the feedlot. The bucket was mechanically dumped and the cylinders were high-pressure type to match the tractor's 3,500-psi system.

The 160 loader was a lower-cost mechanically-dumped loader for the Model One-Sixty tractor. The standard bucket held 9 ft^3 and attachments such as the dozer blade, 80-in. snow bucket, and slip-on snow bucket expanded the loader's versatility.

The 450 for the 6140 tractor and the 460 for the 6060-6070-6080 tractors were easily detached from their running mates when necessary. Good clearance around the loader towers allowed easy operator mounting and dismounting. The 460 loader could be mounted on 6000 Series cab tractors.

all the tractors had the high-pressure hydraulic systems, lift cylinders were small in diameter. Attachments included standard combination bucket, material bucket, 80-inch light material and snow bucket, dozer blade, and grill guard. The Models 8, 9, 14, and 17 were manufactured by the Freeman Loader Corporation from Peru, Indiana.

In 1963 the loader line was redesigned with box-welded booms and sub-frames to provide a much more rugged and eye-pleasing appearance, and was called the 100 Series. Models WD90, D140, D150, and D170 all had full height-lift capacity of 1,500 pounds and break-out force of 3,000 pounds. The D110 and D120 had capacities of 800 pounds to full and 2,000 pounds breakout. As later model tractors such as the D17IV, One-Sixty, One-Seventy, and One-Eighty came out, the 160, 170, and 180 loaders with standard pressure lift cylinders were added to the line. The 100 Series loaders, also built by Freeman, remained in the line for a low-cost loader even after LaCrosse produced the 400 and 500 Series loaders.

The 400 and 500 Series Loaders

A substantial market demand for double-acting lift cylinders to permit digging and hydraulic bucket control precipitated the development of a new line of all-hydraulic loaders. The 400 and 500 Series were announced in 1963, and had standard pressure cylinders only.

The 400 Series fit the One-Sixty, 5040, and 5050 tractors, as well as the 5045 when it arrived in the late 1960s. The 500 Series fit the D17 III and D19 (with auxiliary pump), the D17 IV, and all Hundred Series tractors from the One-Seventy through the Model 200. These two loaders were unique in that the bucket cylinders were located on the underside of the boom which provided two important performance benefits. First, during rollback, the dump cylinders pushed the bucket into the material for a much higher breakout force, 4,100 pounds, compared to the larger 470 loader at 3,500 pounds, and second, the material was more likely to stay in the

The 470 loader on a Model 7000 tractor (back) had top-of-the-line loading capacity. The 400 loader was an ideal workmate for the 5050 two-wheel and front-wheel assist tractors. The 500 Series (not shown) was similar to the 400 but fit the 175, 185, and comparable tractors. The 400 and 500 loaders were first produced at LaCrosse then transferred to LaPorte when the former was closed.

bucket due to the crowding action. With all box-welded boom and subframe construction, these were extremely rugged farm loaders which provided reliable service for their owners. The 400 could lift 2,000 pounds and break out 4,100 pounds, while the 500 could lift 2,300 pounds and break out 4,000 pounds. A 400 loader for a 5050 two-wheel drive tractor included a pump and valve and sold for $3,179. The 500 for a 175 tractor utilized the tractor's hydraulic system and sold for $2,816.

The Model 470 loader, new around 1975, was matched to the 7000 and 7040 tractors, therefore it could have been mounted on any of the 7000 Series tractors. On the 7040 tractor, the 470 could lift 5,800 pounds to 13 feet, had a breakout force of 3,500 pounds and lift cycle time of 6.3 seconds. The 470 was built by the Great Bend Manufacturing Company, Inc., of Great Bend, Kansas.

Two new all-hydraulic loaders were developed at the LaPorte plant. The 460 for the 6060, 6070, and 6080 tractors (1980) and the 450 for the 6140 tractor (1982) shared a new design that emphasized quick detachability and sufficient clearance over the front wheels for use with front-wheel assist tractors. Both used the tractors' hydraulic systems since adequate hydraulic output for loader operation had been design criteria for the tractors. The 460 could lift 1,800 pounds and break out 4,400 pounds. The 450 had a maximum lift of 1,660 pounds and break out force of 4,275 pounds. Prices in 1985 were $3,719 and $3,398, respectively.

Construction was heavy box-welded sections with large pads at all high-stress points. Hydraulic couplers, parking stands, grill guards, and gauge rods were standard, and a complete line of buckets was available. Teamed with their specific tractors, they had good performance and enjoyed good customer acceptance. At some point the loader line was sold to the Lasco Company of Vicksburg, Mississippi, who then manufactured them for A-C on an OEM basis. These loaders were phased out as the West-Allis tractor line disappeared.

The Roper-built posthole digger could dig holes from 6 to 24 in. in diameter. It was made for Snap-Coupler tractors.

Farmstead Tools

One of the better known Oxnard tools was the 6-foot rear-mounted blade, since it was sold all over the United States and Canada. It could be angled and tilted to work in any way it was needed — even ditching when it was equipped with blade points. When reversed, it was good for light dozing and backfilling. The blade extension group added 18 inches on each side for a 9-foot cut and endplates added capacity. The blade with extensions and endplates weighed 745 pounds and had a list price of about $285.

One of the handiest tools on any farm was a rear-mounted scoop. There was always some dirt to be moved, rocks to haul, holes to fill. The Oxnard scoop would hold 13 1/2 cubic feet (one-half yard) and was reversible. With Snap-Coupler, it fit WD45, D14, and D17 tractors. It tipped the scales at 354 pounds and could be hired full time for $139. That's cheap help!

Digging postholes or planting trees was back-breaking work until someone invented the tractor-driven posthole digger. The first A-C digger was built by the Roper Manufacturing Company of Zanesville,

A D14 tractor with an Oxnard 6-ft rear-mounted blade was just the ticket for grading and scraping feedlots.

The D14 with the half-yard rear-mounted scoop made dirt moving easy and avoided shovel handle blisters.

Ohio, and came out about 1960. It was designed to fit Snap-Coupler hitch tractors. With a 12-inch auger, it sold for $165.

The rear-mounted carrier and any Snap-Coupler tractor could save a lot of aching muscles and strained backs while moving all kinds of things around the farmstead. For $54.10 it could eliminate visits to the chiropractor.

If the lift was too much for one man, the farm crane was much more efficient than calling for a neighbor. Mounted on a WD, it could lift 1,400 pounds or 650 pounds on a CA. The WD crane sold for $43.50, the CA model for $25.75. Both the rear carrier and farm crane were made by the Athens Plow Company.

When the No. 10 posthole digger was shown to Canadian dealers, the "demonstration posthole" was found to have been occupied by a slightly bewildered resident.

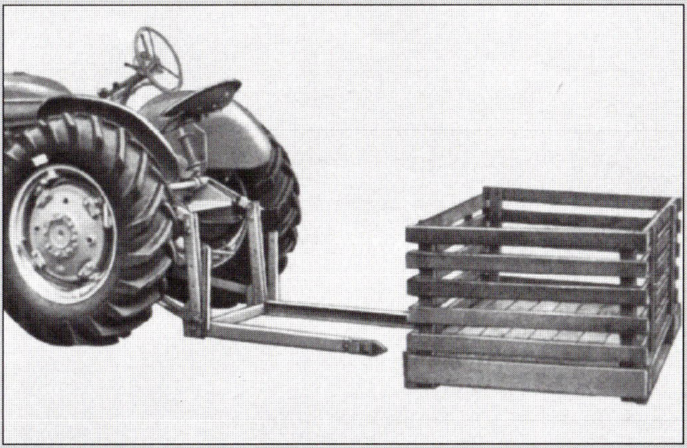

The inexpensive rear carrier hauled 500 to 1,050 lb most anywhere on the farm.

The No. 10 Posthole Digger

LaCrosse engineering designed a three-point hitch posthole digger to replace the Roper-built unit. The No. 10 was shown to dealers at French Lick, Indiana, in the fall of 1960. To show dealers the quality of work done by the No. 10, one of the holes it had made was dug up and displayed during the program. Or so they said. The No. 10 fit category 1 three-point hitch and had 6- to 24-inch augers available. It disappeared when LaCrosse closed.

No wants a posthole digger— they want postholes, and the No. 10 powered by a D15 Series II could put them wherever holes were needed, 6 to 12 in. in diameter.

With the rear crane on a CA one could lift 650 lb with one finger— hydraulically. The WD crane could lift 1,400 lb.

Hay and Forage Equipment

A WC with its 7-ft side-mounted mower could cut 25 to 30 acres per day. Up-front location let the operator see the cutterbar at all times for safe operation.

Mowers, Rakes, and Conditioners
Sickle Bar Mowers

The company's emphasis on the family farm obligated it to provide a line of implements for the WC, RC, C, and B tractors to maximize their utility and versatility. Tractor-mounted equipment fostered the systems approach which was widely embraced by all farm machinery manufacturers starting in the late 1930s. Sickle bar mowers were an obvious choice for tractor-mounting, and even side-delivery rakes would eventually ride on the tractor.

An early reference to mowers, that was found in a 1937 WC tractor catalog, described: "WC mowing attachment. Power-driven by the WC engine through the transmission by means of the power take-off furnished complete with the mowing attachment." It had a 7-foot cutterbar and was side-mounted. The full visibility and greater safety of side-mounting was emphasized. A slip-clutch and cutterbar break-back protected the mower from obstructions. A combination foot pedal/hand lever lifted the cutterbar. Later, as more models were offered, an improved WC side-mounted mower was designated the No. 2. It was built from 1950 through 1952, with approximately 4,600 units being produced. In 1949 a 7-foot WC mower sold for $195.

A side-mounted mower with either a 5- or 6-foot cutterbar was available for the Model B tractor. It was simply called the Model B mower and renamed the B-C mower when the Model C came along in 1940. Hand lift was standard and hydraulic lift was optional. The 6-foot B-C mower had a 1949 list price of $170.

The B-C mower put the Model B to work in the hayfield. It also fit the Model C and there was a choice of 5- or 6-ft cutterbars.

The No. 5 was designed for the Model CA but could be adapted to the B and C tractors. Twin V-belt drive on adjustable sheaves allowed changing sickle speed to match crop conditions. There was a choice of 5-, 6-, or 7-ft cut.

The No. 5 side-mounted mower was designed specifically for the Model CA tractor. Cutterbars in 5-, 6-, and 7-foot lengths were available and hydraulic lift was standard. It replaced the B-C mower, so attaching parts were available for B and C tractors. Production of the No. 5 started in 1951.

The No. 3 semi-mounted rear mower was a versatile machine, having been adapted to the B, C, CA, WC, and WD tractors. An important advantage was its ability to cut square corners by pivoting the tractor on the right rear wheel. This eliminated the need to stop and back the tractor, or make a left-turn loop as was necessary with a side-mounted mower. It also reduced plugging the cutterbar at corners in heavy hay. Furnished with 5-, 6-, or 7-foot bars and hand or hydraulic lift, the No. 3 was the backbone of the A-C mower line until the new Twin-Wheel Drive No. 7 came out in 1956. The No. 3, with a 7-foot cutterbar, had a list price of $294.

At the risk of stating the obvious, there are only two ways to increase the daily capacity of a field machine — go faster or increase width-of-cut. More daily capacity was needed in mowers. As a result, some designers tried 9- or 10-foot cutterbars but side draft was intolerable in heavy legume hay crops. Other companies tried mounting a wide cutterbar supported at both ends on the front of the tractor, but these machines were unwieldy and hard to maneuver, and had operating problems because the cutterbar was so far out front. They met with limited success.

The mower-conditioner, either pulled or self-propelled, became an acceptable solution but they didn't come along until the late 1960s and were expensive if only a mower was needed. As a result, more speed was tried. Increasing forward speed quickly plugged the sickle, so increasing sickle speed was tried. This resulted in the sickle heads being jerked off as reciprocating stresses exceeded sickle strength. If increased sickle speeds were to be had, the old pitman would have to go. It went.

The No. 3 semi-mounted mower fit several different A-C tractors, including the WD. There was hand or hydraulic lift and 5-, 6-, or 7-ft cut. An important advantage was its ability to make non-stop square corners.

Twin-Wheel-Drive No-More Pitmans

Allis-Chalmers engineers invented Twin-Wheel Drive. In this exclusive design, two gear-driven counterweighted flywheels drove the sickle through two short connecting rods and a cross-head. When running, the counterweights were always opposite the sickle to balance reciprocating forces. Operation was incredibly smooth with sickle speeds up to 2,800 strokes per minute, nearly double that of pitman drives. Field speeds were now limited by ground conditions, not cutting capability. Since the relationship of the flywheels and sickle did not change, the cutterbar could operate at any angle from 45 degrees below horizontal to vertical. This was just one more example of the many innovations to come from the fruitful A-C engineering departments.

The unique Twin-Wheel Drive mower went to the field in two configurations of the No. 7 mower. A rear-mounted model hitched to the Snap-Coupler and lift arms of A-C tractors, and a trail-type carried on a pair of 4.00 × 8 air-tired wheels. Cutterbars of 6 and 7 feet were offered. The No. 7 mower was built from 1956 to 1959 in the Gadsden, Alabama, plant. At that time both mower and cotton picker production were transferred to LaPorte. A quantity of 6,000 No. 7 mowers were produced. In 1960 the mounted mower sold for $404 and the trail mower for $469.

The No. 7 Twin-Wheel drive mower could be mounted on Snap-Coupler hitch tractors. A trail-type version was also available.

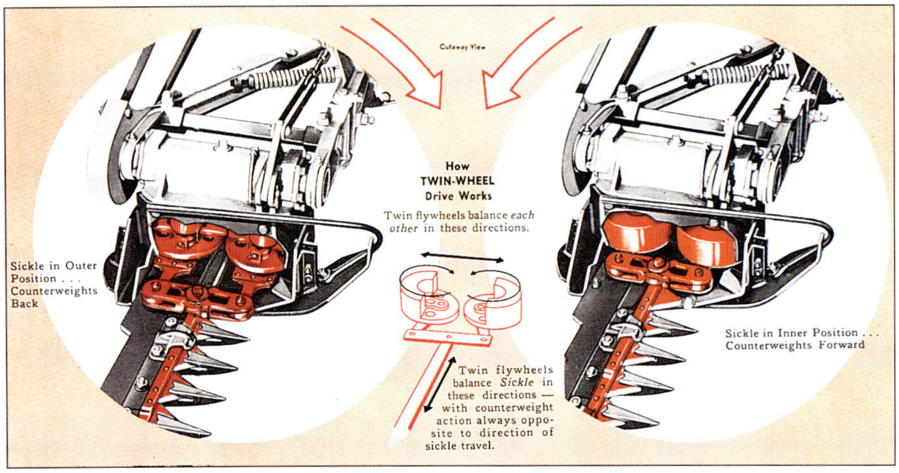

The exclusive A-C Twin-Wheel drive took the vibration, noise, and wear-and-tear out of sickle-bar mowers.

The Twin-Wheel Drive was carried over intact to the 80 Series mowers in 1959. The 80 Series featured the non-removable rotating power take-off shields. This important safety device was invented by Charlie Scranton, LaPorte chief engineer, and became an industry standard. The 80 Series was offered in four models: 80-R rear-mounted for the D Series, WD, and CA tractors; 80-T trail-type for all tractors; 80-S side-mounted mower for D Series and WD-WD45 tractors; and 80-SI, a fully hydraulically controlled side-mounted industrial mower for D Series tractors. Cutterbars of 6 and 7 feet were available for all models, with a 5-foot bar for the 80-SI. Prices in 1960 ranged from $413 for the 80R to $512 for an 80S or an 80T.

As the Snap-Coupler hitch was replaced by the three-point hitch on A-C tractors, the 80-R3 mower was added around 1963. The 82-T, with new frame design and staggered wheels for better balance, replaced the 80-T. The 80-R was replaced by the 82-R, a semi-mounted mower with castering tail wheel and built-in hay conditioner hitch. The 82-S replaced the 80-S. A 9-foot cutterbar was added. The 80-SI was supplanted by the 82-SH. Twin-Wheel mowers remained in production until the hay and forage line was phased out in 1975.

Eighty and 82 Series mower production from a summary dated April 1966:

80-S	3,303	80-T	6,200
80-SI	397	82-S	2,350*
80-R	18,500*	82-SH	50*
80-R3	3,537*	82-T†	

* Still in production.
† Still to come.

The 80-R3 fit most three-point hitch tractors and had a hay conditioner drive.

The 82-T was a trail-type and replaced the 80-T.

The 82-R was semi-mounted on the three-point hitch and had a built-in hitch for a hay conditioner.

(Top left) The 80-T was quick-hitch and go. (Top right) The 80-S side-mounted left tractor available for drawbar work. (Lower) The 80-R rear-mounted mower for A-C tractors was the most maneuverable.

Rotary Mowers

Many mowing and cutting jobs existed where a farmer would not take his sickle bar mower if he wanted to keep it in one piece. The sickle bar mower was made to cut grass and legume hay crops, clip pastures, and perhaps trim roadsides to keep the farmstead looking sharp for the neighbors. Clearing brush was not in its job description. But it was in the job description of the new boy on the block — the rotary mower. In user terms, these rugged machines were called "bush-hogs," a term derived from the name of its innovator, the Bush Hog folks in Selma, Alabama. The rotary mower found a home on nearly every farm — who didn't have brush or heavy weeds growing somewhere on the place?

Starting about 1959, Allis-Chalmers marketed a line of rotary cutters built by the Danuser Machine Works, Inc., of Tulsa, Oklahoma. The 160 cut 5 feet and came in both mounted Snap-Coupler hitch and pull-type. The Model 172 was a 6-foot trail-type. Less well-known was the 166 which was a center-mounted cutter for the D14 wide-front tractor and was suited to highway mowing. Highway mowing equipment was a limited market for A-C, and since farmers preferred rear-mounted or pull-type machines, the 166 was soon deleted from the line. In 1960, list prices were $464 for the 160 and $544 for the 172.

In 1962 rotary mower production started at LaPorte with two models, the 160 and 172. Although model numbers remained the same, these models

This Snap-Coupler hitched 160 rotary mower was an early Danuser-built model.

The 166 center-mounted mower cut 5 1/2 ft and was matched to the D14 for highway mowing.

The LaPorte-built 160 rotary mower cut 5 ft and was well-matched to the D15 tractor.

The 172 6-ft rotary mower had two-position wheels; toward the rear like this one for close cutting or along the sides for better balance.

were entirely different from the purchased mowers. Cutting 5 and 6 feet, respectively, both were available in Snap-Coupler hitch, three-point hitch, and trail-type versions. More models were added during the late 1960s. The 184 pull-type cleared 7 feet at a pass, while the big 284 cut nearly 14 feet to chop four rows of corn or cotton stalks.

Gearboxes were rated from 60 horsepower on the smaller units to 110 horsepower on the big 284. The power lines were protected by slip-clutches and over-running clutches kept the flywheel effect of the heavy blades from pushing the tractor. Spindle speed was 750 rpm. There was a choice of standard air-lift, heavy-duty brush, or flat shredding blades, all of forged heat-treated alloy steel for rugged service and long life. From 1962 through 1965, LaPorte manufactured 3,225 Model 160 and 4,150 Model 172 mowers and production of these and the larger models continued until 1975.

The 284 rotary cutter had a 110-hp gearbox and cut 13 ft 8 in. to shred four rows of corn or cotton stalks.

Side-Delivery Rakes

When the LaPorte engineers presented the Roto-Baler, they had two more machines waiting in the wings to complete the system — a side-delivery rake and a bale loader. Development of the A-C side-delivery rake began in 1944, with first production in 1947. The side delivery was power take-off driven, a new concept in the industry. With one wheel at the front and another at the rear of the basket, front weight was carried on the tractor's drawbar, yet another departure from conventional design. The rear wheel was fixed while the front wheel could castered so the rake trailed straight behind the tractor, again an improvement over others based on old horse-drawn rakes. The rake was more versatile than most with two raking speeds and a reverse tedding speed provided by a shiftable gearbox in the power line. For raking in ridges and rough ground, tandem wheels were available for both front and rear wheels. The 10-foot reel had four bars and 128 teeth, and would rake a 7-foot mower swath. The A-C rake set a new standard in rake design. It sold for $461 in 1951. The No. 7 rake appeared in 1957 and was an improved and simplified version of its predecessor. The front caster wheel was moved to a position behind the basket, improving performance in rough ground. The two-speed gear box was replaced by a gearbox with

The A-C side-delivery rake and tedder was one of the first PTO-driven rakes in the industry. A gearbox provided two raking speeds and one tedding speed.

The No. 7 had the front wheel moved in back of the basket and a simplified drive with one raking speed and one tedding speed.

two input shafts, one for raking and the other for tedding. To change the mode of operation, the drive shaft was moved from one input shaft to the other. Basket and reel were essentially the same as the former model. The No. 7 had a 1960 list price of $520.

Parallel-Bar Rakes

The No. 77M mounted parallel-bar rake was produced for the 1959 haymaking season. Matched to Snap-Coupler hitch tractors at first, it later became available with three-point hitch. Parallel-bar rakes were noted for their extremely smooth, vibrationless operation, even at relatively high power take-off speeds. The 77M had a five-bar reel with 100 teeth. A simple V-belt drove the reel, eliminating the angular drive found in conventional rakes. Parallel-bar rakes were not adaptable to tedding, but that feature was not widely used. After tedding was eliminated, curved teeth with better lifting ability were used on the parallel-bar units.

The mounted rake was highly maneuverable and road transport was limited only by the tractor's speed. Depending on the hitch, the 77M list price varied from $485 to $515. The mounted rakes found early acceptance but as ground-driven pull-type rakes became available, the mounted models quickly phased out. Farmers liked the convenience of pin hitching and adaptability to any tractor with a drawbar.

The reel and basket assembly of the 77M was carried over intact to the drawn 77G. The first 3,250 77-G rakes (1960-1963) had a roller chain drive to a gear box on the rear basket shaft. Beginning about mid-1963 the reel drive was changed to a V-belt. The reel clutch was rope-controlled from the tractor seat and a single crank controlled basket height. The 77-G was furnished with 4 bars and 80 teeth or 5 bars with 100 teeth. A special model for raking heavy bean crops had five bars with 145 closely-spaced teeth. A

This 77-G pull-type rake had V-belt drive but first 3,250 rakes had a chain and gearbox drive.

The mounted No. 77 parallel-bar rake was vibrationless. It was easy to mount on a D-Series Snap-Coupler hitch tractor to become an easy-to-handle, highly-maneuverable outfit.

The No. 77-G bean rake had five bars with 145 closely-spaced teeth for raking heavy crops of edible beans.

total of 508 77-G and 30 77-G bean rakes were equipped with rubber-mounted teeth. The 77-G remained in production until the hay and forage line was dropped. The 77G sold for $615 in 1960.

The 78 rake, with offset wheels, was an improved version of the 77. The right wheel was moved forward so that it was as close to the basket fore-and-aft as the left wheel, which improved gauging and operation in rough ground conditions. An axle with universal joints connected the wheels so both drove the reel. Other than this major change, the 77 and 78 were essentially the same. The 78 was also offered in 4-bar, 80-tooth and 5-bar, 100-tooth versions. The 78 came out in the mid-1960s and continued until the hay line was phased out.

Hay Conditioner and Mower-Conditions

Ever since farmers discovered that legumes made the highest quality, most nutritious hay, they have faced a frustrating dilemma. As the hay cured under a summer sun, the leaves dried first, some fell off and were lost, yet the stems could still be too wet and green to keep in the haymow. This was especially true of heavy, rank first cuttings — those most likely to suffer the early summer rains. Some process to speed the drying of the stems was badly needed. Someone discovered that breaking or crushing the stems slightly allowed the moisture to escape, thus greatly reducing the drying time.

The industry responded with crimpers, crushers, and conditioners, which the hay growers were quick to snatch up. Anyone who made legume hay had to have a conditioner if he wanted to improve the odds of having quality hay for his livestock or to sell in the commercial market. A-C jumped on the bandwagon with its version of the hay conditioner in 1960. It had two 6-foot rolls; the upper was 8.75 inches in diameter and rubber-covered, while the lower was 6.5 inches with staggered steel bars. This combination provided the right amount of crushing and crimping without cutting the stems. The rubber-against-steel rolls provided long life and quiet operation. If the mower had been equipped with a divider board, the conditioner could handle a 7-foot mower swath.

The obvious next step was to combine the mowing and conditioning into one operation and hitches to accomplish this task were soon made available by all manufacturers. A-C provided a wheeled arched frame to pull the conditioner behind an 80-R or 80-T, and of course the 80-S didn't need one. The hay conditioner was built for about 10 years, from 1960 to 1970. By the end of 1965, 7,987 units had been built and it was still going strong. An early price was $806.

Mower-Conditioners

An even better solution to combining the mowing and conditioning operations was to build the mower and conditioner into one high-capacity machine with

The No. 78 rake with staggered wheels had improved gauging in rough conditions and a smooth V-belt drive.

A special hitch pulled the hay conditioner behind an 80-R3 mower on a D15 Series II tractor. This was a once-over outfit with up to 50 acres per day capacity.

The 390 mower-conditioner replaced the 904. The biggest change was a greatly-simplified main drive which eliminated the long cross shaft to the center of the conditioner.

The 904 mower-conditioner cut, conditioned, and windrowed a 9-ft swath. With a 35-minimum hp requirement it was no load for this One-Seventy tractor.

a wider swath than the typical power mower. Although some 7- and 12-foot units were offered by the industry, the 9-foot size was the most popular. A-C introduced the 829 mower conditioner in 1966. The 829 cut and conditioned a 9-foot swath and with deflectors in place, formed a windrow. The header could float 10 inches above or below center at each end to follow ground contours for complete cutting. A four-bar reel fed the hay into the sickle and the conditioner rolls. Sickle speed was 730 strokes per minute at 540-PTO rpm. The upper steel conditioner roll was 6.25 inches and the lower rubber-covered roll was 6.75 inches in diameter and had a peripheral speed of 20 mph for high capacity conditioning. The power take-off delivered power through a right-angle gearbox to the power band main drive belt. Power requirement was 35 minimum horsepower.

The Model 904 replaced the 829 in early 1969. The overall appearance was similar, but it was geared to run slighter slower at 618-sickle strokes per minute and conditioning roll peripheral speed was reduced to 18 mph. A 60-horsepower wobblebox provided smooth sickle operation.

The 904 was replaced by the improved 390 in 1972. The most noticeable difference was the 390's simplified main drive which eliminated the long cross shaft to the center of the machine. It was replaced with a much a shorter drive on the left side of the header which reduced the number of parts substantially. Other changes were minimal. The 390 was discontinued in 1975 with the rest of the hay and forage line.

Experimental Self-Propelled Windrowers

During the late 1960s and early '70s the self-propelled windrower replaced the pull-type unit, much the same as the self-propelled combine displaced the drawn combine. Allis-Chalmers wanted a share of the self-propelled windrower market and developed an entry. It was ready for production in 1971. Like too many other products, the tooling costs proved to be prohibitive in light of estimated sales. Furthermore, the market was primarily in the western United States

The prototype self-propelled windrower (right) was lined up with other new products at LaPorte in 1971. No production occurred when feasibility studies concluded that there would be insufficient sales to produce a profit.

and dominated by at least two other well-established manufacturers. The project was cancelled.

Pull-Type Swather

The spring wheat growers of the northern United States and the prairie provinces of western Canada have always had to cope with a short growing season. One of our Canadian friends described their climate as "nine months of winter and three months of poor sledding." If spring were a little late and snow a bit early, it was easy to lose a crop of wheat.

When binders and threshers ruled the harvest, the grain could dry and cure in the shock. Combines required dry grain and straw so to rush the wheat harvest as much as possible, it was (and still is) customary to cut and swath the crop as soon as the grain was mature but the straw was still green. Hopefully, the straw would dry enough in the windrow to allow combining the crop before the snows came. Wheat farmers in that part of the world spent a lot of time looking toward the northwest, hoping to outguess the early snows.The swather was a simple machine consisting of a reel, sickle, canvas-conveyor platform, and the necessary parts to make it work. Very early swathers were converted from grain binders when farmers deserted threshing for combining.

This prototype pull-type swather was a 1950 attempt to move into a new market, but it was never produced.

As combines took over, combine manufacturers were quick to build pickup attachments for their combines and companion swathers to round out their line in the spring wheat country. Advance Rumely built a bull-wheel-driven swather to complement their prairie combines. It had about a 12-foot cut. The swather appeared in early 1930s catalogs but none were built after A-C acquired Rumely.

Swathers soon were equipped with PTO drive and sized so that the following combine could handle the windrow. Therefore as combine capacity grew, so did the width-of-cut of the swathers. LaPorte engineering built a prototype swather in 1950, no doubt to help sell the All-Crop harvester in the spring wheat country. The sickle and platform conveyor were power take-off driven, while the reel was ground-driven. It appeared to cut 6 or 7 feet, about right for an All-Crop harvester to handle the windrow. This prototype was never produced.

In 1959 the Independence plant built 15, 12-foot and 35, 15-foot swathers. Whether the machines were too expensive to manufacture or did not perform up to expectations is not known, but no more were built. During the winter of 1959-60 the Farm Equipment Division entered into an agreement with Kilberry Industries of Winnipeg, Manitoba, Canada, to buy disker-drills and swathers for resale with the

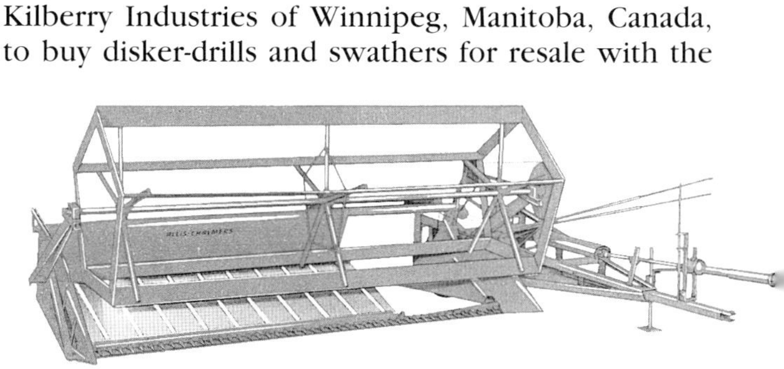

The Kilberry-built swather came aboard in 1960. Built in 12- and 16-ft sizes, it was of conventional design with a canvas platform draper. It was sold only in the Canada's prairie provinces.

The heavy tubular frame is apparent in this rear view. The wheels could be rotated 90 degrees for endwise transport.

Allis-Chalmers label. The company distributed these products only in western Canada on special quotation to the Calgary, Regina, Saskatoon, and Winnipeg branches.

The swather was offered in 12- and 16-foot sizes and was of conventional design. The frame was of heavy tubular steel construction and carried on two rear wheels, which could be rotated ninety degrees for endwise transport. An optional third wheel carried the main hitch member while the hitch extension was moved to the right end of the header. Transport width then was reduced to 9 feet for both models. The swather was power take-off driven by V-belts to the draper, reel, and sickle drives. The draper was treated to withstand the weather and provide long service. The reel was a five-bat, enclosed-end design. Both header height and reel height were hydraulically controlled by the tractor's hydraulic system or an available power pack for tractors without hydraulics.

	List price (Canadian Funds)
Model 12 swather, 12-foot cut	$ 975.60
Model 16 swather, 16-foot cut	1,097.60
Hydraulic power pack	144.40
Transport group	41.95

Allis-Chalmers sold the Kilberry products for a very short time.

Make Hay the Fluid Way — Forage Harvesters

During the 1930s when agricultural engineers were searching for ways to mechanize crop handling, they knew that the most important factor might be the ability of the crop to flow. As a result the term "fluidize" was used. Small grain and soybeans were naturally fluidized and mechanized handling presented no special challenge. Ear corn and roughage crops were obviously another matter. After the corn was shelled, it was fluidized but that was well after the hard work of handling it in the ear was finished. Field shelling was the operation that fluidized corn.

Corn silage could be fluidized to a degree, again after all the manual labor had been expended to bring the corn bundles to the ensilage cutter. The writer grew up on a central Indiana dairy farm where the silo was filled with corn silage. Both alfalfa and soybean were chopped and blown into the hay mow by the same Papec chopper, after being brought from the field by a buckrake. With the Papec always located on the south side of the barn, forking hay, especially soybean hay, into the chopper could qualify as the hottest, driest, dustiest, itchiest job this side of Hades. The evening shower was the nearest thing to heaven at that time. The payoff was that feeding chopped hay was many times easier than handling 60-pound bales and the cows seemed to thrive on it without wasting as much. The ultimate answer to that dirty job was field chopping, which allowed the roughage to be blown or conveyed mechanically (fluidized). Sure would have been welcome in pre-World War II Indiana. In fact, at that time in northern Indiana, A-C engineers at LaPorte were working on a machine which would field-chop forage so it could be handled by machine from start to finish.

Corn silage was bad enough, but grass silage with its high tonnage of heavy wet green material was worse, and grass silage was gaining popularity in America's Dairyland at that time. If there was ever a crop that called for mechanization, it was grass silage.

In 1940 Allis-Chalmers demonstrated their new forage harvester at a field day in Wisconsin where F. W. Duffee, head of agricultural engineering at the University of Wisconsin, hailed it as the "last link in mechanized agriculture." The first A-C forage harvester was a relatively simple machine which had a pickup header and draper that fed the crop to a wide chopping cylinder. The cylinder, which was equipped with knives made from angle iron, did not have a blowing action so the chopped forage dropped to an enclosed conveyor which elevated the

One of the first Allis-Chalmers forage harvesters was demonstrated in Wisconsin in 1940 and hailed as "The last link in mechanized agriculture".

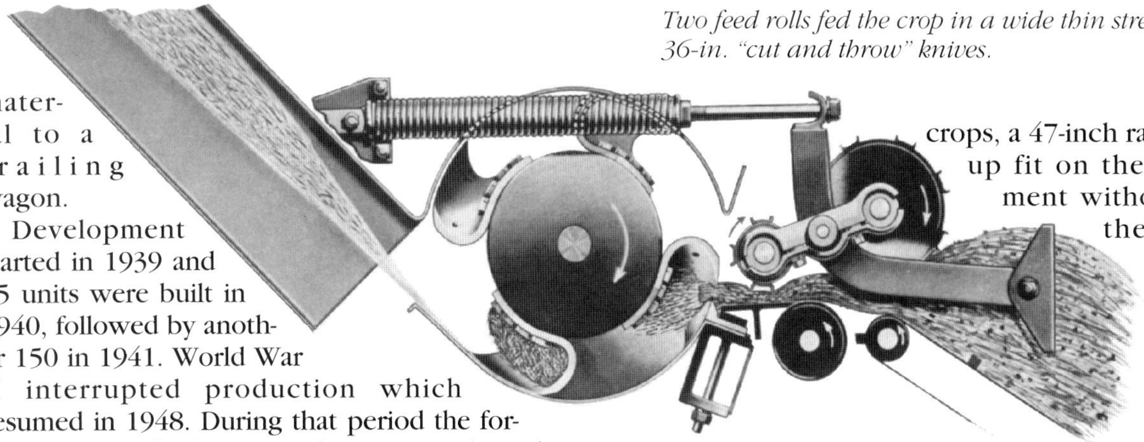

Two feed rolls fed the crop in a wide thin stream to the four 36-in. "cut and throw" knives.

material to a trailing wagon.

Development started in 1939 and 25 units were built in 1940, followed by another 150 in 1941. World War II interrupted production which resumed in 1948. During that period the forage harvester had major redesign to replace the conveyor. Starting at Serial No. 276, the chopping cylinder was equipped with "cut and throw" curved knives which blew the forage into the trailing wagon, eliminating dozens of parts. Two powered feed rolls fed the crop into the cylinder. They had a stone-stopper clutch which stopped the rolls when an object became wedged, thus eliminating much knife and shear bar damage and downtime. The four-knife cylinder was 36-inches wide so the crop was fed in a wide thin layer which was easy to cut and also kept the power requirement low. A two-plow tractor could power the unit or it could be equipped with a B125 auxiliary engine for use with a one-plow tractor. The built-in knife sharpener allowed the operator to sharpen the knives without removing them from the machine.

Several different attachments fit the forage harvester. The direct-cut grass attachment had a 48-inch sickle and gathered a 52-inch swath. For windrowed crops, a 47-inch raddle-type pick-up fit on the grass attachment without disturbing the sickle. The one-row row-crop attachment adapted the forage harvester to sorghum, corn, and other row crops and retired the old ensilage cutter at the silo. In 1956 a new heavy-duty grass attachment with many improvements, including a new rubber draper pickup, replaced the earlier design. A total of 32,681 forage harvesters of the first model were manufactured. In 1956 a forage harvester with power take-off and row-crop attachment sold for $1,490.

A WC tractor with a direct-cut forage harvester making grass silage. Starting at Serial No. 276, the forage harvester gained the "cut and throw" curved knives which allowed replacing the rear conveyor with a blower hood, greatly reducing the number of parts.

The Model 50 Forage Harvester

The Model 50 forage harvester replaced its predecessor in 1958, beginning with Serial No. 40001. Introduced with a four-knife cylinder, a six-knife version was soon made available. This allowed finer cutting of silage crops. Feed roll speed was adjustable from 165 to 425 feet per minute to provide theoretical cutting lengths of 3/8 to 2 1/4 inches with a four-knife, 1/4 to 1 3/8 inches with a six-knife cylinder. The new shear bar was similar to an oversized razor

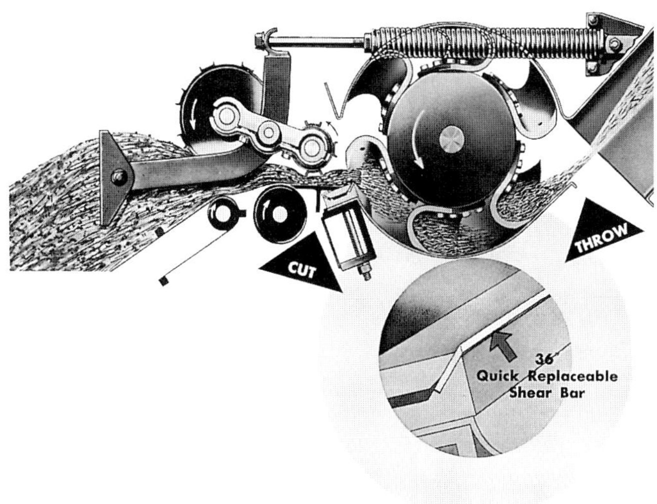

The Model 50 chopping cylinder had four knives like the original model. This was later changed to six knives for a finer chop. A unique replaceable shear bar glooked like a large razor blade and helped maintain chopping efficiency.

A D17 with a Model 50 and one-row row-crop attachment making corn silage. Optional extension and rotating hood permitted side delivery for loading trucks.

This Model 50 with a 5-ft direct-cut attachment had increased capacity for making grass silage.

A D17 and 50 forage harvester chopping hay. The Model 50 had a 60-in. pickup which was 12 in. wider than the earlier model.

blade in that it was made of high grade steel with a keen cutting edge, and was quickly and easily replaced. The knife-saving stone-stopper clutch and built-in sharpener were retained in the Model 50. A two-way rotating delivery hood could be rope-controlled from the tractor seat, allowing the operator to load the wagon more evenly.

The one-row row-crop attachment changed very little. The big news was a new grass attachment with a 60-inch sickle bar that gathered a 65-inch swath. A four-bar Hume pickup reel was standard equipment. The Model 50 was built until 1963 when it was replaced by the Model 780. Production quantities included 1,181 four-knife and 988 six-knife machines for a total of 2,169 units.

The 780, 782, and 720 Forage Harvesters

The 780 forage harvester was a big change in design — it was a right-side cut machine. More important was the location of the cylinder, which was just above the main axle. The lower position simplified feeding the crop into the cylinder. Five feed rolls controlled the crop right up to the replaceable shear bar. Using six, three, or two knives on the cylinder, the cutting length could be varied from 1/4 to 2 1/4 inches. The proven knife sharpener and stone-stopper clutch were continued in the new chopper. A rope-controlled feed reverser allowed the operator to unclog the feed rolls from the tractor seat — a great safety feature as well as a time saver.

The 6 1/2-foot pickup attachment could handle the largest windrows, and its 64 steel pickup fingers didn't miss much. The full-width auger, with retractable-finger feeder, carried the crop into the feed rolls in a smooth even flow for uniform chopping. The 6 1/2-foot direct-cut head for standing crops had a sickle speed of 1,188 strokes per minute and the four-bat pickup reel speed could be adjusted from 64 to 110 rpm. The 20-inch cross auger and retractable-finger feeder were similar to those in the pickup attachment.

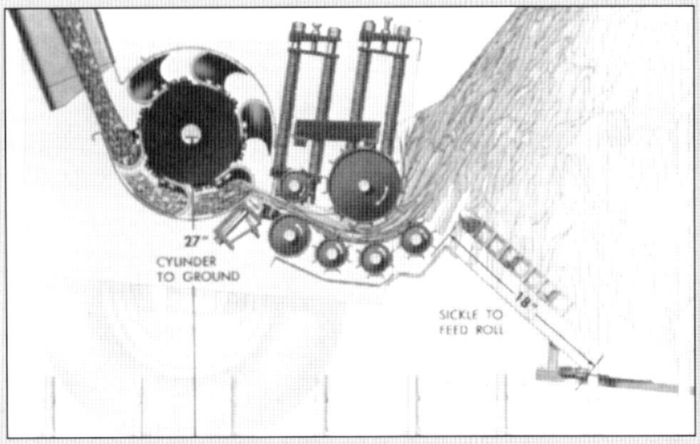

The 780 and 782 forage harvesters had two upper and four lower feed-rolls to feed the crop into the cylinder which improved handling of heavy wet silage.

A One-Ninety XT, 782 forage harvester with a row-crop attachment, and a forage wagon were taking corn silage out of the field at a fast clip. There were one-row, two-row, and three-narrow-row attachments for the 782.

A One-Ninety XT-powered 780 forage harvester with a 6 1/2-ft direct cut attachment could put grass silage in the wagon in a hurry.

A 6 1/2-ft pickup attachment equipped the 780 or 782 for high-capacity hay chopping and the One-Ninety XT had the power to utilize that capacity.

The 720 Forager was a smaller, lower-cost machine with two-row harvesting capacity. A one-row was also available. The 720 had an 18 1/2-in. wide by 20-in. diameter cylinder, completely unlike prior models.

A pickup attachment adapted the 720 Forager to chopping dry hay and haylage, and a direct-cut attachment was available for making grass silage.

One-, two-, and three-row row-crop attachments were available for the 780. There were three gathering chains per row to deliver the crop a short 18 inches from the sickle to the feed rolls. Capacity of the 780 was rated at up to 60 tons per hour and although power requirement was said to be a three-plow tractor or larger, the emphasis would have to be on larger. It was usually operated by a D19 or One-Ninety XT tractor with 70 to 95 horsepower.

The 780 was built from 1963 through 1967, then was replaced by the Model 782. The 782 was an improved 780 and was very similar in appearance. Attachments for the 782 included the 6 1/2-foot direct cut, 6 1/2-foot pickup, and row-crop units for one row, two 30-inch rows, two 40-inch rows, and three 20-inch rows.

The Model 720 Forager was added to the line in 1972, but did not replace the Model 782. The 720 was smaller than the 782 and the only A-C forage harvester to use other than the original 36-inch wide, 15-inch diameter cylinder dimensions. The 720 cylinder was 18 1/2 inches wide and 20 inches in diameter, which was comparable to most other medium capacity forage harvesters on the market at that time. Direct cut, pickup, and row-crop attachments were available. Forage harvesters were discontinued in 1975.

Taking the Pasture to the Cows — The 56F Forage Harvester

The 56F flail-type forage harvester was the company's answer to a need for a relatively low-cost unit used primarily for green-chopping forage for confinement feeding. Dairy cows were being kept confined close to the milking parlor rather than being allowed to roam the fields. The same fields could produce more forage if the cows did not trample and contaminate the crop. The farmer needed a machine that was compact, easily maneuverable, and required only moderate power. It was often left hitched to a bunk-feeder wagon. The operator could go to the field and chop the day's needs into the feeder wagon, return to the feed lot, and leave the whole outfit there for feeding until green-chop was needed. The system kept the cows out of the field, didn't require much labor, and eliminated the need to keep the fences repaired.

Unless it rained for days on end, everything was fine. Flail harvesters were also used to make grass silage, but they produced a more coarsely chopped silage than the conventional forage harvester, which was undesirable because it would not pack as well in the silo. Flail machines could not be used successfully in tall row crops. Since flail-type harvesters lacked the versatility of interchangeable-attachment harvesters,

Powered by a D17 Series II, the 56F flail-type forage harvester was geared to either green-chop or windrow work. It took a 6-ft swath.

they found only a special-purpose market with limited sales potential.

The LaPorte-built 56F was similar to most other flail-type harvesters on the market around 1960. The chopping rotor turned backward in relation to forward travel, while the 36 knives cut a 6-foot swath of forage, shearing the crop against a full-width shear bar at the front of the rotor housing, and threw it over and back into a cross auger. The auger carried the forage to the large diameter fan which blew it into the wagon. The 56F was a very uncomplicated machine. With hydraulic lift, tires, and rotor shield, it had a list price of about $1,332.

Forage Blowers

High field chopping capacity called for even higher capacity at the silo if the field equipment were to be fully utilized. A-C's first attempt at designing a blower utilized a "C" shaped fan blade similar to the cut-and-throw knives of the forage harvester. It would throw, not blow, and it wouldn't throw very high — back to the drawing board.

In 1949 A-C launched a high-capacity forage blower that quickly earned a good reputation for filling tall silos with minimum power input. Two features made this possible: a larger 9-inch pipe diameter and a fan with six cupped and curved blades which kept the crop material away from the fan housing to reduce friction. The long hopper had a chain and raddle conveyor to carry the crop to the fan. About 1953, the conveyor was changed to a smooth-running belt. Three star-wheel retarders leveled the incoming forage for even feeding to the blower fan.

The early models were belt-driven but tractor belt pulleys were rapidly disappearing. By the late 1950s, a right-angle PTO drive was offered. The PTO drive had a gearbox, two pulleys, and multiple V-belts to change input speed and to turn the drive 90 degrees. The list price of $228 for the optional PTO drive increased the blower's total price about 40 percent.

The 30L was the original long conveyor unit with a new name, while the 30S had a short auger conveyor for front-unloading forage wagons. Both of these blowers were belt, power take-off, or auxiliary engine driven. The D14 149-cubic inch engine was optional auxiliary power. List prices were $803 for a 30L and $683 for a 30S. A D14 engine added $574. From 1949 through 1964, 31,841 standard and 30L blowers were manufactured. Between 1959 and 1961, 753 No. 30S blowers were produced.

A 1949 blower with engine drive. Early models had a chain-raddle conveyor which was later replaced by a rubberized belt.

This 30S blower is equipped with a sodium bisulphite attachment and PTO drive. The right-angle PTO drive was expensive but effective.

The 480 blowers had a 48-in. fan for increased capacity. This 480S short-conveyor blower worked with forage wagons that were equipped with a front cross-conveyor.

The 480 Blowers

The 30L and 30S blowers were replaced by the 480 blowers. The 480S and 480H came along in 1964, followed by the 480L a year a later. The 480's boasted capacity of up to 1 ton per minute. The fan was 48 inches in diameter with six straight blades. The PTO drive was direct to the fan shaft, eliminating the expensive gearbox and V-belt. Normal input speed was 550 rpm with 850-rpm maximum. The high capacity 9-inch diameter pipe was retained. The angle of delivery could be changed by rotating the fan wrapper.

All models had a short 12-inch diameter auger to feed material into the fan. The 480H short hopper model eliminated the conveyor which resulted in a lower cost machine. It depended on gravity feed so it was best adapted to grain and finely chopped silage. The 480S had a short chain and raddle conveyor with star-wheel retarders and was adapted to self-unloading forage wagons with front cross-conveyors. The 480L had a long, folding chain and raddle conveyor with star-wheel retarders for use with rear-unloading forage wagons. In about 1970, as front-unloading wagons replaced rear unloaders, the long hopper blower was dropped from the line. The 480H and 480S were replaced by the new 660 blower in early 1973.

The 660 Blower

The 660 was a high-capacity blower with a 60-inch six-blade fan and 9-inch pipe. It was a short conveyor type with a chain raddle and an auger feeding the fan. Star-wheel retarders and the rotatable fan wrapper were retained in this design. The 660 had PTO drive direct to the fan shaft. Small transport wheels lowered the fan housing and drive shaft so that it was aligned with the tractor power take-off. The 660 blower was discontinued in 1975.

Wagons and Power Feed Box

No forage system would be complete without transportation from the field to the blower. To complete its forage system Allis-Chalmers turned to an outside supplier, the Calhoun Company of Cedar Falls, Iowa, for a self-unloading box. It was a 14-foot box with 52-inch sides, two beaters, and cross-conveyor, and was extendible to 74 inches with a third beater. An elevating conveyor for bunk feeding was available. For the customer who wanted to build his own box, the base mechanism with one beater was offered, along with a 2-foot platform raddle extension for a 16-foot box.

A D15 Series II drove a power feed box to unload forage into a 480H blower powered by a D17 Series III. All the operator had to do was control the unloading rate.

The 660 blower of 1973 had a 60-in. fan for maximum capacity. It only came in the short conveyor type.

The first A-C forage box was was a purchased-finished product from the Calhoun Co. of Cedar Falls, Iowa. Base unit was a 14-ft box with 52-in. sides.

The LaPorte forage box came as a complete factory-built unit or as a mechanical unit for a home-built wagon box.

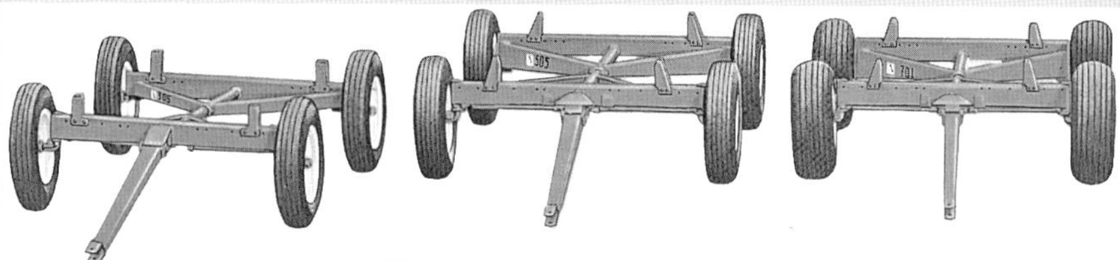

Wagon gear with capacities from 3 1/2 to 10 tons covered a broad span of farm hauling needs.

The purchased-finished forage box was marketed from about 1959 until 1962, when LaPorte began building a power feed box. It was similar to the purchased unit and was also available as a complete factory-built box or as a kit for a "field-built" box. In 1964, LaPorte started building a rear self-unloading box, which was essentially the same except that it didn't have the front beaters and cross-conveyor.

Wagons, flare boxes, and barge boxes were purchased from Electric Wheel, a division of Firestone Tire and Rubber Company, Quincy, Illinois. Due to their limited capacity the flare and barge boxes were soon dropped, but the wagons remained available. Six auto-steer wagons were offered, though not all at the same time:

Model 305 — 3 1/2-ton economy model.
Model 405 — 4 1/2 ton.
Model 505 — 5 1/2-ton for heavier loads.
Model 605 — 6 1/2-ton.
Model 701 — 7 ton for power feed boxes.
Model 1001 — 10 ton for hydraulic hoists.

The Model 500 was a fifth-wheel steer wagon with telescoping reach for a 16- to 19-foot wheelbase. This wagon was ideal for cotton boxes or other high-bulk low-weight loads.

A Better Way To Package Hay The Roto-Baler

Like the All-Crop harvester, A-C didn't invent the rolled bale or the machine to make it, but they did develop the concept into a workable baler (far beyond anything the actual inventor might imagine). The rolled-bale baling press was invented by Ummo F. Luebben of Lincoln, Nebraska, and he was granted Patent No. 972,884 on October 18, 1910. A-C built a baler far different from the device pictured in the patent drawings. Harry Merritt was instrumental in acquiring rights to the Luebben patent in 1940.

Work on the A-C round baler began in December of that year, with the first unit built in 1941. Six experimental machines were built by the LaPorte engineering department, who retained two while the other four were placed with users who returned them at the end of the season. The same program was repeated the following year. In 1943, a pilot run of 25 balers was built, most of which were placed

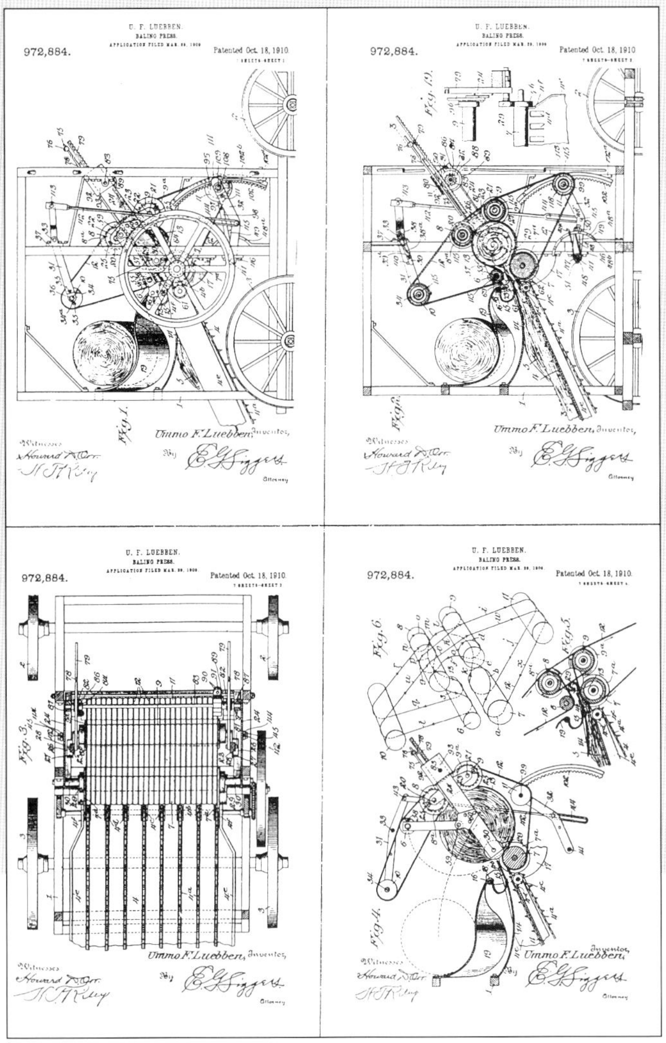

The Luebben patent of 1910 defined the rolled bale and a system of belts to make it. Allis-Chalmers engineers took this crude, basic design and developed the successful Roto-Baler.

Round bales often spent the winter right where they dropped from the baler. Water on the top was no problem but they would soak up water from the bottom, so well-drained land was important for outside storage.

with farmers and custom operators in seven midwestern states. For the next year several of these balers were rebuilt and all placed with users. This extensive test program produced the proven Roto-Baler.

Also like the All-Crop, the Roto-Baler was unlike any other of its kind. Most unconventional machines take considerable selling, and the round baler was no exception. Farmers were accustomed to seeing hay baled in relatively large, usually heavy, rectangular packages held together by two or three wires, although the new-fangled twine-tie balers were getting started. Now here was a bale that wasn't rectangular, wasn't really very heavy, and was wrapped, not tied, with binder twine. Can't be! Won't work!

But that strange-looking machine, which certainly didn't look like a baler, had a lot going for it. It was simple, lightweight, and low-priced, compared to the conventional balers of the late 1940s. At only $985 (1947), a farmer could afford to own one and avoid the custom baler who didn't always come to bale before the next rain which was sure to come. That had appeal. But who wants a round bale?

A Roto-Baler working on a large double windrow that made full, well-shaped round bales. The Model CA had two-clutch power control and plenty of power for convenient, high-capacity baling.

So the A-C sales crew had to learn all about "Roll-Up Compression" and how it formed the bale without mangling the hay. They had to learn how to convince farmers that this was a better bale. They had to show that the leaves were still on the stems because of gentler handling and rolled inside for protection. They had to unroll bales to demonstrate easier feeding. They had to soak bales with water and then unroll them to show that the hay inside was still dry. They talked their first customers into leaving some bales in the field all winter, so they could open them in the spring and show the non-believers the bright

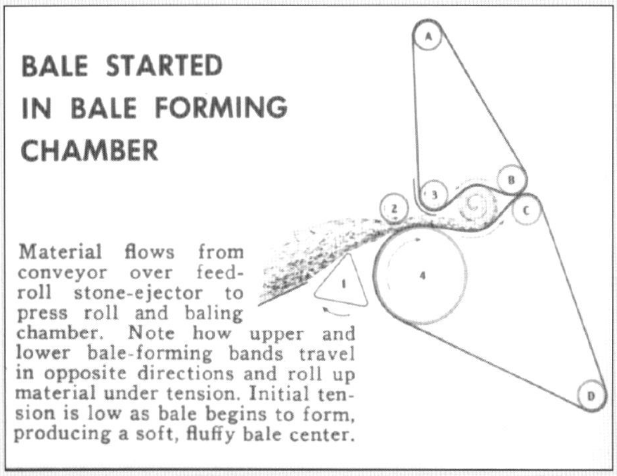

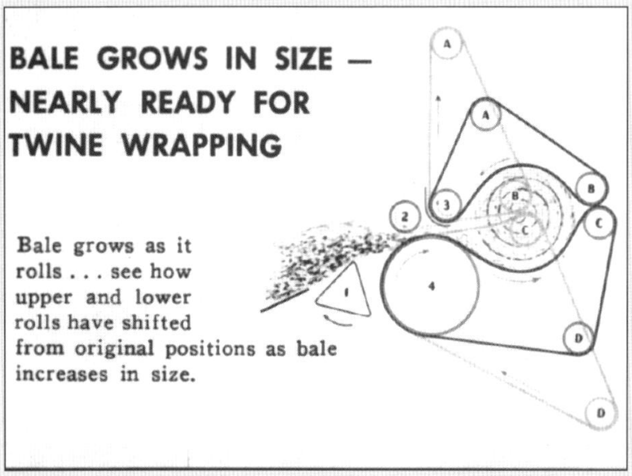

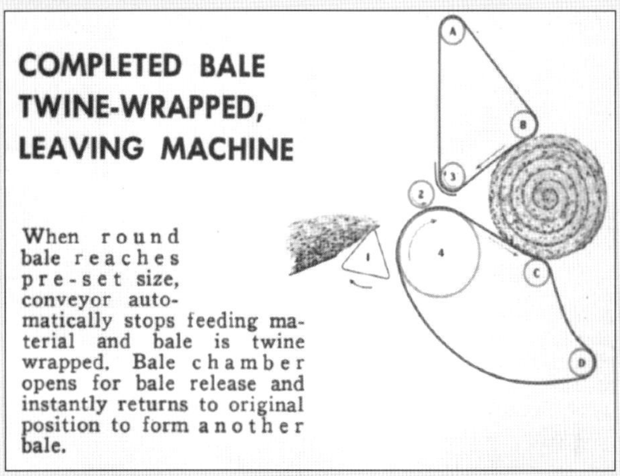

The baling cycle of the Roto-Baler was a relatively simple operation. Only genuine A-C bands would make well-shaped bales due to their unique fabric construction.

green, dry hay. Like the thatched-roofed dwellings in high rainfall areas, those bales were dry inside. Even the cattle became part of the selling team–they liked the rolled hay better because there were no sharp ends to jab their mouths.

It worked! The first year's (1947) production was 3,000 units. Not bad, but the proof was in the second year's production, just a few units short of 10,000, followed by a like number the third year. No one builds 23,000 of anything in three years if it doesn't satisfy the customers.

The rolling mechanism was simplicity itself, consisting of six upper and six lower bands, which formed an expandable diameter chamber 36-inches wide. The bands were made with a special cross-ply design in the fabric which made them run straight. Ordinary belts would all run to one side resulting in poorly formed bales. Rolled-bale size was adjustable from 14 to 22 inches in diameter, weighing 40 to 100 pounds.

Operators soon became aware that successful Roto-Baler baling required the right windrow, so users were advised to rake a windrow, then turn around and rake another windrow adjacent to the first, forming a wide double windrow. This allowed the baler to form a full-width bale, something nearly impossible to accomplish with small windrows. This also reduced baler travel and increased capacity, so it worked out well.

It was necessary to stop forward travel while the bale was wrapped with binder twine. The stopping was probably a bit tough on the operator of a WC tractor, which required disengaging the clutch, putting the transmission in neutral, engaging the clutch to wrap the bale, disengaging the clutch, shifting into gear, and engaging the clutch for each bale. The new 1948 WD with its two-clutch power control was a godsend to the Roto-Baler operator. However, not every farmer who used a Roto-Baler saw fit to purchase a new WD so the company began a development to eliminate the stopping.

The No. 10 Roto-Baler Non-Stop Baling

The development program to eliminate the stopping resulted in the No. 10 Roto-Baler, which hit the market about mid-1955. The No. 10 had a bypass conveyor which rerouted hay back to the front end of the pickup while the bale was being wrapped. This allowed continuous forward travel. An overdrive (fast wrap) mechanism reduced wrapping time to about 3 seconds to minimize the amount of hay which had to be bypassed during the wrap cycle. The No. 10 was about 500 pounds heavier and $300 higher than the

A No. 10 Roto-Baler in the bypass mode as the bale was wrapped in a brief 3 seconds.

standard Roto-Baler, which might have been acceptable if it had worked successfully in more conditions. Unfortunately it did not, and many of the No. 10's were eventually converted to standard operation.

However, the fast-wrap mechanism was retained in the converted units so baling capacity was gained. As a result, the Roto-Baler was offered in two models — standard and overdrive (fast-wrap). The overdrive model was about 100 pounds heavier and $175 higher in price.

Another outcome of the No. 10 design was an enclosed wing-type rotary feeder just ahead of the main roller and above the feed conveyor which improved feeding in adverse conditions. This feature was added to the standard Roto-Baler in 1956. Over the production life of the Roto-Baler, many other changes and improvements were made, including a new conveyor drive, soft bale-center attachment, automatic twine-tube trip, and several others.

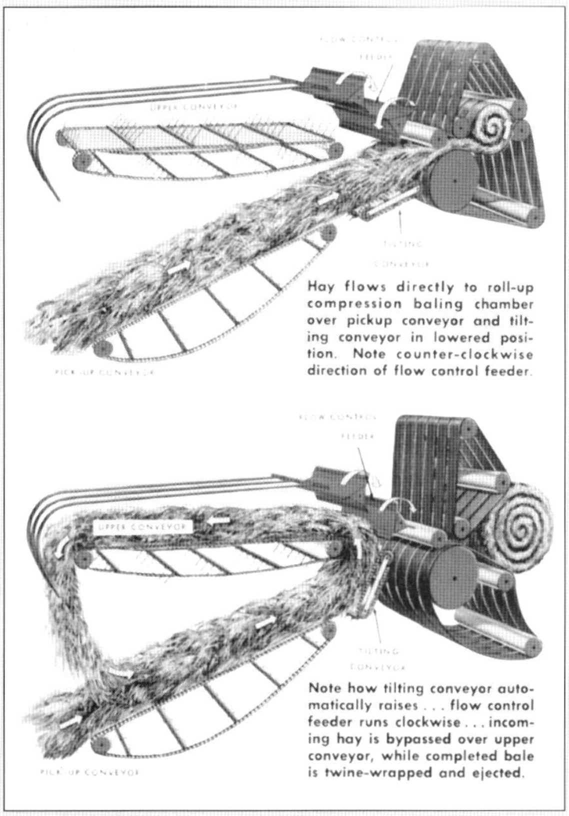

(Top) In the No. 10 baler normal feeding and bale forming was just like the standard baler. (Lower) During the 3-second wrap cycle, the upper conveyor carried the hay back to the front of the feeder. A smaller windrow worked better in the No. 10.

The standard baler with overdrive (fast-wrap) with its 3-second wrapping cycle had improved performance and greater field capacity.

The rotary feeder became standard equipment in 1956, while the soft-center attachment was added the previous year.

A Model 185 tractor pulling the 1972 version of the Roto-Baler. It was equipped with the full shielding which was added when the Roto-Baler was reinstated.

All three versions of the No. 10 appeared in the 1960 price book:

Model	Weight (lb)	List Price ($)
No. 10 with PTO drive	2,918	1,950
Standard with overdrive	2,488	1,845
Standard	2,410	1,670

The Roto-Baler remained in production from 1947 through 1964. Due to popular demand there were three small production runs of 800 each in 1971, 1973, and 1974. Altogether 77,200 standard and fast-wrap Roto-Balers and 1,500 No. 10 Roto-Balers were built. Not bad for something that didn't even look like a baler.

The bale loader was designed specifically to load round bales and would fit the C, CA, WC, WD, and WD45 tractors. It sold for $150 in 1960. This 1945 photo was two years ahead of Roto-Baler production.

Big Round Bales at Allis-Chalmers

The success of the Roto-Baler would seem to have set the stage for Allis-Chalmers to follow with another success in the form of a large round baler. The large round bales were becoming popular for the same reasons their smaller A-C cousins had changed hay harvesting because of the ability to shed rain and snow, provide a nutritious product, and ease of storage — just leave them outside. The huge capacity of the large round baler was an important factor in their broad acceptance. Certainly the task of gaining dealer and customer acceptance would have been an entirely different matter from that of the first Roto-Baler.

Feasibility studies indicated that purchasing a unit for resale was a better alternative than attempting to

The Hawk-Bilt large round baler was tested in 1971 but performance did not meet performance criteria. Its ground drive could not make a tight bale.

At first Roto-Baler advertising had a huge job to do in convincing farmers that round bales would work at all, let alone produce a higher-quality product.

design and tool a proprietary model. Discussions with Hawkbilt of Vinton, Iowa, led to the acquisition in 1971 of a test unit which Hawkbilt had imported from Australia. It was an unusual concept in that it did not have a PTO drive, but a ground-drive arrangement which rolled up a large loose round bale as the unit was driven across the field. It was found to be lower cost, but performance was not competitive to large balers of domestic manufacture. The project was dropped. Thus the story of why the Roto-Baler didn't grow up.

Square Balers

Although the round bales were widely accepted by livestock farmers, they never caught on in the commercial hay market. Experienced haulers could handle round bales, but the majority of truckers didn't want to try to load them. As fewer Roto-Balers were sold, the company had a definite need for a square baler.

The Model 200 Baler

A search for a suitable baler resulted in the acquisition of the Jones baler, built in Mold, Wales, United Kingdom. This baler was launched into the market in early 1962 as the Model 200, a PTO driven twine-tie unit. It produced a conventional 14×18-inch bale 18- to 36-inches long. The 12×19-inch (228 in.^{2}) feed opening was slightly smaller than subsequent models. The 28-inch stroke plunger delivered 78 strokes per minute to form bales up to 80 pounds. The pickup was 54 inches wide and had 55 coil-spring teeth on five bars. A rod-type crop guard kept the hay in contact with pickup teeth in windy conditions.

The Twin-Feed rake system provided smooth, even feeding to the bale chamber but was gentle enough to minimize leaf loss. This feeding system, with modifications, was used on all the following domestically-produced balers. The knotter was of conventional design with disk-type twine holders that assured positive tying. Needles were protected by a positive ram stop, if for some reason they were to remain in the bale chamber.

The chicks in their round-bale coop often appeared in early Roto-Baler brochures and advertising.

The 200 was A-C's first square baler (1962). It was a PTO-driven twine-tie unit that made conventional 14x18-inch bales of up to 80 lb. It was manufactured by the Jones Baler Co. of Mold, Wales, U.K. which the company acquired and made a part of Allis-Chalmers — Great Britain.

Power was transmitted through a three-joint power line to a heavy flywheel which smoothed shock loads. Roller chains were used throughout the baler. An overrunning clutch protected the tractor power take-off and power line, while slip clutches and shear pins at strategic points protected all operating components.

The Model 200 set the stage for a number of square balers to follow — all of which were produced at the LaPorte plant. The 200 was offered only during 1962 and early 1963, giving LaPorte engineering time to develop the Model 303 which appeared in 1963.

The LaPorte-Built Square Balers

Although there were many detail changes, the specifications of the 303 were much like those of the 200. The proven Twin-Feed rake system was somewhat redesigned but its function was basically unchanged. The pickup was 60 inches wide with 48 teeth on four bars. The feed opening size was increased to 264 square inches and ram speed was reduced to 70 strokes per minute. Bale size was unchanged and only twine-tie was offered. The three-joint power line was standard and a Wisconsin V4 air-cooled engine was available.

About 1964 the 303 was redesignated the Model 303 Bale Chief with some important new features.

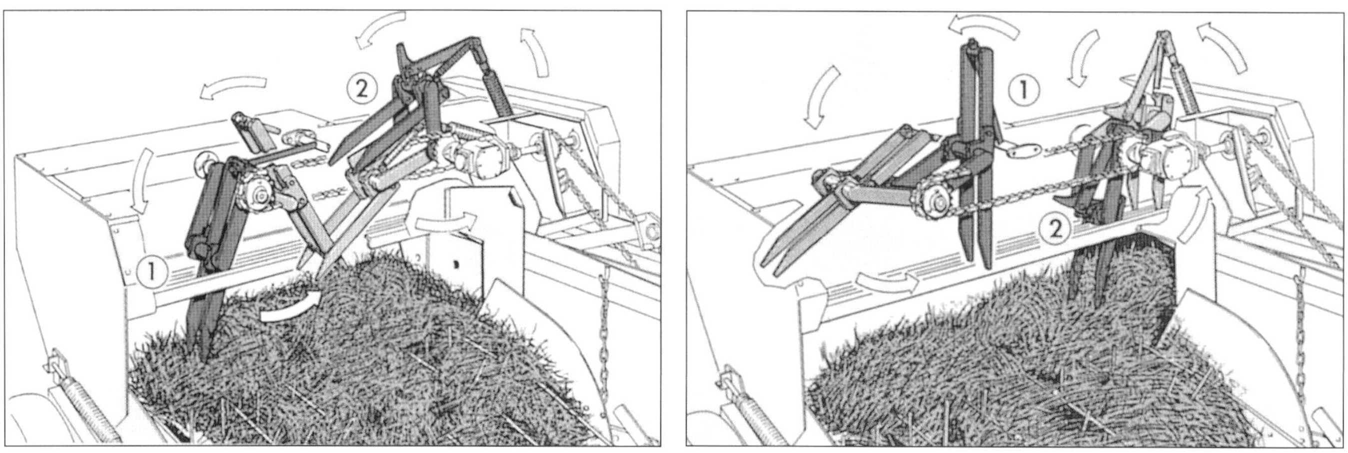

The Twin-feed rake system. Synchronized double forks moved the hay into the bale chamber. A full charge was delivered on each ram stroke. Rake No. 1 reached into the crop at the corner of the feeder housing and carried it toward the chamber. As the crop reached the center of the housing, rake No. 2 picked it up and moved it into the bale chamber. Rakes retracted up and out of the hay on each back stroke. One rake was always moving hay toward the chamber.

An engine-driven bale thrower was available for the 303 Bale Chief. The LaPorte-built 303 replaced the 200 in 1963.

Basic specifications remained unchanged, but a new bale thrower was added. It had six corrugated rubber rollers, and was driven by a 5.75-horsepower air-cooled engine. Engine speed could be controlled from the tractor to change the throwing distance for loading both the front and rear of the wagon. A wire-twist model became available and 450 were produced that year. The wire-twist unit could make bales up to 110 pounds.

In 1966 the Model 302 baler was added. With a two-joint power line and other changes, it was a lower-priced baler and was available in twine-tie or wire-twist versions. Major functional components were the same as the more deluxe 303. The roller-type, engine-driven bale thrower was replaced by the new Model 44-PTO-driven belt-type thrower. The advantages included improved operation and lower cost.

In 1970 the Model 444 baler replaced the 303. Upgraded from the 303, the bale chamber, knotters,

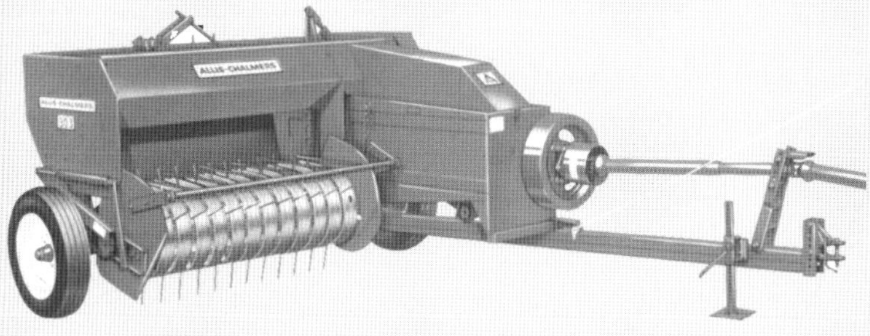

The 303 was offered in both twine-tie and wire-twist models. A three-joint PTO driveline was standard.

The 444 was the top baler starting in 1970. A five-bar 90-tooth pickup improved performance in short-stemmed and fine hay. It came in twine-tie and wire-twist models and had a three-joint driveline.

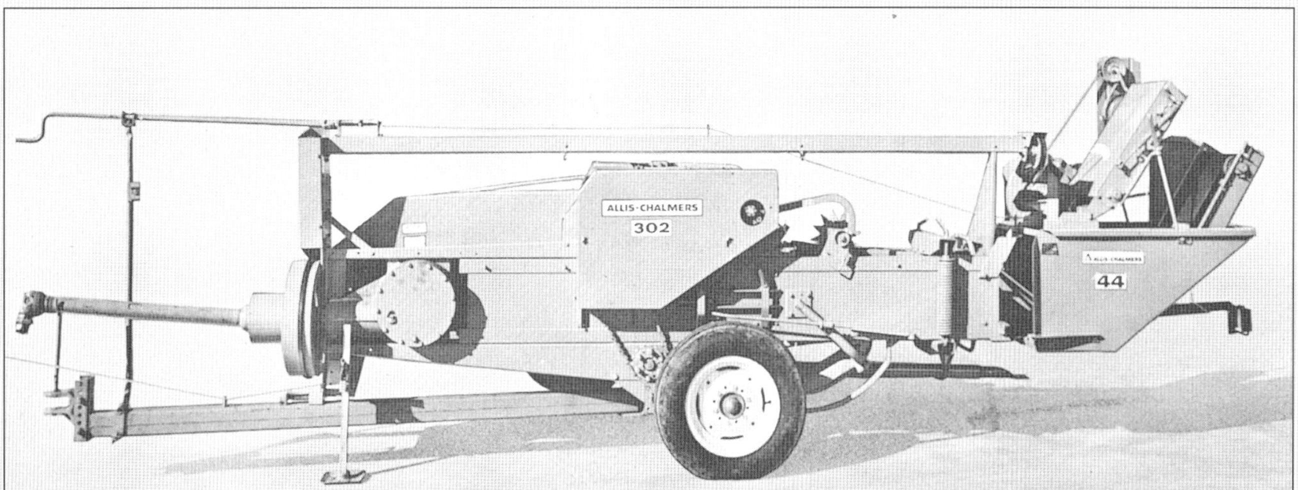

The 302 had a two-joint driveline and was lower-priced than the 303. This one is equipped with the 44 PTO-driven bale thrower. Thrower speed could be changed with the crank near the operator to control throwing distance.

The 442 was lower-priced than the 444, but had a four-bar 48-tooth pickup and a slightly smaller feed opening.

The 443 replaced the 442 in 1973. It was mechanically the same as a 444 except that the three-joint powerline was optional.

The bale conveyor was available as a portable unit in sizes from 25 to 49 ft. Stationary systems for in-barn installations could be straight, L, or T arrangements in any size barn. Engine or electric motor drive was offered. The bale conveyor complemented the square baler and thrower to provide a mechanized system of bale-handling from field to storage. The conveyor was a purchased-finished product.

and feeding mechanism were similar, but the ram speed was increased to 80 strokes per minute which increased baling capacity. The 60-inch pickup had 90 teeth on five bars for improved performance in short or fine-stemmed hay. The 444 had a three-joint power line. The 44 PTO-driven bale thrower was carried over from the previous model.

The Model 442 replaced the 302 as the lower-priced companion to the 444 and was in the line from 1970 through 1972.

It had a slightly smaller feed opening at 253 square inches and a four-bar, 48-tooth pickup. Twine-tie only, it made bales up to 80 pounds. The ram speed was 80 strokes per minute. Power was transmitted by a three-joint power line.

The Model 443 replaced the 442 at the beginning of 1973. The 443 shared the larger 264-cubic inch feed opening and five-bar, 90-tooth pickup with the 444. To hold the price down a two-joint power line was standard, which left the three-joint type as optional. Optional equipment, including the 44 bale thrower, was the same as the 444. The 443 and 444 were produced until the entire hay line was discontinued in 1975.

Specifications for Square Balers

	200	303	302	442	443	444
Years built	1961-62	1963-69	1966-69	1970-72	1972-75	1970-75
Type	Twine	Twine	Twine	Twine	Twine	Twine
	---	Wire	Wire	---	Wire	Wire
Bale size (in.)	14 × 18	14 × 18	14 × 18	14 × 18	14 × 18	14 × 18
Bale length (in.)	18 to 36	12 to 42	12 to 42	12 to 42	12 to 42	12 to 42
Feed opening (in.^{2})	228	264	264	252	264	264
Strokes per min.	78	70	70	80	80	80
Stroke (in.)	28	28	28	28	28	28
Bale weight (lb)	*	*	*	*	*	*
Pickup width (in.)	54	60	60	60	60	60
Bars	5	4	4	4	5	5
Teeth	55	48	48	48	90	90
Power line						
U-joints	3	3	2	3	2 or 3	3
Drive	†	†	†	†	†	†
Optional	---	§	§	---	---	---
Bale thrower	None	Engine-roller	ll	ll	ll	ll

* Up to 80-lb twine, 110-lb wire.
† PTO standard.
§ Wisconsin engine.
ll PTO-driven belt-type.

Roto-Baler Serial Numbers and Production Data

Year	Serial No.-Serial No.	Production
1947	101-3100	3,000
1948	3101-13043	9,943
1949	13044-23880	10,837
1950	23881-33706	9,826
1951	33707-42427	8,721
1952	42428-53903	11,476
1953	53904-61536	7,633
1954	61537-65151	3,615
1955	65152-68401	3,250
1956	68402-71401	3,000
1957	71402-72401	1,000
1958	72402-72701	300
1959	72702-73525	824
1960	73526-74601	1,076*
1961	---	0
1962	---	0
1963	74602-74901	300
1964	74902-75401	500†
Discontinued 4/68, Reinstated 9/71		
1972	74902-75700	799
1973	75701-76500	800
1974	76501-77300	800
No. 10 Roto-Baler		
1958	C101-C1600	1,500

* Includes 500 with overdrive.
† Includes 250 with overdrvie.

Square Baler Serial Numbers

	200	303	302	442	443	444
1962	INA*	---	---	---	---	---
1963	INA*	101-	---	---	---	---
1964	---	2601-	---	---	---	---
1965	---	5405-	---	---	---	---
1966	---	6101-	30001-	---	---	---
1967	---	INA*	INA*	---	---	---
1968	---	INA*	INA*	---	---	---
1969	---	INA*	INA*	---	---	1501-
1970	---	---	---	INA*	---	2101-
1971	----	---	---	INA*	---	INA*
1972	---	---	---	INA*	101-	2501-
1973	---	---	---	INA*	350-	2523-
1974	---	---	---	---	INA*	INA*
1975	---	---	---	---	INA*	INA*

* Information not available.

Forage Harvester Prices (1960)

	Standard	Model 50
Base unit	$1,162.80	$1,350.00
PTO	146.30	180.00
Row-crop attachment	570.00	570.00
Grass attachment	432.50	600.00
Hume reel	90.00	Std.
Pickup attachment	166.40	180.00

Serial Numbers and Production Data for the 56F

	Serial Numbers	Production
1960	101-600	500
1961	601-1600	1,000
1962	---	0
1963	1601-1850	250
Total		1,750

The Gleaners — Gleaner-Baldwin Combines

"The Gleaners" by Jean Francois Millet. This 1857 painting of women gathering the grain left by the reapers inspired the Baldwin brothers to use it to exemplify their harvesting machines.

The "stingy Scotch" (Curtis Baldwin's own words) Baldwin family produced three sons: Curtis, the eldest, George, and Ernest. Each of whom made significant contributions to the development of grain combines. Curtis grew up with the dawn-to-dark toil of feeding Tom Dugan's threshing machine, "so tired at night we could hardly make the bed, usually in the straw pile". He realized that the right machine could eliminate most of the hard work of handling the crop and was "fired with ambition to build such a machine" (Baldwin, 1931).

Baldwin learned to run the threshing engine, then acquired a threshing rig, later several outfits, and then did custom threshing throughout the wheat belt. He gained more experience than money and soon saw that handling the straw twice or more, only to later burn it, was gross inefficiency that no one could afford. He conceived a "standing grain harvester" which would leave the straw in the field, and applied for a patent. On September 26, 1911, at age 23, he was granted patent No. 1,004,134.

The Baldwin brothers: (top) Curtis, (left) Earnest, and George.

The Standing Grain Harvester was a pusher-type machine with four horses hitched to a rear draft pole. A small gasoline engine powered the harvester. A large fan directed a blast of air through a forward-mounted hood which blew the standing grain back against an overshot or rearward-turning toothed cylinder. It stripped the grain heads from the standing straw and delivered them to another cylinder for rethreshing. Baldwin claimed that the grain passed the cylinder at the horses' pace, about 3 1/2 mph, therefore the cylinder struck the grain heads 17 times while a conventional cylinder machine would contact them only once. One man drove the horses while another tied the grain sacks and attended the engine. This rig could harvest 25 to 30 acres in a 10-hour day.

To pay the building expenses of the machine, Baldwin took a job selling and "experting" threshers. After two years, the Standing Grain Harvester was ready to test. Acceptance was slow but he eventually sold about 100 units throughout the wheat states. He claimed his harvester could reduce the cost of harvest from 14 to 2 cents a bushel.

To improve performance in lodged crops and humid conditions, Baldwin replaced the air-blast arrangement with a slotted mechanical gathering reel. These changes were covered by patent No. 1,290,484, which was granted on January 7, 1919. The Baldwin Standing Grain Harvester was first manufactured in Nickerson, Kansas. The Baldwin Manufacturing Co. of Hutchinson, Kansas, was incor-

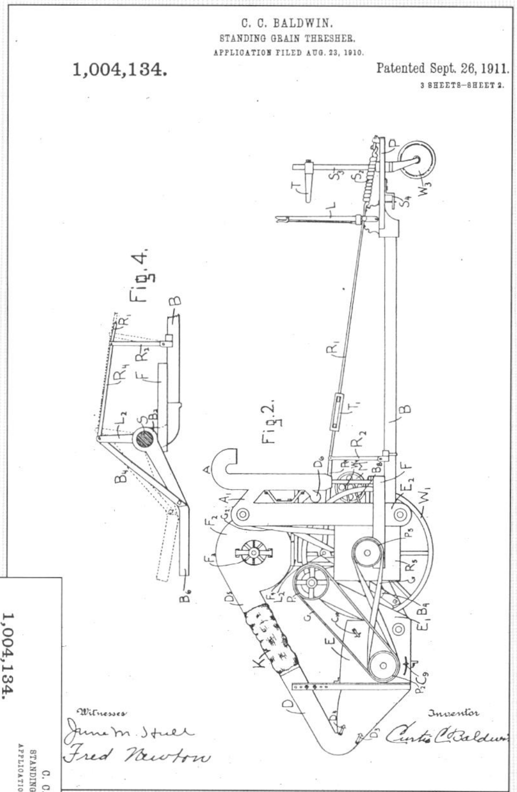

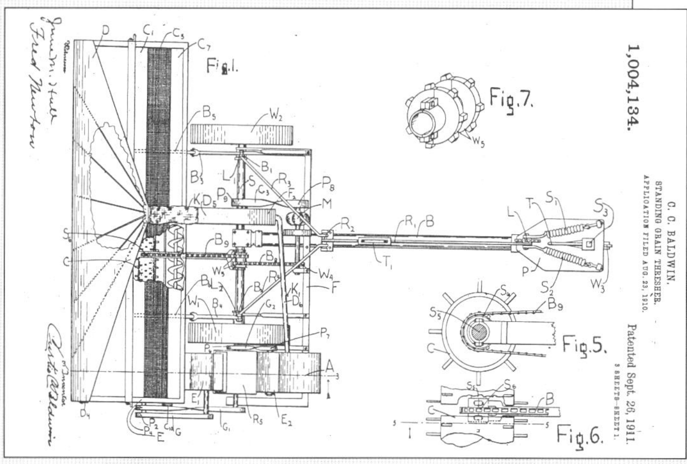

Patent No. 1,004,134 for the standing grain harvester was issued to Curt Baldwin when he was 23 years old.

The Baldwin standing grain harvester (1910). Four horses pushed the machine and two men operated it to harvest 20 to 30 acres per day.

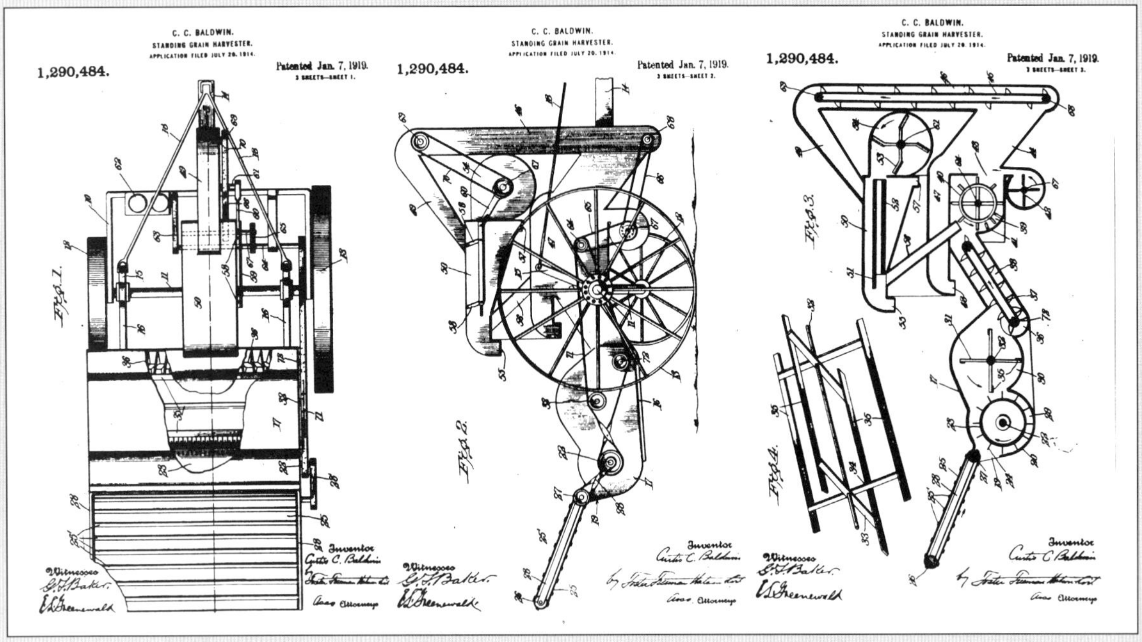

Patent No. 1,290,484 replaced the blower with a slotted mechanical gathering reel.

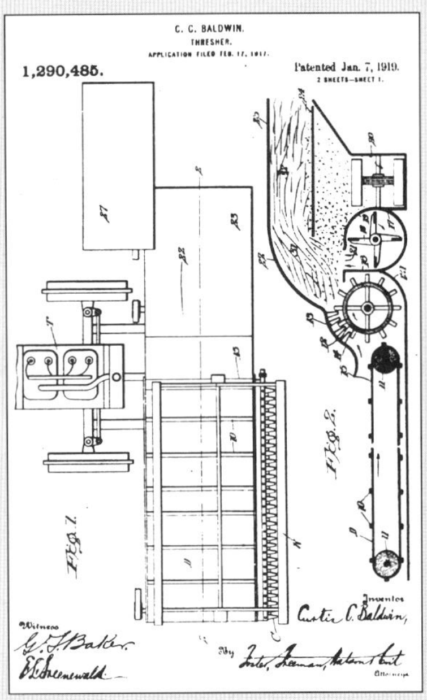

Baldwin's patent No. 1,290,485 was for a harvester mounted on the front of a tractor.

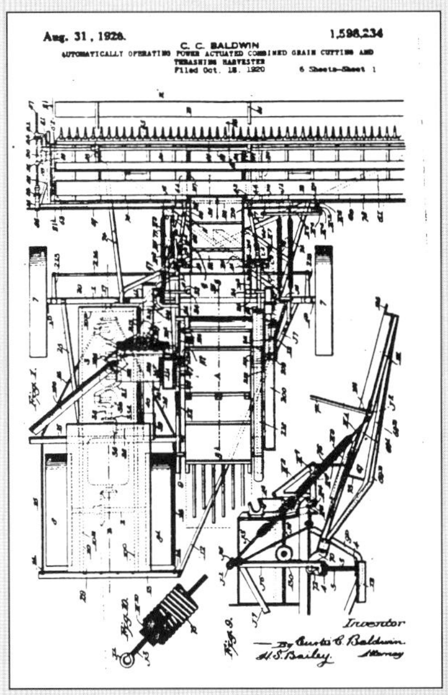

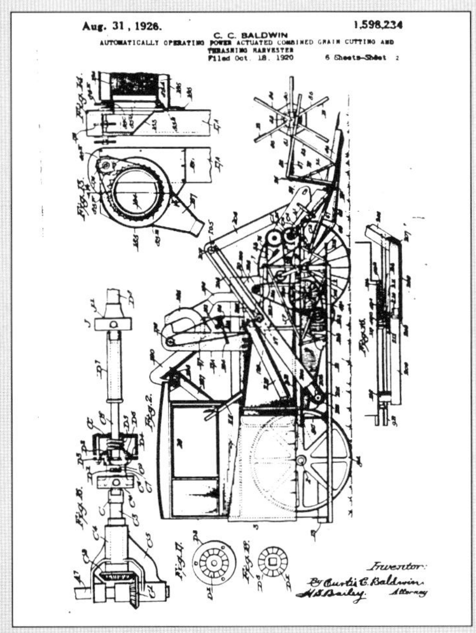

Baldwin's Savage harvester had steerable wheels which also carried the front end of the 44-hp Savage tractor.

porated in Kansas in 1915 and capitalized at $275,000. The prospectus went into great detail about the corporation's new factory in Sparta, near Grand Rapids, Michigan. Curtis Baldwin was the general manager of the corporation.

The Standing Grain Thresher Co. closed in 1918. Clarence A. Stevens replaced Baldwin as general manager when he later took off alone to develop the Savage combine in Denver, Colorado. Curt Baldwin began work on his tractor-mounted combine in 1921, having found in Savage a place to work, good help, and financing. He hired Norman E. "Pop" Bunting who came from International Harvester and had developed the Deering Reaper and Thresher.

The Savage combine was built around the 44-horsepower Savage tractor. Cutting a 16 1/2-foot swath, it could harvest 2 acres per mile of travel, 5 acres per hour at 2 1/2 miles per hour. Patent No. 1,598,234 (1926) covered the Savage combine. It featured the Baldwin patented chaff cutterbar which

A very early version of the Fordson-mounted Gleaner.

claimed to reduce the size and weight of the separator by 50 percent. The Savage enterprise failed and the patent was assigned to the Advance-Rumely Thresher Co., of LaPorte, Indiana. When Advance-Rumely acquired the Savage patent, Curt Baldwin went to LaPorte, and worked for Rumely from 1922 to 1924.

Concurrently, his brothers George and Ernest, Cal Stevens, and George Michael were in Wichita, Kansas, designing the Fordson-mounted Gleaner. The new harvester was built during 1922 and tested the next year near Winfield, Kansas. Tests were successful and the group was able to get financing. During 1924, four combines were built, of which three were sold and the fourth retained for demonstrations. The demonstrator traveled over 1,000 miles during the harvest, then earned more than its cost doing custom work. During this 1,000 mile trek, it was operated without lugs on the drive wheels to demonstrate its low traction requirements.

Curt Baldwin was sending his ideas for the Gleaner to his brothers and Cal Stevens. New patents were filed in Stevens' name.

Baldwin left Rumely and returned to be vice-president of the Gleaner Manufacturing Co. In 1925, Baldwin began development of his pull-type combine. He saw the limitations of the Fordson-mounted design, among which were dirty operator position, poor visibility, and the dependence on a specific tractor. A pull-type machine would serve a much larger potential market.

The Gleaner Manufacturing Co. was in business but it didn't have a factory. The Butler Co., Kansas City, was contracted to build 100 Gleaners for the 1925 season. At the Wichita implement show those

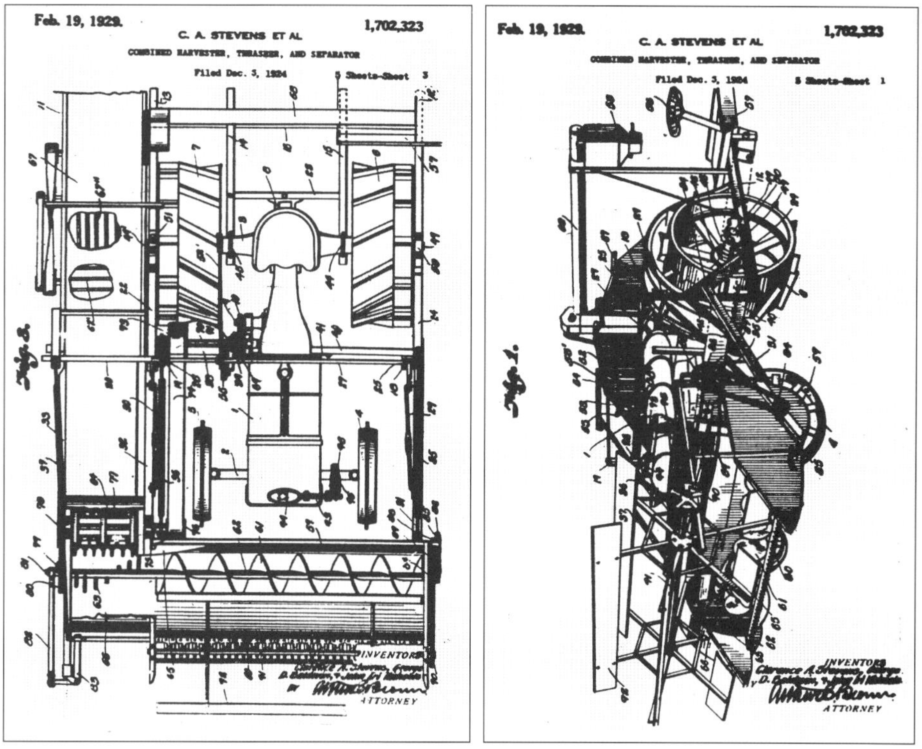

Patent 1,702,323 for Fordson-mounted Gleaner was issued to C. A. Stevens, although much of the design was done by Curt Baldwin.

100 units were quickly sold with $100 deposits. Deposits were accepted for a second hundred, which Butler was able to produce. All of them were sold, which indicated good performance for a new company with an unproven product.

In 1925 the company purchased the factory and machine tools of the Baird Pneumatic Tool Co. in Independence, Missouri, and moved the Gleaner Manufacturing Co. headquarters there. One account said that 641 "pushers" were built in 1926, with a loss of $33,619.11. S. H. Hale advanced $160,000 and asked for a reorganization. The Gleaner Manufacturing Co. changed its name to the Gleaner Harvester Corp. on December 30, 1926.

Curt Baldwin completed the design and testing of the 10-foot Pull Model in 1926 and insisted that the Gleaner Manufacturing Co. build it, but the directors declined. He again left, formed the Baldwin Harvester Co., and contracted with Standard Steel of North Kansas City to build his new machine. Sales of the new Baldwin Pull Model combine quickly surpassed the Gleaner and Standard Steel was unable to produce enough machines in time for the 1927 harvest. In March of that year the Gleaner Harvester Corp. assumed all the inventory, work-in-process, manufacture, and sales for the Baldwin. By the end of 1927, the corporation had produced 500 Gleaners and 1,050 10-foot pull-types. Henry Ford's decision to discontinue the Fordson tractor meant the demise of the Gleaner mounted combine. As a result, the 1927 run was the last production of this unique machine.

Shortly after Curt Baldwin sold his rights to the Pull Model to the Gleaner Harvester Corp., he formed the Baldwin Harvester Corp. to build the new Baldwin Pull Model. The Gleaner Corp. filed suit to protect their trademark "Gleaner-Baldwin." The suit was settled and the name was changed to Curtis Harvesters, Incorporated.

The Curtis Harvester was built by the Ottawa Manufacturing Co. in Ottawa, Kansas, under Curt Baldwin's supervision. His Pull Model 30 Curtis Combine was sold directly to the customer, eliminating the dealer which Curtis contended did little or nothing to merit his commission. The Curtis Harvester did not survive the Great Depression. However, Curt Baldwin continued his work on threshing and separation using rotary principles.

Meanwhile back at Gleaner, the 1928 12-foot model was more popular than the 10-foot, but it failed in Canada due to lack of separator capacity and

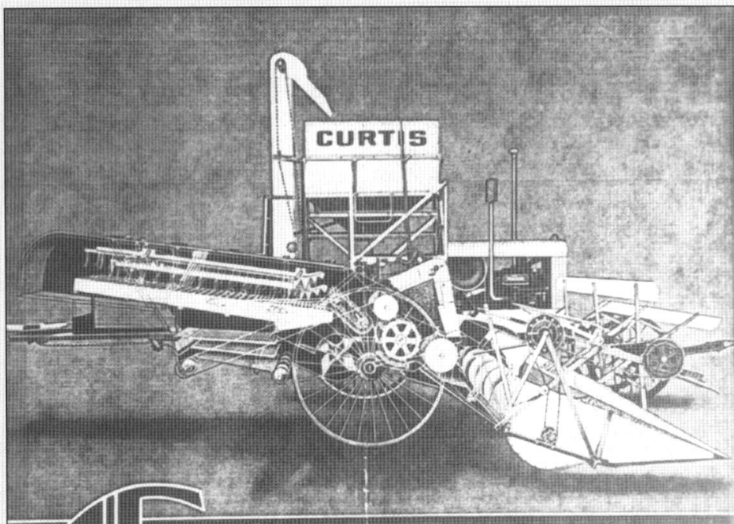

Curt Baldwin's Pull Model 30 had rake-and-rack separation. Baldwin sold them directly to the user, eliminating the dealer and his commission.

The 1927 Gleaner cut an 8-ft 3 -in. swath, harvesting 1 acre per mile of travel.

In 1928 the Gleaner Harvester Co. built 223 Gleaco grinders. No more were ever built.

power. Larger engines from Continental, Waukesha, and LeRoi helped overcome the lack of power.

For 1929, the separator width was increased by 4 inches and the Model T engine was replaced with a Continental W-9. Inadequate filtering allowed dirt to enter the engines and 2,500 engines had to be rebuilt. In 1930, Model A Ford engines powered the combines.

Major improvements included using cut-steel sprockets and roller chain, which replaced the detachable link chain and cast gray iron sprockets. Two models were offered — the Model R 12-foot raddle machine, and for the first time, the Model A 12-foot straw walker unit.

Gleaner-Baldwin pioneered corn combining. Material to build 1,500 corn attachments was purchased for the 1930 season and 500 had been shipped by August 31. Although the combines could shell corn, the system lacked grain dryers and the shelled corn didn't keep in storage. Gleaner's field shelling was 30 years ahead of its time.

The company had profits of $1,406,646 in fiscal 1930, but these were to be the last for a while. Both Gleaner and its dealers had high inventories. Low crop prices had caused heavy defaulting by earlier buyers. Production workers were laid off and those left saw their pay cut by 30 percent. The company was broke and went into receivership in February 1931. It was sold at auction in April. Many hoped that Curtis Baldwin would buy the ailing corporation, but he did not bid. The Gleaner Harvester Corp. was purchased by the Gleaner Harvester Corp., a new compa-

The horse-drawn Gleaner was for the farmer who didn't believe in "them new fangle' tractors," but didn't believe in threshing machines, either.

The 1930 12-ft combine was powered by a Ford Model A engine which was equipped with a starter and lights.

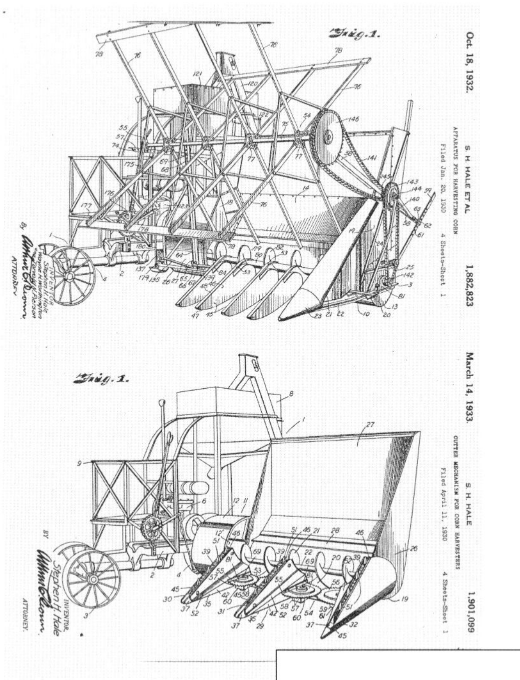

The Hale patents covered the Gleaner cornheaders which were whole-stalk machines.

ny made up of the Commercial Credit Co., some Kansas City banks, and stockholders and creditors of the former company. The new corporation, which elected to use the established name, paid $270,000 for the assets and assumed liabilities of $81,086.

New management wanted to replace the Model R and began development of the Model NR, a raddle-type machine with a 32-inch separator. Concurrently, an updated Model A straw walker machine was being tested in Illinois soybeans. Meanwhile, sales were caught in the doldrums of the depression. Only one Gleaner was sold in Kansas during 1932, and it was a carryover 1931 model. In April 1936, the Model NR was shown at the Wichita implement show and advance orders were taken. The NR continued the unique Gleaner features of auger platform conveyor and down-front cylinder.

The All-Crop harvester's overwhelming success caused every farm equipment manufacturer to rush a small combine to market. During 1936, Gleaner's 6-foot pull-type combines, the Model S walker and Model T raddle, were developed. These were powered by a two-cylinder Cushman engine, although later it changed to a Wisconsin four-cylinder engine. In 1937, fifteen of each model were built by the engineering department and sold.

The Model T 6-foot raddle combine went into production in 1938. Inadequate testing of the earlier 30

Gleaner experimented with several concepts for whole-stalk corn harvesting and built 500 cornheaders in 1930.

resulted in "a graveyard of hundreds of combines that were taken back, were rebuilt, and many of these did not stay sold." It appears that "no small fix" was required to save the Models S and T. The situation with the NR (Model E) was much better. A new Model J 9-foot pull-type raddle machine was designed. The tribulations with the S and T combines caused engineering to design an all new 6-foot machine which was designated the S42. The Model S was tested in southern California to take advantage of the earlier harvest.

The 1935 prototype of the NR combine became the Model E (raddle) and Model F (walker) 12-ft combines.

In 1942 the Model S walker machine supplanted the Model T raddle unit in the production schedule. From this time there were fewer raddle machines and more walker models in production quantities. This signified the movement of the smaller Gleaner combines into soybean country where walkers were preferred. For Gleaner, ever strong in the wheat belt, larger pull-type raddle machines would remain strong for many years but standardization eventually eliminated raddle separators altogether. In 1943 the 9-foot Model J raddle was replaced by the 9-foot Model H walker, further evidence of a changing market for Gleaner.

Farm equipment was rationed during World War II, but there was plenty of equipment available for planting and harvesting food crops, especially wheat. During 1943, Gleaner built 1,295 combines, 20 of which were a special allocation to the state of Kansas. The state War Board developed a program by which these combines would be available to custom operators who agreed to follow the harvest from Texas and Oklahoma to the Canadian border, to report on their custom operations, and to keep the machine for custom harvest during the next year, 1944. Sixteen custom operators filed reports at the season's end. With the16 new and 9 old machines, they harvested 33,198 acres of wheat which yielded 682,038 bushels. The top operator harvested 4,457 acres and 113,706 bushels. This first year of operation was extremely successful and saved hundreds of acres in the Montana and North Dakota which would have been lost to early snows otherwise. Gleaner's fine record and contribution led to the formation of the "Harvest Brigade" by Massey-Harris for 1944 (Donald J. Martin to Lee Oberholtz, 1967).

In 1949, 499 12-foot windrowers (swathers) were built. Little information is available concerning the windrowers and no more were built for 10 years.

The Model S 6-foot pull-type combine was built in small numbers until 1953, when it was replaced by the Model P-80. The "80" denoted the 80-inch cutter-bar which was extended to better handle two rows

The Gleaner Six was a variation of the Model T raddle machine. No production quantities are shown but it may have been included in the Model T numbers.

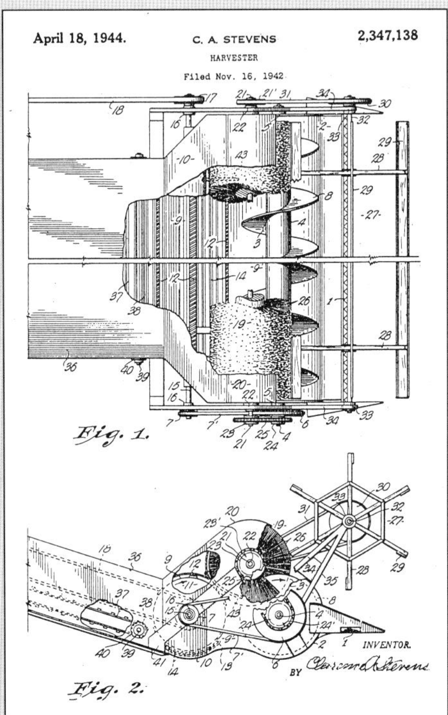

This interesting design for a straight-through combine apparently had a large cylindrical brush to convey the crop to the down-front cylinder. Stevens patented the Fordson-mounted Gleaner.

This is a bagger version of the 1951 Model S walker combine.

The P-80 was the last of the Gleaner pull-type combines. Only 287 were built in 1954 and 1955. When A-C acquired the company in 1955, the P-80 was immediately dropped in favor of the proven All-Crop harvester.

of soybeans. The P-80 was discontinued in 1955 when Allis-Chalmers acquired the Gleaner Harvester Corp. Carryover P-80's had prices of $1,632 for a PTO-driven machine and $1,936 for an engine-driven model. Swept away by the orange flood of All-Crop harvesters flowing from LaPorte, there was no room in the line for the P-80. The curtain closed on Gleaner pull-type combines.

The Great Silver Fleet Gleaner Self-Propelled Combines

By the late 1940s it was apparent that big changes in combines were coming. Gleaner pull-type combines had served their owners faithfully, but the handwriting was on the wall — and it said that self-propelled combines would replace their pull-type counterparts.

Models A, R, and B Combines

Market studies and development had been under way for some time and during 1950, four prototypes were built. These included both 32- and 36-inch separators, both raddle and straw walker separators, as well as 12- and 14-foot headers (one lost its grain bin in a low underpass).

Five hundred of the new Model A (walker) and Model R (raddle) combines were built in 1951. The news release expounded on several important features, among which were center-line design with equal weight distribution, down-front cylinder, and two-fan cleaning system. A new stone-ejection concave door was a Gleaner exclusive. Although the propulsion components were new, the thresher and separator had been proven in the pull-type machines all over North America. The down-front cylinder was 30 inches and the separator was 36 inches wide.

The safety concave door which opened when a stone or stump entered the cylinder was a Gleaner exclusive. This feature could save a Gleaner owner hundreds of dollars over the life of his combine.

A 226-cubic inch Ford engine developed 70 horsepower at 2,000 rpm. A variable-speed V-belt drove a three-forward, one-reverse speed transmission. Forward speeds ranged from 1.5 to 12 mph, and 10-, 12-, and 14-foot headers were offered. The grain bin held 50 bushels and could be unloaded in 90 seconds, standing still or on-the-go. The operator enjoyed a high, centered platform with hydraulic controls for the header lift and variable-speed drive. In 1953 the power source was changed to the Hercules six-cylinder 237-cubic inch engine, then to the A-C six-cylinder 230-cubic inch 6B230 in 1957.

In 1953 the Model AB rice combine (soon called the Model B) was produced. It was a walker type machine with a 14-foot header. In 1955 the Models A and R had prices ranging from $5,536 to $5,586. In 1960 the Model A sold for $7,173 and the Model B retailed for $7,196.

The Models A, R, and B were the first of a long line of self-propelled Gleaner combines.

The Model B rice combine was similar to the Model A except that it had a spike-tooth cylinder, pickup reel, and R2 rice tires.

The Model T came as the T2 (two rows) and T3 (three rows of soybeans). It was produced only in 1954. When A-C purchased Gleaner in 1955 they elected to continue production of the SP100 self-propelled All-Crop harvester and the Model T was dropped.

Model T Combine

In the early 1950s, a market for a smaller, self-propelled combine materialized, primarily in the eastern Corn Belt and other areas with medium-sized diversified farms having both corn and small grain crops. A rash of "baby" self-propelled combines were rushed to fill this niche, and Gleaner offered the Model T in 1954. The T2 (two rows of soybeans) had a 7-foot header and the T3 (three rows of beans) had a 10-foot header.

With a 23-inch rasp cylinder and 28-inch walker separator, the Model T was about three-fourths the size of the Model A. General arrangement was also similar except that the T had only one fan. Grain bin capacity was a moderate 30 bushels. A Ford four-cylinder 172-cubic inch engine drove the unit through a variable V-belt and three-speed transmission.

The Model T combine was only built in 1954. After Allis-Chalmers acquired the Gleaner Harvester Corp. on February 1, 1955, both the Model T and P-80 combines were dropped. Relatively unsuccessful in heavy small grain, the Model T proved to be a winner in edible beans. The unsold inventory acquired by A-C found a home in the edible bean country of Michigan's thumb. In April 1955, the T2 had a list price of $4,459 and the T3 sold for $4,560.

Gleaner Heads For the Hills — The AH Combine

Allis-Chalmers was well represented in the wheat-producing steep slopes of the Pacific Northwest by dealers who were selling Springfield crawlers and other A-C equipment, but couldn't reach the combine market unless they also had a contract to sell the Harris self-propelled hillside combine. A-C dealers wanted a hillside machine and company management was sympathetic to their needs. As a result the folks at Gleaner designed the self-leveling 14-foot Model AH which was added to the line in 1959 with a pilot run of 12 units.

Models C and CH Combines

Larger farming operations, and particularly custom operators, had outgrown the Model A and were demanding more capacity. The Model C with 14- to 20-foot grain headers and four-row cornheads had the capacity they wanted. In 1960 the Model C was brought to market with production of 706 machines. The Model C had a 40-inch cylinder and 48-inch separator (four walkers instead of three) and was exactly one-third larger than the A. Grain bin capacity grew to 60 bushels and the machine was powered by the A-C six-cylinder 262-cubic inch gas engine. In 1960,

The Model AH hillside Gleaner combine was well-adapted to the steep slopes of the Palouse wheat country. It had a 14-ft header.

Gleaner built two-row cornheads for the Model A in Independence from 1957 through 1959. Cornhead production was transferred to LaPorte for the 1960 run.

a C with a 16-foot header sold for $8,615 and four-row cornhead was priced at $2,926.

The Model CH Hillside with 16- or 18-foot headers replaced the AH, and the Models CR Rice and CT Track special models were offered. From 1960 through 1963 when it was replaced by the CII, 4,000 Model C, CR Rice, CT Track, and CH hillside combines were built. The Model C did much to help establish Gleaner as a leader in self-propelled combines.

Model E Combine

Discontinuing both the Model T Gleaner and the SP100 All-Crop left a void in the line. Small diversified farms were still out there and their operators wanted self-propelled harvesting. Field shelling was rapidly gaining acceptance, creating more demand for a small combine. The Model E fit the specifications.

The Model CH hillside Gleaner combine had 16- or 18-ft headers and the unique Esca ladder which moved up or down with the left drive wheel as the machine self-levelled.

The Model CT tracked combine was used in rice fields. A few were sold for corn harvesting when wet weather delayed field work and marooned wheeled machines.

The Model C with headers up to 20 ft and four-row cornheads gave Gleaner the big capacity needed by custom operators and larger farms. It was introduced in the fall of 1960.

The experimental Model E had only one fan but was changed to two before production began.

The Model E with a two-row cornhead and grain headers up to 12 ft was ideal for family farms. It filled a market slot vacated by the Model T and the Super 100.

Originally developed with only one fan, it was redesigned with two fans before it went into production in 1962. Eight-, 10-, and 12-foot grain headers and two-row cornheads were offered. With a 27-inch cylinder and 32-inch separator it was sized between the Models A and T. The grain bin carried 40 bushels and the E was powered by the workhorse four-cylinder 226-cubic inch gasoline engine also used in the D-17 tractor.

The Model E proved to be a high volume unit for Gleaner, with nearly 21,000 E and EIII combines built from 1962 through 1968. The EIII was built only in 1968. It had a 66-bushel grain bin, longer separating area, and a better cab. Capacity on the EIII was increased slightly with a 13-foot grain header and three-row cornhead.

AII and CII Combines

In 1964 the AII and CII replaced their predecessors. Major changes included larger separating and cleaning areas and improved cabs. Bin capacity went from 50 to 55 bushels on the AII and from 60 to 85 bushels on the CII. Both larger grain headers and cornheads were added to the II Series line. Four, six, and eight 20-inch row cornheads were furnished during the 1960s when there was a great deal of experimentation with 20-inch rows for corn. From 1964 through 1967, over 8,500 AII and 12,700 CII and their various special models were built.

The CII and AII combines had larger grain bins and more separating and cleaning areas. Wider grain headers and 20-in. row cornheads were offered.

The larger Models F and G (center and right) replaced the AII and CII in 1968, and the Model K (left) supplanted the Model EIII in 1969.

The 1968 Models EIII, F, and G in their open platform versions. Hydrostatic steering was standard and the Tilt-away cab for reduced transport height was optional.

Models F and G Combines

In 1968 the new Models F and G replaced the AII and CII, with some very major changes. All vital specifications such as bin capacity, cylinder width, separating, and cleaning dimensions were increased to handle the higher crop yields which were becoming commonplace during the 1960s. The Model F was about 20 percent larger than the AII, and the Model G about 13 percent larger than the CII. Greater emphasis was placed on operator comfort and convenience with more modern cabs.

Grain headers ranged from 13 to 20 feet on the F and 13 to 24 feet for the G. Model F cornheads were two-, three-, and four-row, while the G offered four-, six-, and eight-row sizes, with more emphasis on narrow-row heads as the trend to 30-inch rows swept the Corn Belt.

For the first time, Gleaners were offered with diesel engines. The F diesel was the A-C 2800, six-cylinder 301-cubic inch naturally aspirated engine at 84 horsepower, while the G diesel was the 2900, the same engine except it was equipped with a turbocharger which boosted output to 109 horsepower. The standard gasoline engines were the A-C G262 six-cylinder 262-cubic inch 93 horsepower engine on the F and the A-C G2800 six-cylinder 301-cubic inch power plant with 105 horsepower on the G. These gasoline engines were relatively expensive to build since they were essentially modified diesel engines with all the strength required for diesel operation. Uncompetitive prices were the downfall of the Harvey-built gasoline and LPG engines and they were discontinued around 1970.

As a result, the F and G gasoline-powered combines were fitted with General Motors (Chevrolet) automotive gasoline engines beginning that year. The Model F received a six-cylinder 292-cubic inch 101-horsepower engine, while the Model G got a V8 with 350 cubic inches and 120 horsepower. Chevy

Hillside combines had an additional feeder beater immediately behind the header auger which rotated with the header and stayed parallel to the ground. The header swiveled between the first and second feeder beaters as the combine self-levelled.

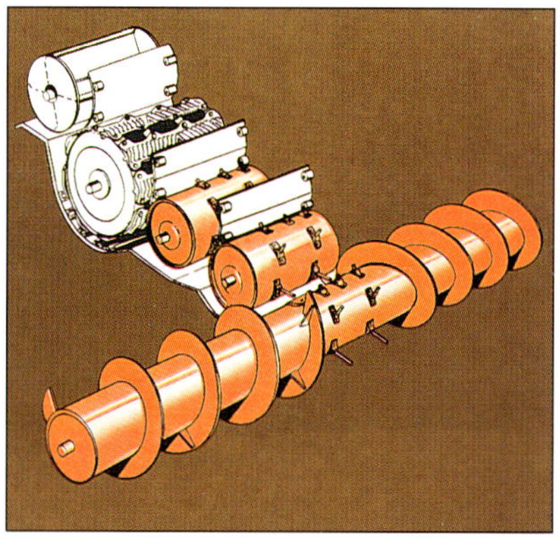

The Model GH Hillside Special was claimed to be able to harvest any slope that a crawler tractor could seed. Headers were 16 to 20 ft.

engines were used on the K, F, G, M, and L combines until the gas-burners were dropped about 1977.

In 1971 two new and important features were added to the F and G combines. A new header-attaching device allowed the operator to mount a grain head or cornhead from the seat by merely engaging a pair of top hooks and lifting the thresher housing, then dismounting to engage a pair of lower wedge pins and connect the drive. The other feature was the spring-loaded auger latch which would release to permit the unloading auger to swing back without damage when an obstruction, such as a pole or tree, was struck. This certainly saved a lot of sweat and tears for unfortunate and forgetful operators.

During late 1969 the F-Hy and G-Hy combines with optional hydrostatic drive appeared in the production figures with about three dozen units being built. Although this feature was a failure in farm tractors, it quickly found its rightful application in combines. A relatively expensive option at first, it improved field performance so much that it eventually became standard equipment on the top-of-the-line machines.

The Model K Combine

The Model K replaced the Model EIII in 1969. The most obvious change was the use of the GM Chevy six-cylinder 250-cubic inch 78-horsepower engine which replaced the venerable West Allis four-cylinder 226-cubic inch gas engine. Except for the engine, K specifications were essentially the same as the EIII. No diesel engine option was offered until the K2.

The K and F were upgraded to the K2 and F2 in 1977. They received the transverse flow fan which was already proven in the L and M combines. Grain bin capacities were increased slightly and extensions were available to increase the K2 bin capacity from 68 to 96 bushels, and the F2 bin from 80 to 120 bushels. The cab and seat were improved for more operator comfort. The optional diesel engine for the F2 was changed to the A-C four cylinder 200-cubic inch turbocharged and intercooled diesel with 95 horsepower. For the first time an optional diesel was offered for the K2, which was the turbocharged version of the same engine at 72 horsepower. The Chevrolet gasoline engines were carried over from the K and F combines.

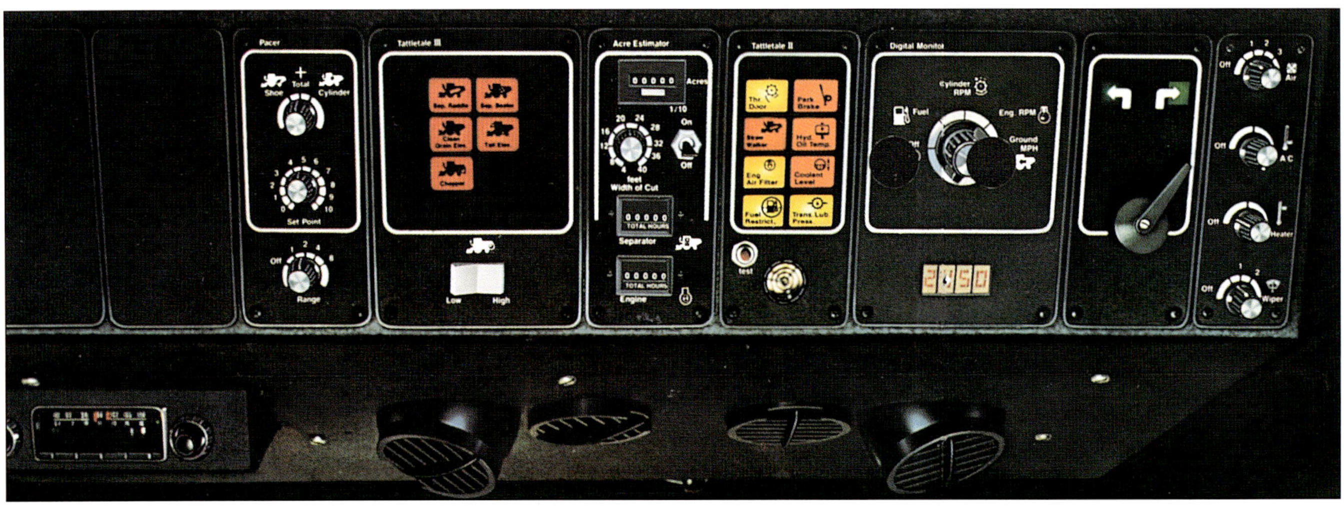

The "Tattletale" electronic monitoring system gave K2 and F2 operators the low-down on many combine functions and warned of stoppages.

The Models K2 and F2 working in Kansas maize. They had transverse-flow fans and larger grain bins, Chevy gas engines and A-C diesels.

A Gleaner F3 with floating cutterbar and pickup reel saving low-hanging soybeans. The Model F3 had the electro-hydraulic control system carried down from the larger Model M and L Series combines.

The electronic monitoring systems, later called the "Tattletale" systems, made their appearance on the 1977 models and were upgraded in 1978 to include monitoring a greater number of functions. As combine cabs became quieter and more isolated, the use of effective monitoring systems became more important, since the sounds of malfunctions could not be heard in the cab.

For 1978 both the K2 and F2 received longer walkers for a 183-inch separating length with total separating area increased to 5,837 and 7,818 cubic inches, respectively. Total cleaning area was also increased to 2,887 (K2) and 3,892 cubic inches (F2). Specially-equipped versions such as Grain, Corn-Soybean, and Corn-Plus were offered in both models. The K2 combine and the F2 gasoline engine model were discontinued at the end of the 1981 production run. The diesel-powered F2 continued for another year.

Gleaner celebrated its 60th anniversary in 1983, calling attention to its 60 years of building fine combines, service to customers, and great contribution to the production of world food. That year the F2 became the F3. Basic specifications remained unchanged from the F2, but the F3 gained features found on the larger M3 and L3, which included electro-hydraulic controls and expanded Tattletale electronic monitoring. Hydrostatic-drive rear-wheel assist and tracks in place of the main drive wheels were offered for operation in rice fields and other poor traction conditions. The three-speed transmission was changed to a five-speed similar to that in the M and L combines.

The Model F3 was continued through 1986, wearing the Deutz-Allis name after May 1985. The last Allis-Chalmers price list showed the Model F3 base price as $53,412, which rose to $64,415 with common accessories and a 13-foot grain header.

The Electronic Age Arrives

When the Model L rolled off the assembly line in 1972, it brought a complete list of new features and systems. With a 48-inch cylinder and 53-inch separator, it would prove to be the largest conventional combine ever produced in the Independence plant. It

The Model L brought many changes to the combine line, among them open concave for more separation at the cylinder, single transverse-flow fan for improved air distribution over the shoe, and electro-hydraulic controls for greater operator convenience.

was the first to carry a 24-foot grain header or eight-row cornhead. It was also the first Gleaner with an open concave for better separation at the cylinder. The Model L had a transverse-flow fan with two outlets that delivered more air with better control. The straw walkers were longer, the cleaning shoe was larger, and the grain bin held 120 bushels. The 216-inch separating length provided a whopping 11,634 square inches of separating area,and there were 5,525 square inches of cleaning area. For its time, it was a big combine.

Perhaps the most interesting feature on the Model L was the new electric-over-hydraulic controls. Electric rocker switches activated solenoids which in turn opened or closed hydraulic valves to control nearly every function on the machine which could require changing on-the-go. It truly was a push-button machine. The new cab, with ergonomically designed controls and better seating, eased the operator's long hours of harvesting.

Engine power was either the Chevy 120-horsepower 350-cubic inch V8 or the A-C six-cylinder 301-cubic inch turbocharged diesel with 109 horsepower. The main drive was a torque-sensing variable V-belt to a five-forward, one-reverse speed transmission. The torque-sensing mechanism in the driven pulley was a cam arrangement which squeezed the pulley halves together to tighten the drive belt as the load increased. Hydro-Traction drive was optional.

Besides the new features, the Model L continued all of Gleaner's well-established concepts and principles which included center-line design with equal weight distribution, down-front cylinder, two- or three-stage separation, and two-stage cleaning. As ever, the safety-concave door protected the cylinder from rocks, roots, or stumps when they were ingested.

The L was followed by its smaller brother, the M, in 1973. Smaller, yes, but not by much. With a 40-inch cylinder and 45-inch separator, it had Model G capacity and therefore replaced that veteran. The M shared the L engines and power train components, cab, and controls. In fact, most everything was the same as the L except width, so total separating and cleaning areas were reduced accordingly. The grain bin carried 105 bushels. Grain headers ranged from 13 to 22 feet and four-, six-, and eight-row cornheads were available.

The MH Hillside Special joined the lineup in 1974. The MH had similar specifications to the M, except that Hydro-Traction drive was standard equipment. A

The MH Hillside Special could self-level up to 23 degrees or a 43 percent slope. It had Model M specifications but with standard Hydro-Traction drive. Headers from 18 to 22 ft were offered.

The Model M came out in 1973. Smaller than the Model L, it had a 40-in. cylinder and 45-in. separating width. The grain bin held 105 bushels.

unique feature of this machine was the self-leveling axle which caused the uphill main wheel to move closer to the combine, shifting more weight to that wheel and improving stability. The leveling system could level to 23 degrees (43 percent). Header choices were 18, 20, or 22 feet. The basic harvesting mechanism was the same as the Model M except that the hillside unit had another feeder beater immediately behind the platform auger which rotated with the header as the combine self-leveled.

The M, MH, and L evolved into the M2, MH2, and L2 in 1977 and remained in the line through 1982. With the 2 series, gasoline engines were no longer offered, and the standard engine was the six-cylinder 301-cubic inch turbocharged diesel with horsepower increased to 130, or the turbo-intercooled version with 145 horsepower, depending on the model. The six-cylinder 426-cubic inch turbocharged diesel with 158 horsepower was optional on the L2 and MH2.

Grain bin capacities were unchanged in 1977 with the M2 holding 105 (165 with extensions) bushels, and the L2 carrying 120 (185 with extensions). In 1978 the standard bin capacities were increased to 180 on the M2 and 200 bushels on the L2, while the MH2 was raised slightly to 115 bushels. The shaft-monitoring system evolved into the comprehensive "Tattletale" system which gave the operator the necessary information for efficient operation or spotting trouble. Both the L2 and M2 were offered as special models such as Grain, Windrow Plus, Corn-Soybeans, Corn Plus, Edible Bean Plus, and Rice-Soybean, each of which was equipped at the factory for that particular crop. Hydrostatic rear-wheel assist was a popular option for poor operating conditions and tracks were offered for rice harvesting. The 1982 models received an attractive new light tan cab interior and seat covers, a pleasant touch after many years of black.

Gleaner's 60th anniversary in 1983 also brought the L3, M3, and MH3 combines and Series 3 Flex headers. The Series 3 Flex headers had many

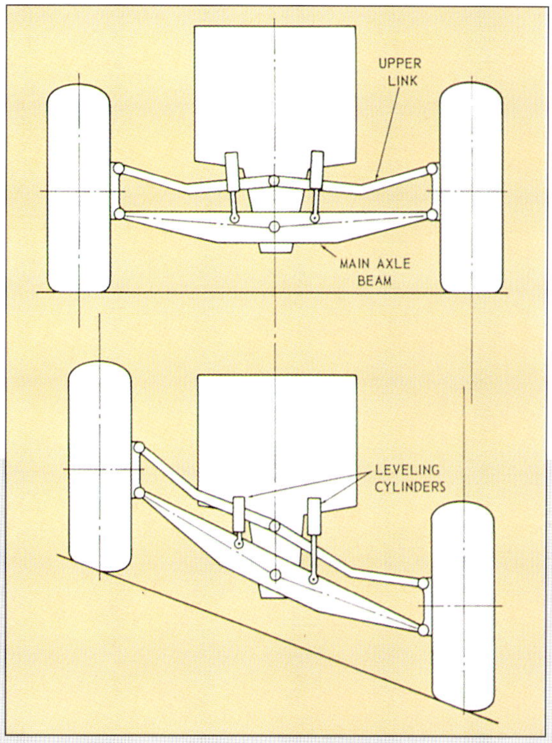

To maximize stability on steep slopes, the MH main axle beam center pivot was lower than its end pins so that the combine's center of gravity moved closed to the uphill wheel when the combine self-levelled.

This Model L2 had both tracks and hydrostatic rear-wheel assist — a nearly unstoppable combination in the worst of field conditions.

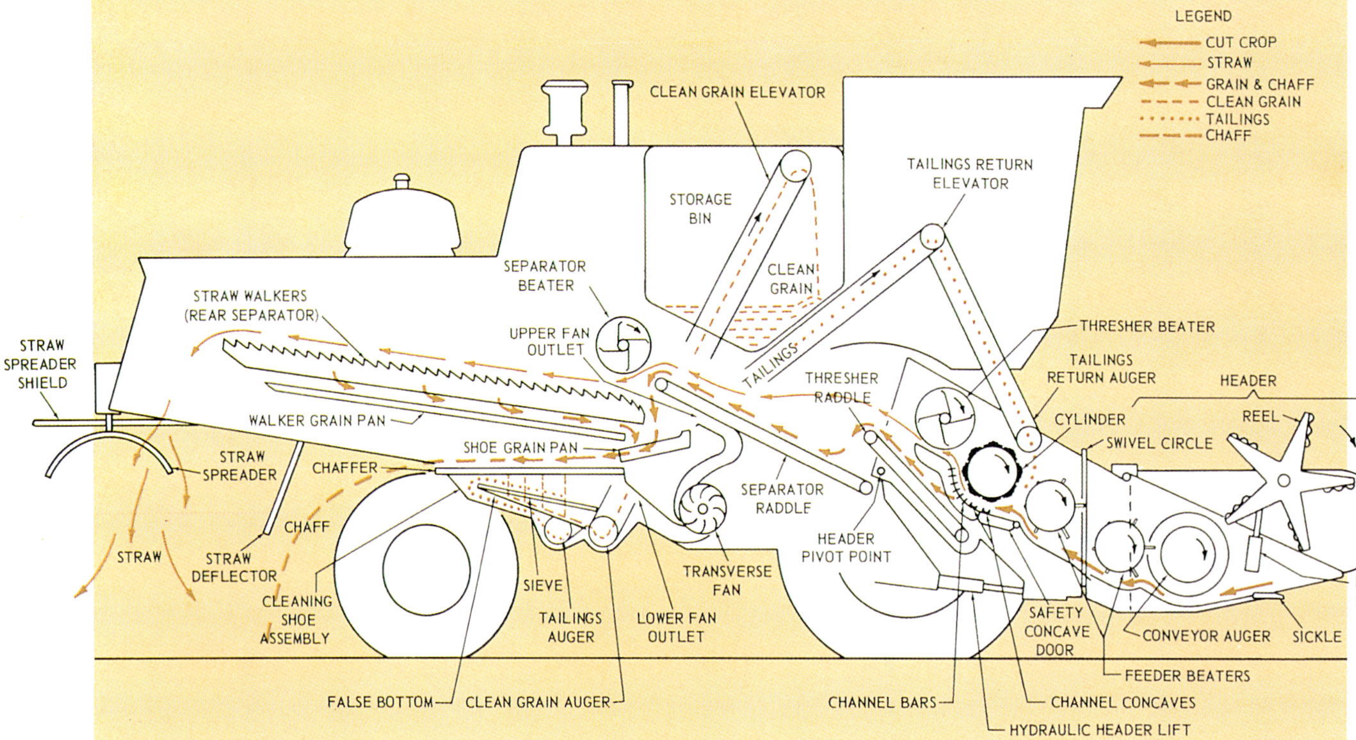

WHAT A COMBINE DOES (Figure 1)

A combine performs five grain harvesting functions—cutting, feeding, shelling, separating and cleaning—while moving through the field. Various headers are made for standard combines to suit the crop such as a grain header, a corn header and a soybean header. The schematic shows a grain header on a hillside combine. With its guards and sickle, the cutter bar, which is the leading edge of the header, cuts the crop. The reel feeds it evenly into the header. The conveyor auger then moves the material to the center of the header. At the center, retractable fingers on the conveyor and on the feeder beaters comb and feed the crop directly into the cylinder. A standard combine uses only one feeder beater, bringing the auger conveyor closer to the cylinder. An additional feeder beater on the hillside combine assures proper feed control through the swivel circle to the second feeder beater.

Shelling, the process of breaking the seed loose from the rest of the plant, is achieved in a combination open and closed concave system. The rotating cylinder, which has eight bars with deep rasps, breaks the grain heads against the concave system to free the seed. A pair of channel concaves at the front of the system does most of the shelling, then a seven-bar open section finishes the job. Rocks and other potentially damaging objects entering the cylinder and concave area are automatically ejected through the safety concave door.

The separating process is based on the difference between the densities of the seed and the chaff and straw. When tossed in an air blast, the seed drops a short distance away while the chaff and straw are carried farther. The first stage in separating occurs as the seed drops through the open section of the concave system to the thresher raddle, a conveyor which carries it to the separator raddle. The second stage of separation begins as the remainder of the crop leaves the cylinder. There, the thresher beater fluffs and agitates the crop as it delivers it to the separator raddle. Lighter material is kept in suspension by the air blast from the cylinder and the beater, while the grain falls to the raddle below and is delivered to the cleaning shoe assembly. Most of the separation takes place in these first two stages. The separator beater and the rotary straw walkers complete the separating by agitating and fluffing the material while conveying it to the rear of the combine. The remaining grain falls to the grain pan and is delivered to the front of the cleaning shoe assembly while the chaff and straw drop out the rear of the combine.

A heavy blast of air through the upper fan outlet cleans most of the grain as it comes off the separator raddle and grain pan. The lower fan outlet directs a controlled volume of air through the end-shaker-type shoe assembly. The shaking action of the shoe causes the chaff to drop off the back of the chaffer while grain and small unthreshed particles of heads fall through onto the adjustable sieve. Threshed grain falls through the sieve and goes up the clean grain elevator into the storage bin. The sieve conveys the unthreshed particles to the tailings elevator, which returns them to the cylinder.

This article from a 1975 A-C Engineering Review is a very concise description of a combine's function.

The Model L3 was the largest of Gleaner's conventional self-propelled combines with grain headers up to 24 ft, flex-heads up to 22 ft, and cornheads from four to eight rows.

improvements, among which was an increased float range of 5 inches for improved harvesting in soybeans. They were offered in 13- to 22-foot sizes. Basic specifications of the 3 series combines were unchanged, but a full schedule of improvements kept these conventional machines competitive. By this time the primary emphasis of combine engineering was on the rotary line. Production of the conventional combines continued until the end of 1986 and bore the name Deutz-Allis after May 1985. That was the end of the Gleaner conventional combine line. These were fine machines that were replaced by more modern technology.

The Model M3 carried flex-heads up to 22 ft. It was powered by a 130-hp 301-in.^{3} turbocharged diesel or an optional 145-hp turbocharged and intercooled diesel.

An All New Concept — The N Series Rotary Combines

Gleaner engineers built a prototype of a larger conventional combine, the 635 (63.5-in. separator), in an effort to provide greater capacity with proven principles and components. The 635 was based on the Model G design with closed concave and open-bottom walkers, both of which were changed in the Model L to increase capacity. The 635 had six walkers and the cylinder was 54 inches wide, 10 more than the Model G. It turned out

The Tattletale electronic monitoring systems for the M3, MH3, and L3 provided visual and audible warnings for about 20 functions, including shaft rotation. The Pacer efficiency monitor measured separator efficiency at the walkers and shoe. An acre estimator tallied the ground covered.

to be a very wide machine. Considering the larger tires required, the 635 was about 24 inches wider than a Model L, which was 123 inches wide on standard tires. It was too wide for easy truck transport, an essential for the custom operator. The extra width also reduced maneuverability in the field. To alleviate the hauling problem, the machine had steering controls on the back and special transmission gears so that it could be driven backward at speeds up to 40 mph. Power came from by a 350-cubic inch Chevrolet engine of about 120 horsepower. In short, the conventional combine could no longer be expanded for more capacity without becoming an unwieldy monster.

Since conventional threshing, separation, and cleaning relied on the force of gravity, more capacity also required more size. The converse, therefore, is that greater capacity without greater size requires more force, a force greater than gravity. Innovative thresher engineers had turned to the rotary concept to utilize centrifugal force as the increased force in the equation. Curt Baldwin was issued three patents concerning rotary threshing in the 1930s. Charles Scranton, chief engineer at LaPorte, was issued a patent on a rotary concept in 1960. Although not new, the rotary principle had yet to be proven.

During 1967, Gleaner engineering began development of a rotary combine with the following objectives: greater capacity without increasing size, higher productivity per unit of cost, less sensitivity to slopes, reduced grain damage, and lower vibration and noise. After laboratory testing produced encouraging results, a prototype was built on a Model G chassis for field testing. Tests were successful and resulted in the development of the N6 combine.

The transverse arrangement of the cylinder was selected over the longitudinal type to gain the advantages of tangential feeding, later named "Natural

Three views of the 1968 experimental N1 rotary processor mounted on a Model G combine. Successful tests led to the development of the Model N6 rotary Gleaner combine.

The experimental Model 635 was Gleaner's largest conventional combine, with a 63.5-in. separator. Its excessive width precluded truck transport so it had special gears and rear steering controls which allowed it to be driven backward at speeds up to 40 mph.

Flow." This proven principle did not require reorienting the crop flow to the cylinder and would reduce feeding problems. The crop was threshed in the conventional manner by rasp bars over an adjustable concave section in the cage. The transverse design also permitted the installation of the Gleaner hallmark — a stone ejection door in front of the cylinder.

As the crop made its first revolution, it was guided to the left past the feeder opening by helical bars, and spiraled around the inside of the cage while high centrifugal forces threw the threshed grain through the cage openings. A cage sweep kept the top of the cage free of chaff or other crop materials. The straw continued through the cage to be thrown out the rear discharge by paddles on the cylinder shaft. When the grain left the cage it was distributed by augers and dropped into the accelerator rolls which accelerated the falling grain and chaff downward at about 1,500 feet per minute. This resulted in a relatively thin flow of material through which the air blast from one of transverse flow fan outlets could easily clean much of the grain before it ever reached the shoe. The other fan outlet supplied air to the shoe which cleaned the grain in a conventional manner.

Although the feeder housing and conveyors were offset slightly to the right in this design, the weight distribution on each drive wheel differed by only 200 pounds. Variable weight elements such as grain bin, fuel tank, and header size were concentrated on the drive wheels, while the weight on the steering axle remained nearly constant for easier handling. The transverse cylinder design effectively combined rotary principles with Gleaner features such as centerline design, concave stone door, tangential feeding, and two-stage cleaning.

The N Series Arrives

A pilot run of 50 machines had been built in the last half of 1978, so there had been plenty of scuttlebutt and everyone was primed to see the new flagship of the Silver Fleet. The Model N6 made its debut to the Allis-Chalmers North American dealer organization in February 1979, when it appeared out of a big cloud of smoke at center stage in the Las Vegas Convention Center. It must have been what they wanted — it brought the house down.

It was big, but still transportable.

	Model L	N6
Height	139 in.	132 in.
Length-less header	266 in.	235 in.
Width-standard tires	123 in.	146 in.

Not bad for a machine with at least 50 percent more capacity. The design objectives had been met with room to spare.

The N6 carried grain headers up to 30 ft and cornheads up to 12 rows. The A-C 426-in.^{3} turbo-intercooled diesel developed 220 hp and the bin held 245 bushels.

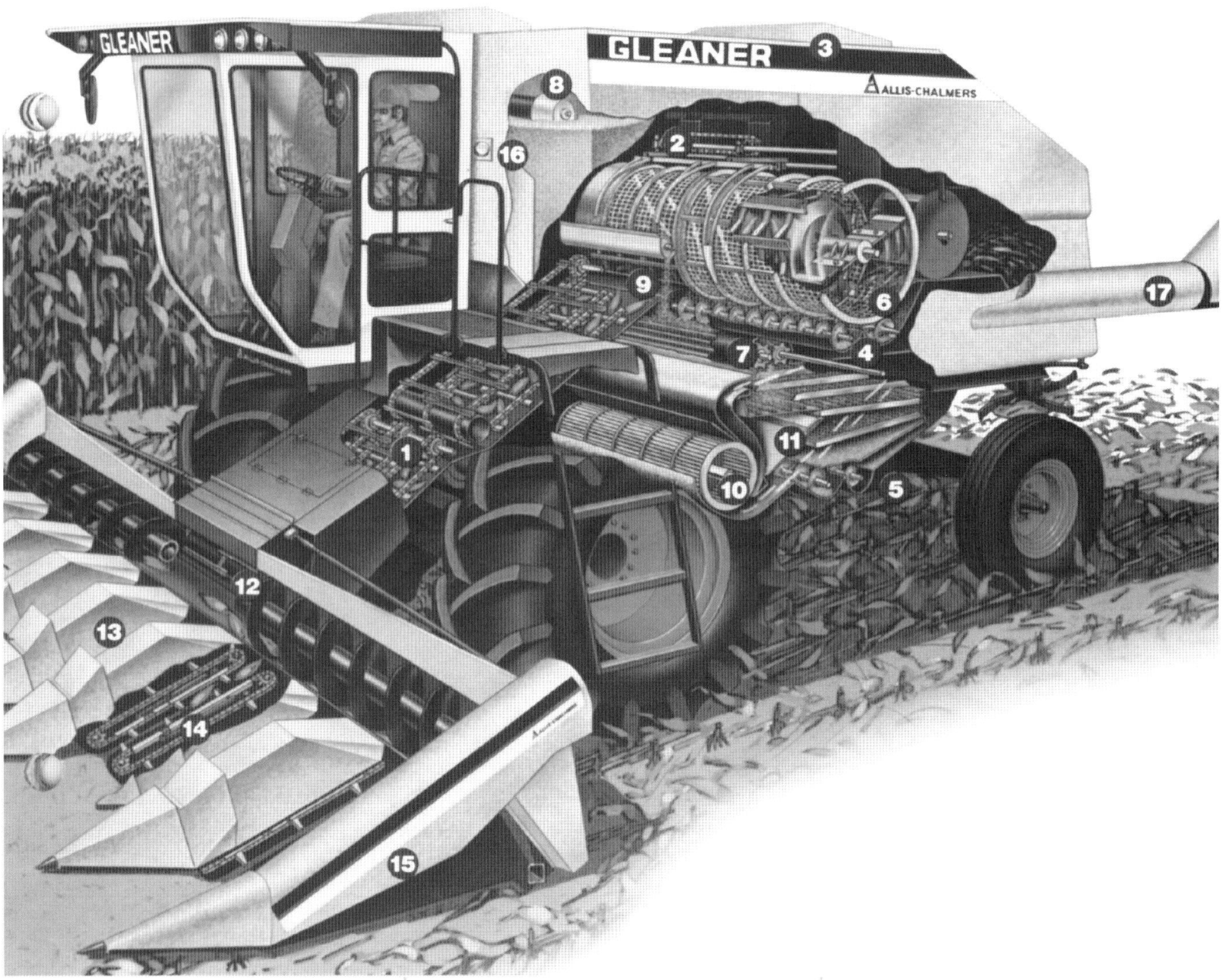

1. Sure-feed system. — Center-pivot system keeps crop at a constant angle right into the cylinder. Straight-line feed.

2. Exclusive cage sweep. — Oscillating action keeps cage free of straw build-up.

3. 315-bu. bin on the N7—World's largest. Center-mounted for balanced weight distribution. N6: 245 bushels; N5: 200 bushels.

4. Distribution augers. — Provide uniform delivery of material, after separation, to accelerator rolls.

5. Tailing auger. —Tailings are returned to the accelerator rolls for recleaning, or to the cylinder (optional) for rethreshing.

6. 360° separating action. — Exclusive Natural Flow. Entire cage is perforated so separation occurs around its full circumference. 6,680 sq. inches of separating on the N7 and N6; 5,441 sq. inches on the N5.

7. Accelerator rolls. — N-Series exclusive. They accelerate grain at high speed through high-velocity air blast for maximum cleaning and better side-hill operation.

8. Clean grain elevator. — Moves grain at up to 50 bu./min.

9. Concave door stone protection. — Automatically ejects rocks and other damaging objects before they can enter the cylinder. Another exclusive.

10. Exclusive transverse flow cleaning fan. — All steel fan with straight-through air flow. Two outlets provide air for pre-clean and finish cleaning at shoe. Cab-controlled velocity.

11. 2-stage high-velocity air cleaning. — Handles most of the cleaning before crop reaches shoe area, reducing shoe load. Adjustable chaffer and sieve for final cleaning.

12. 14-inch auger. — Maintains a steady, even crop flow for high-speed, multi-row harvesting.

13. Sloped gather sheets. — Minimize hairpinning and reduce whipping of downed stalks.

14. Individually mounted row units. — Units feature extra-long gathering chain lugs for positive feeding.

15. Extra-low profile dividers. — Gently skim under down and tangled crop.

16. 150 gal. fuel tank. — Constructed of Marlex plastic. Biggest anywhere for more harvesting time between fuel stops.

17. Pushbutton activated, swivel unloading tube. — Unloads the N7's entire 315 bu. load; the N6's 245 bushels and the N5's 200 bushels in approximately two minutes.

The Gleaner rotary combine was unique in that the processor was oriented transversely. This diagram tells why it was successful.

The N5 had grain headers from 15 to 30 ft and four-, six-, or eight-row cornheads. The grain bin held 200 bushels and the engine was a 190-hp 426-in.³ turbocharged and intercooled diesel.

The grain bin held 245 bushels, and could unload at a rate of 2.1 bushels per second. The fuel tank carried 150 gallons. The cab, similar to the L-M cab, was mounted 7 inches lower and had a visor roof to better shade the operator and carry field lights. Electrohydraulic controls were also similar to the L-M system, except that the hydraulic pump was a variable-displacement type. "Tattletale" electronic monitoring kept the operator well informed about operating systems and warned of trouble. The Pacer Efficiency Monitor reported on conditions in the separator and cleaning shoe.

The N6 was powered by an A-C six-cylinder direct-injection turbocharged and intercooled diesel of 426-cubic inch displacement rated at 2,600 rpm and 220 horsepower. Drive was hydrostatic, coupled to a four-speed constant-mesh transmission.

Early in 1979 the Model N5 joined the line and a few months later the even bigger N7 took over as the flagship of the Great Silver Fleet. The N5 had somewhat less standard equipment, no visors on the cab, and reduced open area in the cage. The engine was the same as the N6, but derated to 2,400 rpm and 190 horsepower. The Model N7 had similar threshing and cleaning specifications to the N6, but because a rotary combine responds to greater power input with greater capacity, the N7 could handle more crop. The N7 was powered by an A-C six-cylinder 516-cubic inch turbo-intercooled

The leviathan N7 had high daily capacity with up to 30-ft grain headers and 12-row cornheads. The enormous 315-bushel bin meant fewer unloading stops even with record yields and the A-C 516-in.³ 270-hp turbocharged and intercooled diesel had the power to make it happen. It was the largest combine of its day.

diesel rated at 270 horsepower at 2,200 rpm. At 315 bushels, its grain bin was the biggest in the business.

With the N-Series, productivity was the name of the game. Grain headers came in 15- to 30-foot sizes and flex-headers from 15- to 24-foot sizes. Four- to eight-row cornheads were available for the N5, while 6- to 12-row heads matched the N6 and N7. The N-Series was built through 1985, and was continued in 1986 as the R-Series by Deutz-Allis.

Gleaner Cornheads

Gleaner was at the forefront of corn combining with several cornhead concepts, all for whole stalk harvesting. In 1930 the company built 500 cornheads for pull-type combines. The unsuccessful results were not related to the field machine, but to the lack of grain-drying equipment on farms or country elevators. After that 500, none were produced until 1957 when 105 cornheads were built, followed by 501 in 1958, and 751 in 1959 — all were built for the Model A.

The Gleaner-built cornheads were two-row models with 22-inch snapping rolls for the Model A combine. Since LaPorte was still building corn harvesters and corn pickers and there was a great deal of commonality, all cornhead production for Gleaners was transferred to LaPorte after the 1959 production run. Snapping-roll length was standardized at LaPorte's 21.25 inches for all cornheads, and this length would remain in production until 1970. Gleaner cornheads had adjustable stripper-plates which kept the butts of the ears from contacting the snapping rolls which reduced grain loss there.

The 1957 Gleaner two-row cornhead for the Model A was A-C's first cornhead for a self-propelled combine.

The 1959 Model A two-row cornhead had different sheet metal from prior production.

In 1960, LaPorte produced 1,500 A-240 cornheads and a pilot run of 12 C-440 cornheads. Production of the E-240 began in 1962, the A-330 in 1963, and the C-430 in 1965. By the end of 1965, the plant had manufactured 17,162 cornheads, only 500 of which were for 30-inch rows. The big move to narrow rows was just around the corner.

The company was aggressively promoting the Gleaners for corn combining and supported farmers who entered corn combining and corn picking contests with active service assistance. It paid off with four national championships in five years, with a second place only .002 of a point out of first the other year. A Model A with two-row cornhead won in 1962, 1963, and placed second in 1964. A Model E with a two-row head won in 1965, and another E two-row won in 1966. Also in 1963, an A-C 170 cornpicker on a D17 tractor won the cornpicking contest. The clean sweep was great.

In the late 1960's narrow rows began to take over in the Corn Belt, and engineering responded with an expanded line of cornheads for the A, C, and E combines to harvest 20- and 30-inch rows. The E-330, A-330, and C-430 (combine-rows, row spacing) units became available and these were followed by the C-630 and the E-420, A-420, C-620, and C-820. All these sizes were continued

The C-440 cornhead for the Model C was built at Laporte after cornhead production was transferred from Independence. Twelve units were built in 1960.

and a new 435 (for 34- and 36-in. rows) added for the Models F and G combines. When the Model K replaced the E, a 335 became available for it and the F.

The Models F and G featured the new header mounting system which permitted the operator to mount the header or cornhead by engaging the mounting hooks and raising the thresher housing, then locking the wedge pins and connecting the drives.

A Model G combine with a four-wide row cornhead at work. Longer 30.75-in. snapping rolls were more aggressive.

The Model C combine with a C-620 six 20-in. row cornhead. Other 20-in. row cornheads were the C-820, A-420, and E-420. The 20-in. cornheads had a short production life.

Gleaner pioneered quick-change headers to accommodate the needs of the corn-soybean grower. With this and quick combine adjustments, it was possible to harvest corn on damp mornings, then change to soybeans in the dry afternoons.

The Low-Profile cornheads were introduced in 1970 and brought many fundamental changes. True to their name, they were much lower with snapping rolls at a reduced angle to the ground to lift lodged cornstalks. The snapping-roll length was increased to 30.75 inches. Roller gathering chains promised quieter operation, reduced maintenance, and longer life. The 20-inch row sizes were dropped so the line consisted of the six cornheads: KL-238, KL-330, FL-430, GL-438, GL-440, and GL-630.

When the Model L and M combines came out, six-row wide and eight-row narrow heads were offered, as well as a new size, a five-row wide model.

A line of adjustable row-spacing cornheads was launched with the K2, F2, M2, and L2 combines for 1977. Of the low-profile design, they featured a toolbar frame upon which the individual row units could be moved 2 inches to either side (only narrower on the eight 30-in.) of their nominal row spacing. Wide row (nominal 38-in. spacing) were available in two- to six-row sizes. Narrow row (nominal 30-in. spacing) came in three-, four-, six-, and eight-row models. Snapping rolls were slightly longer at 31.1 inches and larger in diameter at 4.3 inches.

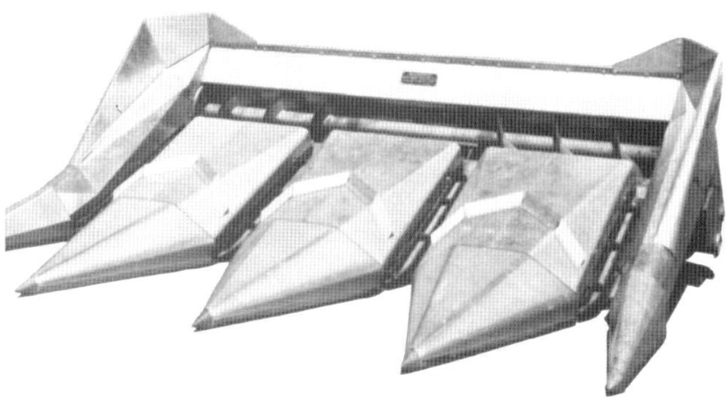

The low-profile cornheads had reduced snapping-roll angle for better work in down corn. Six models for 30-, 38-, and 40-in. rows were produced.

When the N-Series combines came out in 1979, corn harvesting capacity leaped by 50 percent with the 12-row cornhead. The N-Series cornheads were of the low profile-adjustable row-spacing design much like the conventional line, but could not be interchanged because of the offset feeder housing of the N combines. Six or eight narrow-row sizes matched the N5, N6, and N7, with the twelve-30-inch row size reserved for the N6 and N7. A four wide-row model fit the N5, with six and eight wide-row sizes for all three combines. The N-Series cornheads continued in

The adjustable cornheads allowed moving the snapping units up to 2 in. either way to center on the rows for better performance and reduced ear loss.

The twelve 30-in. row cornhead on an N6 or N7 combine had enormous capacity and could keep the operator calling for more trucks to haul the corn away.

production with new model designations to match the succeeding R-Series Deutz-Allis combines.

After the Industrial Tractor Division was closed, grain header production at Independence and cornhead manufacturing at LaPorte were shifted to the Topeka plant. Further consolidation brought both back to the main combine plant in 1984.

Some late 1984 cornhead base prices:

Combine	Cornhead Size	Base Price
K2-F3	3-30-in. rows	$9,500
M3-L3	6-30-in. rows	17,200
N5-N6	8-30-in. rows	22,400
N6-N7	12-30-in. rows	35,800

Of course, that was without the combine. That's a lot of bushels.

Production Data from the 1928-1954 Gleaner Pull-Type Combines

	PT	A	R	NR	E	F	J	H	T	S	P-80
1928	2,965	---	---	---	---	---	---	---	---	---	---
1929	---	1,039	3,663	---	---	---	---	---	---	---	---
1930	---	---	---	---	---	---	---	---	---	---	---
1933	---	1,837	3,207	---	---	---	---	---	---	---	---
1934	---	---	197	---	---	---	---	---	---	---	---
1935	---	---	523	---	---	---	---	---	---	---	---
1936	---	---	---	1,056	---	---	---	---	---	---	---
1937	---	---	---	2,042	---	---	---	---	31	---	---
1938	---	---	---	2,557	---	---	---	---	702	---	---
1939	---	---	---	---	269	181	160	---	61	---	---
1940	---	---	---	---	204	---	---	---	122	---	---
1941	---	---	---	---	899	---	241	---	636	---	---
1942	---	---	---	---	596	---	---	---	---	100	---
1943	---	---	---	---	1,100	128	---	67	---	---	---
1944	---	---	---	---	1,152	323	---	128	---	160	---
1945	---	---	---	---	666	335	---	---	---	860	---
1946	---	---	---	---	1,425	326	---	---	---	728	---
1947	---	---	---	---	1,500	500	---	---	---	1,750	---
1948	---	---	---	---	1,250	750	---	---	---	1,009	---
1949	---	---	---	---	1,500	---	---	416	---	2,000	---
1950	---	---	---	---	---	---	---	162	---	1,029	---
1951	---	---	---	---	33	---	---	---	---	1,179	---
1952	---	---	---	---	---	---	---	---	---	251	---
1953	---	---	---	---	---	---	---	---	---	225	---
1954	---	---	---	---	---	---	---	---	---	---	148
1955	---	---	---	---	---	---	---	---	---	---	139
Total	2,965	2,876	7,590	5,655	10,594	2,543	401	773	1,552	9,291	287

Specifications for Small Pull-Type Gleaner Combines

	T	S	Six-S	P-80
Years built	1937-41	1937-41	1942-52	1953-55
Width of cut (in.)	72	72	74	80
Cylinder type	Rasp	Rasp	Rasp	Rasp
Diameter (in.)	19.25	19.25	19.25	19.25
Width (in.)	18	18	22	22
Separation	Raddle	Walker	Walker	Walker
Width (in.)	24	24	24	24
Length (in.)	120	126	132	132
Total sep. area (in.^{2})	2,880	3,024	3,168	3,168
Tot cleaning area (in.^{2})	1,464	1,464	1,891	1,891
Bin capacity (bu)	20	20	22	22
Engine (original)	Wisconsin			
Cyl. (in.^{3}-hp)	4-70.4-12.5 hp			
1939	4-70.4-16.0 hp			
1942			Wisc. 4-92-21.5 hp	
1953				Wisc. 4-108-25 hp

Specifications for Large Pull-Type Gleaner Combines

	R-A	NR-NA	E-F	J-H
Years built	1929-35	1935-38	1939-51	1939-50
Width of cut (ft)	12	12	12	9
Cylinder type	RASP			
Diameter (in.)	19.25	19.25	19.25	19.25
Width (in.)	26.0	26.0	26.0	26.0
Separation	R-W*	R-W*	R-W*	R-W*
Width (in.)	32.0	32.0	32.0	32.0
Length (in.)	Raddle- 156 Walker- 164			
Total sep. area (in.^{2})	Raddle- 4992 Walker- 5248			
Tot. cleaning area (in.^{2})	2,464			
Bin capacity (bu)	50	50	50	50
1942:			40	40
1946:			45	45
Engine (original)	Ford A , 4 Cylinder, 200 in.^{3}, 32 hp			
1944:			LeRoi, 4-177, 32.5 hp	
1947:			Hercules, 4-126, 35 hp	
1948:			Ford, 6-226, 48 hp	

* Raddle separation (R) or walker separation (A).

Specifications for Gleaner Models A, R, B, T, C, and E

	A/R/B*	A/R/B*	T	AH	C	E	*CR Rice *CT Track	CH Hillside
Years built	1950-57	1957-63	1954	1959-61	1960-63	1962-67	1962-67	1963-67
Grain headers (ft)	10-14	10-14	7, 10	14	14-20	8-12	13	16, 18
Cornhead rows	2	2	None	None	4	2	4	None
Bin cap. (bu)	50	50	30	50	60	40	60/75	75
Cylinder type	Rasp *Spike	Rasp *Spike	Rasp	Rasp	Rasp	Rasp	*Spike	Rasp
Diameter (in.)	All rasp-19.25	Spike-19.62						
Width (in.)	30	30	23	30	40	27	40	40
Separator	A,B walker R raddle	Walker	Walker	Walker	Walker	Walker	Walker	Walker
Width (in.)	36	36	28	36	48	32	48	48
Length (in.)	156	156	144	156	156	147-159†	156-162†	156-162†
Total area (in.^{2})	5,006	5,006	3,932	5,074	7,056	4,580-4,968†	7,056-7,344†	7,056-7,344†
Total clean. area (in.^{2})	2,685	2,685	2,067	2,685	3,564	2,680	3,564	3,564
Engine	Ford	A-C	Ford	A-C	A-C	A-C	A-C	A-C
Cyl./in.^{3}	6/226	6/230	4/172	6/230	6/262	4/226	6/262‡	6/262
1953-57	Hercules 6/237							
Drive	Variable V-belt§							
Transmission	3 forward, 1 reverse§							

* Rice combines have spike-tooth cylinders.

† With walker extensions.

‡ CR also had 262-in.^{3} turbocharged diesel and 262-in.^{3} LPG engines available.

§ Applies to all models.

The big CH Hillside Special Gleaner Combines on the steep hills of Oregon.

A Gleaner combine with pick-up attachments handled windrowed barley in North Dakota.

Specifications for Gleaner Models AII, CII, EIII, F, G, and K

	AII	CII	EIII	F	G	K	GH Hillside	FR Rice
Years built	1964-67	1964-67	1968	1968-76	1968-72	1969-72	1969-72	1969-70
Grain headers (ft)	10-16	12-24	10-13	13-20	13-24	10-14	16-20	13-15
Cornhead rows	2,3,4	4,6	2,3	2,3,4	4,6,8	2,3	DNA	3,4,6
Bin cap. (bu)	55	85	66	77	100	66	100	77
Cylinder	Rasp	Rasp	Rasp	Rasp	Rasp	Rasp	Rasp	Spike
Diameter (in.)	19.25	19.25	19.25	19.25	19.25	19.25	19.25	19.62
Width (in.)	30	40	27	37.5	44	27	44	37.5
Separator	Walker	Walker	Walker	Walker	Walker	Walker	Walker	Walker
Width (in.)	40	48	32	42.5	53	32	53	42.5
Length (in.)*	162	162	165	165	162	165	162	165
Total area (in.^{2})*	6000	7344	5448	6875	8109	5448	8109	7344
Total clean. area (in.^{2})	2889	3564	2680	3441	4138	2680	4138	3721
Engine								
Gas	A-C	A-C	A-C	A-C	A-C	Chev.	A-C	A-C
Cyl./in.^{3}	6/230	6/262	4/226	6/262†	6/301‡	6/250	6/301§	6/262
Horsepower	76	93	---	93	105	78	105	93
Diesel	---	---	---	A-C	A-C	---	A-C	A-C
Cyl./in.^{3}	---	---	---	6/301	6/301T	---	6/301T	6/301
Horsepower	---	---	---	84	109	---	109	84
Drive	Variable V-belt§							
Transmission	3 for., 1 rev.	3 for., 1 rev.	3 for., 1 rev.	3 for., 1 rev.	3 for., 1 rev.	3 for., 1 rev.	6 for., 2 rev.	6 for.,2 rev.
Optional	---	---	---	Hydro.II	Hydro.II	---	Hydro.II	Hydro.II

* With Walker extensions.

† F gas, 1970 and up--Chev. 6/292, 101 hp.

‡ G and GH gas, 1970 and up--Chev. V8/350, 120 hp.

§ Applies to all models.

II Hydrostatic.

Several Gleaner L3's at work in the fields.

Specifications for the Gleaner Models M, L, and MH

	M	Hillside	MH	M2/M2HY Hillside	MH2/MH2HY	L2/L2HY	M3 Hillside	MH3	L3
Years built	1973-76	1972-76	1974-75	1977-82	1977-82	1977-82	1983-86	1983-86	1983-86
Grain headers (ft)	13-22	13-24	18-22	13-22	18-22	13-24	13-22	18-22	13-24
Cornhead rows	4,6	4,6,8	none	4,5,6	none	4,5,6,8	4,5,6	none	4,5,6,8
Bin cap. (bu)									
Std/extensions	105	120	105	105/165 180*	105/115	120/185 200*	180	115	200
Cylinder type	Rasp	Rasp	Rasp	Rasp	Rasp	Rasp	Rasp	Rasp	Rasp
Diameter (in.)	19.25	19.25	19.25	19.25	19.25	19.25	19.25	19.25	19.25
Width (in.)	40	48	40	40	40	48	40	40	48
Separator type	Walker	Walker	Walker	Walker	Walker	Walker	Walker	Walker	Walker
Width (in.)	45	53	45	45	45	53	45	45	53
Length (in.)	216	216	216	216	216	216	216	216	216
Total area (in.^{2})	9,850	11,634	9,850	9,964	9,964	11,780	9,964	9,964	11,780
Total cleaning area (in.^{2})	4,667	5,525	4,667	5,252	4,667	6,206	5,252	4,667	6,206
Engine									
Gas	Chev.	Chev.	Chev.	None	None	None	None	None	None
Cyl./in.^{3}	V8/350	V8/350	V8/350	---	---	---	---	---	---
Horsepower	120	120	120	---	---	---	---	---	---
Diesel	A-C	A-C	A-C	A-C	A-C	A-C	A-C	A-C	A-C
Cyl./in.^{3}	6/301T	6/301T	6/301T	6/301T	6/301T	6/301T	6/301T	6/301T	6/301T
Horsepower	109	109	109	130	130	130	130	130	130
Opt. diesel	---	---	---	---	A-C	A-C	A-C	A-C	A-C
Cyl./in.^{3}	---	---	---	---	6/426T	6/426T	6/301TI	6/426T	6/426T
Horsepower	---	---	---	---	158	158	145	158	158
Drive	V. V-b‡	V V-b‡	Hydro.	V.V-b‡	Hydro.	V.V-b‡	V.V-b‡	V.V-b‡	V.V-b‡
Transmission	5 F, 1 R§	5 F, 1 R§	4 F, 4 R§	5 F, 1 R§	4 F, 4 R§	5 F,1R§	5 F,1R§	5 F,1R§	5 F,1R§
Optional	Hy-Tr II	Hy-Tr II	None	Hydro.	None	Hy-Tr II	Hyd-Tr II	Hy-Tr II	Hy-Tr II

* Bin capacity 1978 and later.
§ # F (Forward), # R (Reverse).
II Hydro-Traction.
‡ Variable V-belt.

The Gleaner GH "Special Hillside" combine.

Specifications for Gleaner Models K2, F2, F3

	K2	F2	F3
Years built	1977-81	1977-82	1983-86
Grain headers (ft)	10-15	13-20	13-20
Cornhead rows	2,3	3,4,5,6	3,4,5,6
Bin cap. (bu)			
Std/extensions	68/96*	80/120*	120
Cylinder type	Rasp	Rasp	Rasp
Diameter (in.)	19.25	19.25	19.25
Width (in.)	27	37.5	37.5
Separator	Walker	Walker	Walker
Width (in.)	32	42.5	42.5
Length (in.)	183†	183†	183
Total area (in.^{2})	5,837†	7,818†	7,818
Total cleaning area (in.^{2})	2,887	3,892	3,892
Engine			
Gas	Chev.	Chev.	None
Cyl./in.^{3}	6/250	6/292	---
Horsepower	85	112	---
Diesel	A-C	A-C	A-C
Cyl./in.^{3}	4/200T	4/200TI	4/200TI
Horsepower	72	95	95
Drive	Variable V-belt‡		
Transmission	3 for., 1 rev.	3 for., 1 rev.	5 for., 1 rev.
Optional		Hydro-trac.	Hydro-trac.

* Bin capacity 1978 and later.

† 1978 and later. 1977 K2 and F2 had 165 in. separating length with 5448 and 6875 in.^{2} of total area, respectively, the same as the K and F.

‡ Applies to all models in this category.

Specifications for Gleaner N-Series Rotary Combines

	N5	N6	N7
Years built	1979-85*	1978-85*	1979-85*
Grain headers (ft)	15, 16, 18, 20, 22, 24, 27, 30†		
Flex headers	15, 16, 18, 20, 22, 24†		
Cornheads			
Wide rows	4,6,8	6,8	6,8
Narrow rows	6,8	6,8,12	6,8,12
Bin cap. (bu)	200	245	315
Processor			
Cylinder type	Rasp	Rasp	Rasp
Length (in.)	90	90	90
Diameter (in.)	25	25	25
Total open cage area (in.^{2})	4,950	5,924	5,924
Total cleaning area (in.^{2})	6,339	6,339	6,339
Engine	A-C direct-injection diesel†		
Cyl./in	6/426TI	6/426TI	6/516TI
Horsepower	190	220	270
Drive	Hydro-Traction†		
Transmission	4 forward, 1 reverse†		

* Continued with engine changes as Deutz-Allis R-Series.

† Applies to all models in this category.

N-Series 1984 List Prices*

Model	Base List ($)	Header (ft)	Total List ($)
N5	91,100	18 grain	104,833
N6	102,820	24 flex	133,719
N7	121,154	30 grain	152,582

* The fall of 1984 with typical accessories.

Fall 1984 List Prices for the Models F3, M3,MH3 and L3

Model	Header (ft)	Total List ($)
F3	13 grain	64,415
M3	16 flex	83,500
MH3	22 grain	118,152
L3	20 flex	115,300

Serial Numbers of Gleaner Self-Propelled Combines (1951-1986)

	A	R	B	C/CH	E	K	F
1951	51A9001	51A8001	---	---	---	---	---
1952	52A9006	52R3001	---	---	---	---	---
1953	53A5000	53R5000	53B6001	---	---	---	---
1954	54A7001	54R7001	54B7017	---	---	---	---
1955	55A9001	55R8001	55B8001	---	---	---	---
1956	A11361	R11073	B11693	---	---	---	---
1957	A12965	R13110	B13303	---	---	---	---
1958	A15174*	R15260*	B15348*	---	---	---	---
1959	A19602	R19777	B21692	---			
1960	A23573	R24014	B23768	C1001	---	---	---
1961	A25493	---	B26428	C1707	---	---	---
1962	A27482	---	B28129	C2752	E1001	---	---
1963	A28811	---	B29278	C3870	E3263	---	---
1964	A30001 **AII**	---	---	**CII**-C5001	E7312	---	---
1965	A32061	---	---	C7004	E10897	---	---
1966	A34588	---	---	C10476	E13870	---	---
1967	A37701	---	G/GH	C17201	E18301	---	---
1968	---		G1001	---	**EIII**-E25001	---	1001
1969	---	---	7401	---	---	1001	5001
1970	---	---	12201	---	---	5001	9001
1971	**L**	---	17001	---	---	8201	12201
1972	1001	M/MH	21001	---	---	12201	15401
1973	2601	1001	---	---	---	14601	18601
1974	4201	4201	---	---	---	17001	21801
1975	5801	7401	---	---	---	19401	25801
1976	9001	11401	---	---	---	21801	31401
1977	13001 **L2**	15401 **M2**	---	**N6**	---	**K2** 24201	**F2** 36201
1978	15501	18401	**N5**	1001	**N7**	25201	39201
1979	17501	19901	1001	1501	1001	26201	41701
1980	19501	21901	1501	2001	1201	26701	43701
1981	21501	22101	3001	3501	1701	27201	46201
1982	24001	23801	4501	5101	2501	---	48201
1983	26001 **L3**	24701 **M3**	5401	6201	3501	---	**F3** 49601
1984	28001	25901	6101	7301	4501	---	51001
1985†	30501	27401	7101	8801	5501	---	53001
1986†	33001	29401	---	---	---	---	55001

* Tailings return to cylinder effective Serial No. 16001.

† After May, 1985, Gleaner combines had the Deutz-Allis name plate. At the beginning of 1986 the N-Series became the R-Series combines. The conventional models were discontinued at the end of 1986.

A Revolution in the Grain Field — The All-Crop Harvester

The Advance-Rumely OilPull erection floor in 1911. Note the crated machine destined for Argentina.

The Rumely Era — 1852-1931

One might say that the LaPorte plant was the result of a pistol-whipping, strange as that may seem. Meinrad Rumely was born in Germany on February 9, 1823. While attending a military school as a young man, he made a minor mistake of stepping beyond the company line. The captain in charge saw the error and struck him with a pistol. While recovering, he decided that his future lie in America. In 1848, he made the 64-day trip on a sailing vessel across the ocean.

It probably wasn't a difficult decision since two of his brothers had already immigrated to this country. After working in the East and Midwest as a machinist, Rumely eventually arrived in LaPorte in 1852 where he and his brother, John, opened a foundry and machine shop to serve the railroad industry. They made railroad car and drive wheels, locomotive cylinders, and machinery for sugar cane crushing and making sorghum molasses.

Loading threshers on railcars in 1915. This gives an idea of the size of the Advance Rumely Company's business.

By 1855 they employed about 20 people and produced over a ton of castings per day. They also built a corn sheller which produced 1,500 bushels per day. This led to building a thresher which was driven by horse power by either a treadmill or sweep. In 1859, their threshers were successful and took first place in competition at the Illinois State Fair at Chicago. In 1861 they were awarded medals by the United States Agricultural Society for the best horse power and the best separator. Firmly established in the thresher business, they were among the first to use

A 55-hp Advance-Rumely Universal Compound steam traction engine — a locomotive for the farm.

The test area for the OilPulls was a busy place.

steam power for threshers. Portable steam engines, steam traction engines, and clover hullers were added to the line. Since Rumely had become one of the largest manufacturers of engines and threshers at that time, LaPorte was on its way to becoming the "Harvester Capitol of the World."

Meinrad Rumley bought his brother's share of the business in 1882, and incorporated the M. Rumely Company in 1887. As president and general manager, he ran the company until he died in 1904. The firm carried on, managed by his four sons. In 1909, the M. Rumely Company, with the help of engineers brought in from the outside, developed the kerosene burning traction engine which was named the OilPull. It was a forerunner of the modern farm tractor and as such made a tremendous impact on mechanized agriculture. Thousands of acres of good prairie soil first saw the light of day as the OilPulls, with their wide prairie breakers, passed by.

One of the famous Advance-Rumely OilPulls which was developed in 1909 and built until 1930.

During 1911-12, M. Rumely consolidated with three major equipment producers: Gaar-Scott and Company, Richmond, Indiana (1836); Advance Thresher Company, Battle Creek, Michigan (1881); and Northwest Thresher Company, Stillwater, Minnesota (1874). This produced an immense capability which was reorganized into the Advance-Rumely Company in 1915. Eight years later the Aultman-Taylor Company, Mansfield, Ohio, was added to form one of the largest firms of its type in the world. Today this merger would probably bring the federal trust-busters to town in full force.

Advance-Rumely grew and prospered for many years but, like so many others, was beginning to feel the economic pressures of the forthcoming depression and began looking for a buyer. Correspondence dated January 1930, between Dr. Edwin Rumely and General Otto Falk indicated that negotiations occurred over many months until the sale was finally consummated in May 1931. Finley P. Mount and J. J. Daniels represented Advance-Rumely, while Max W. Babb and Harry C. Merritt represented Allis-Chalmers during the final negotiations. A-C acquired the LaPorte plant, several branch houses, a viable line of equipment, and access to hundreds of dealers and thousands of customers. History confirms this as an extremely wise decision on the part of Allis-Chalmers. The next year, A-C purchased the Birdsell Mfg. Company, of South Bend, Indiana, (clover hullers), and the Ryan Mfg. Company, Hegewisch, Illinois (road graders), and moved both operations to LaPorte.

The huge OilPulls with their prairie-breakers turned thousands of acres over to face the sun for the first time.

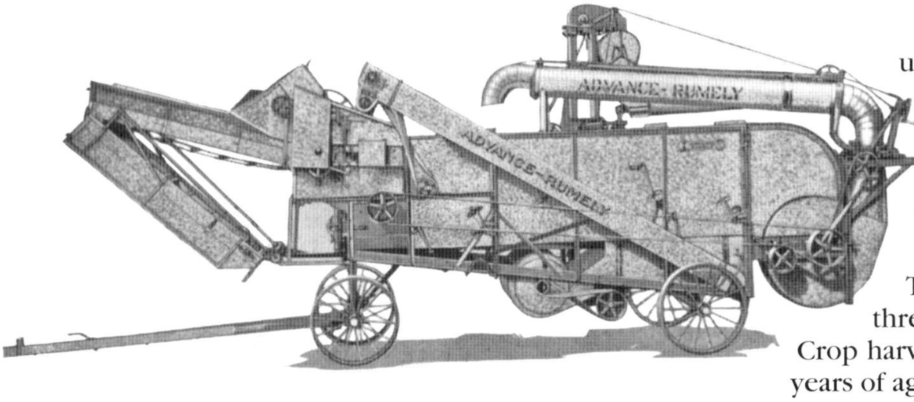

The Rumely Machines

When Allis-Chalmers acquired Advance-Rumely they also received a premium line of equipment. Products included the OilPull, DoAll, and 6A tractors, the Nos. 2 and 3 prairie combines, the No. 3 hillside combine, a swather, and the famous 22 × 36 and 28 × 46 threshers. All of these units appeared in the 1932 to 1935 Allis-Chalmers catalogs.

Only a few of these machines were produced after the May 1931, buy out. These included the threshers and the No. 3 hillside combine, which were built as late as 1936. Production of the well-known OilPull ended in 1930, as did the No. 3 prairie combine and the huge 36 × 60 thresher. Other A-R products such as the silo filler, corn sheller, husker-shredder, Nos. 1, 2, and 4 prairie combines, swather, and the big 32 × 52 thresher, had been discontinued during the 1920s. All the other units in the A-C catalogs were just existing inventory that was to be cleared as soon as possible. No doubt the depression of the early 1930s considerably lengthened the sell-off.

The All-Crop Harvester

In what has to be the most outstanding success story in the farm equipment industry, the Allis-Chalmers All-Crop harvester brought changes to North American agriculture and farm families' lives far beyond anything that could be imagined. This unimpressive little machine challenged the autocracy of the binder-steam engine-thresher and relegated them to the museums. The All-Crop did for the Corn Belt and eastern mid-sized farms what the Gleaner and its contemporaries did for the wheat growers of the plains.

The writer remembers when the local thresherman traded his big rig off for two All-Crop harvesters. That was the year that he, about 16 years of age, was decreed of sufficient maturity to run the bundle wagon through the threshing ring. The disintegration of the threshing ring brought a new combine to the home place. That John Deere 12A was beautiful! At least to a 16-year-old who could not face the prospect of weeks on that tyrannical bundle wagon. What happened to threshing in central Indiana was happening all over the country. Threshing rings were washed away by a wave of small combines from all equipment companies, more than half of which were orange and manufactured in LaPorte, Indiana. It made farm wives happy as those huge threshermen's meals were no longer needed, and it must have made thousands of teenage farm boys absolutely ecstatic.

This revolution in grain harvesting for the small and medium-sized farm started in 1931 when Allis-Chalmers announced that it was offering the Allis-Chalmers Combine, a handy, compact, and light machine which would allow "anyone who could drive a tractor to cut and thresh their small grains, clovers, peas, and beans alone, with no more effort than it took to disk those fields" (From a 1931 sales brochure). Allis-Chalmers did not invent this machine, but they eventually reinvented it. Actually invented by Bob Fleming and promoted by Guy Hall, the Fleming-Hall, also known as the brush combine, had been demonstrated in Racine County, Wisconsin, to industry representatives in an attempt to find a manufacturer. A fairly simple machine, it used a 5-foot wire brush for a cylinder and an overgrown fanning mill for a separator.

The brush combine met with a mixed reception by the observers, but Harry Merritt saw the possibilities in this crude beginning and purchased a shop license to build the machine for $25,000. About this same time, Allis-Chalmers acquired Advance-Rumely and the brush combine was sent there for development. John Mainland, the Rumely inventor of threshers and combines, and Walter Dray, whom Merritt brought from Hart-Parr, inherited the combine development project. They, with Walter Charley and Nelson Hazzard, went to work on the brush design, and for all practical purposes, reinvented it.

As a result of their efforts, the brush was abandoned and a steel bar cylinder working against rasp-

The 22x36, 28x46, and 28x50 threshers and the No. 3 hillside combine were the only Rumely machines produced after the A-C buyout and production continued until 1936.

bar concaves was substituted. Still a 5-foot cut, the 5-foot cylinder received the crop in a thin, straight stream which was unlikely to bunch and plug the cylinder. The separator, now much more sophisticated than the earlier fanning mill, set crosswise behind the cylinder, and provided over 9 feet of separating length without adding much overall length. The drives were strong, lightweight V-belts in place of chains and sprockets. The cylinder-drive pulleys were two-piece, "like two soup plates set back to back," which could be varied with spacers to provide adjustable diameters and different cylinder speeds. It weighed around 2,500 pounds, was mounted on rubber tires, and when pulled by a rubber-tired tractor could harvest at 5 mph. It even had a new name, the Allis-Chalmers Corn Belt Combine (*Farm Implement News*, August 1933).

The Brush Combine

"Cut Crops and Costs with this Handy, Compact, Light Allis-Chalmers Combine" was the lead line on a 1931 folder describing this marvelous new

The improved wire reel swept the grain back to the draper.

The new Allis-Chalmers combine and an All-Crop tractor combining soybeans in Illinois.

The brush which acted as a cylinder had 750,000 teeth and turned 1,000 rpm.

The Allis-Chalmers combine working in a Wisconsin field. It was powered by a Model U tractor (1931).

machine, the brush combine. Harvesting was now a one man job.

"Narrow enough to go through a 12-foot gate and light enough to be operated by the power-take-off of a two-three-plow tractor."

"It has only half as many working parts and half the weight of the conventional combine, yet plenty of strength to withstand the strain. It has a 5-foot cut and a 5-foot cylinder. It is 10 1/2 feet wide and weighs slightly over 3,000 pounds. It is capable of cutting and threshing 15 acres a day at plow speed."

"It is a complete unit which cuts, threshes, cleans, recleans and tanks the grain as it goes through the field. Its unique design and full-width brush-type cylinder permits early and late operation without lost time due to clogging by tough straw."

When it was time to harvest, get-ready time was minimal. There were no belts to put on or concaves to adjust — simply a matter of oiling the machine, tightening the draper, hitching the tractor, and heading for the field. The draper traveled about twice ground speed, spreading the crop, so it went over the brush in a thin layer so the brush could easily sweep the grain out of the heads. The brush was 15 inches in diameter, turned 1,000 rpm, and had 750,000 steel teeth. After the grain was swept out of the heads, it fell to an auger which delivered it to the fanning mill where it was cleaned and then elevated into the tank. An elevator emptied the grain tank into a wagon. The straw passed down over long wire fingers which were right below the brush and dropped to the ground. To change crops it was only necessary to change the screen in the fanning mill. It was unique. Nothing like it had ever been seen before. In fact, it started a revolution.

The Corn Belt combine with its original dog-leg right-angle V-belt cylinder drive before the straight V-belt with adjustable pulleys was used.

The Allis-Chalmers 5-ft Corn Belt combine and a Model U tractor, both on Goodyear tires, harvesting wheat.

Barney Oldfield (on seat) who raced the Model U air-tired tractor said he would not have left the farm if they had had such good equipment when he was a boy.

The air-tired Model U and A-C High-Speed combine were on the leading edge of new age in agriculture. Both exemplified all-new technology which changed an entire industry.

The High-Speed Combine

In the 1935 catalog it was called the Allis-Chalmers High-Speed Combine and 550 units were built that year. Both cylinder bars and concaves were now rubber-faced for rubber-on-rubber shelling, an exclusive feature which was responsible for the combine's ability to successfully thresh delicate grass and legume seeds and soybeans with minimum damage.

It was billed as the original, and only, lightweight one-man combine for two-plow tractor power. The high-speed combine was ideal for small or medium sized farms with diversified crops and wheat farmers with small acreages. Handling wheat, oats, rye, barley, soybeans and other similar crops was possible with this model. It could combine 10 to 20 acres per day and one man from the tractor seat could handle the entire outfit.

The High-Speed combine would soon change its name to the All-Crop harvester.

Like the Corn Belt combine, the crop was delivered from the 5-foot cutterbar to a 5-foot cylinder in a smooth, thin, even flow with no bunching around corners for complete shelling with minimum power requirements. The large separating and cleaning areas were equal to that in a conventional 10-foot combine to permit faster field operation. It was mounted on rubber tires for lighter draft in the field and higher speeds on the road.

Successor to the Binder

Production skyrocketed to over 7,650 in 1936 when the little orange combine finally received its real name, the All-Crop harvester. With an advertised price of $595 fob factory, it cost little more than a big PTO-driven binder. The production numbers started climbing. The peak year prior to World War II was 1938, when over 15,300 combines were built.

The All-Crop design was dynamic with improvements in every production run. The combination clean grain-unloading elevator was replaced with a separate clean grain elevator and an unloading auger at Serial No. 23801 in 1938. During 1943, at Serial No. 86337, the quick adjustable cylinder speed was added. Now the operator could change cylinder

A price of only $595 was little more than a PTO-driven binder but it could do the entire harvesting job.

The factory built 17,904 and 27,639 All-Crop harvesters in its peak years of 1950 and 1951. Daily pace was as high as 125 units.

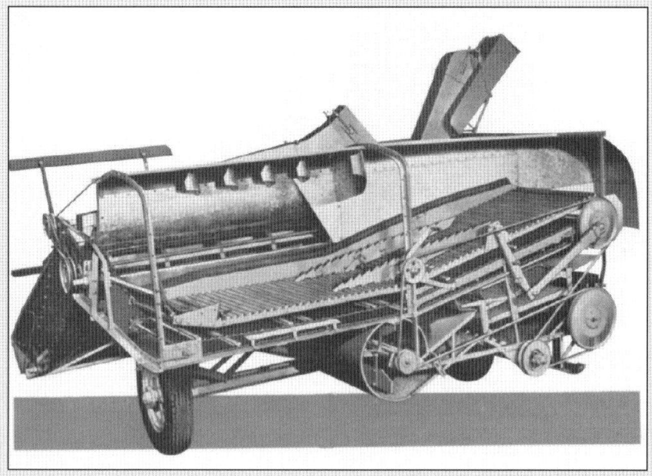

The early All-Crop harvester had a common clean grain and unloading elevator. In the plant it was known as the "No. 10 stove-bolt special" because the separator housing was assembled with stove bolts instead of rivets as later machines were. The simplicity of the All-Crop is evident in this cutaway.

The WC tractor and an All-Crop harvester working in heavy wheat. This machine had no straw spreader, so straw was left in a windrow.

The All-Crop harvester delivered outstanding performance in soybeans because its rubber-on-rubber shelling didn't crack the beans. Effective Serial No. 23801, the All-Crop got a separate clean-grain elevator and unloading auger which was a big improvement.

The 60 All-Crop bagger model was popular wherever clover, alfalfa, and other high-value seed crops were grown. The bagger operator filled and tied bags then slid them down the unloading chute near previously dropped bags for handy pick up. All of the All-Crop harvesters were offered in both bin and bagger models.

The WD tractor and a 66 All-Crop taking a 6-ft swath in small grain. The extra cutting width made the All-Crop more competitive with other small combines. The 66 could be recognized by its flared header, six-bat reel, and unloading auger on the front of the grain bin.

speeds with a crank instead of dismantling the pulleys to change the number of spacers. The original All-Crop was built from 1935 to 1949, with a total of 176,916 units produced. The 60A began rolling off the line in 1949, and 75,000 units were built through 1952, when it was replaced by the B-Series combines.

The B Series started in 1953, at Serial No. B101, with the Model 66. The cutterbar width was increased to 66 inches, and with its 6-inch gather, the 66 could cut 6 feet, which was a needed change. Early production had the same 18-bushel bin as the earlier A-Series. A total of 30,500 Model 66 combines were manufactured with the 18-bushel grain tank. The 66 had a six-bat reel. In late 1954, at Serial No. 30601, a new 25-bushel grain bin was installed and the name changed to Big Bin 66. The unloading auger could be folded for transport or stored on the separator housing, and the control for the bin unloader was extended to the tractor seat so the operator could unload on-the-go. The larger bin was a 39 percent increase. Between 1954 and 1959, a total of 42,755 Big Bin 66's were produced.

In 1959 the Big Bin 66 evolved into the Model 72, effectively replacing both the 66 and the Model 90. An all-new header had a 72-inch cutterbar and cut 78 inches including a 6-inch gather. The canvas conveyors were replaced by an auger with retractable-finger feed to an undershot raddle which fed the cylinder. A six-bat power-driven reel provided aggressive action even when forward speed was reduced for heavy crop conditions. The straw rack had a new step-up design similar to those in the Models 90 and SP 100.

The company claimed the All-Crop could thresh over one hundred crops and it supplied screens and other accessories to accomplish that ambitious objective. Harold Fisher, a veteran of the Harvester Sales Department, would on occasion state that the All-Crop could separate the seeds from pumpkins. Worse yet, he could even say it with a straight face.

The 60, 66, and 72 All-Crop harvesters were built for 25 years, each version better than its predecessor. Slightly fewer than 338,000 units were built. Imagine a field of 1,365 acres covered with All-Crop harvesters parked side-by-side or a chain of combines 1,024 miles long, stretching from LaPorte, Indiana, to Denver, Colorado. At any rate, it was a lot of combines. Somewhere along the line, LaPorte become known as the "Harvester Capitol of the World," a title well earned and most appropriate.

A WD45 and Big Bin 66 gathering a heavy crop of oats. The Big Bin 66 looked like the 66 except that the unloading auger came out the side of the 25-bushel bin.

A D17 diesel puts the Model 72 All-Crop through its paces in Indiana oats. The Model 72 was recognized by its enclosed feeder housing, auger platform, and smaller-diameter six-bat power-driven reel. It was the last of the All-Crop harvesters with production ending in 1969 after nearly 338,000 were built. Must be some kind of record.

This 1946 prototype Model 66H hillside combine was never produced.

The Combines that Didn't Make the Team

When a corporation holds a substantial share of any market, there must be within that organization a constant flow of ideas and new concepts, innovation and improvements, research and development. Certainly such was the case at the LaPorte plant with regard to combines. There seemed to be no shortage of new ideas and innovation.

Many new product concepts never reached the prototype stage for one or more reasons. Many of them did, however, but failed to make the big step into production. There were many reasons why a prototype machine was not placed in production. Perhaps a feasibility study indicated that the machine would never be profitable, regardless of how good it was. Others may not have performed up to expectations and were abandoned for another more promising solution to the problem. Whatever the reason, it is always interesting to see these prototypes and speculate on what might have been.

The Model 40 All-Crop Harvester

By 1938 the 60 All-Crop was a runaway success. So was the Model B tractor, with the Model C waiting in the wings to create the same kind of sales record. A great combine, great tractors, but they weren't quite matched. A Model B or C couldn't really handle a 60 All-Crop unless it was equipped with an auxiliary engine and that ran the price up. What was needed for the small farmer was a small PTO-driven combine, and the 40 All-Crop arrived near the end 1938 to fill that niche. With a 40 All-Crop, the small farmer could enjoy the same independence and freedom from harvesting drudgery as the mid-size farmer with his 60.

Almost perfectly matched to the Model B, the 40 could cut a 45-inch swath with its 40-inch cutterbar and 5-inch gather. The cutterbar fed the crop through a 3-foot cylinder and to a 3-foot rack, straight through. No turns or bottle-necks for the crop. It was a concept that would spread throughout the industry to determine every other small combine's configuration except the master of them all, the 60 All-Crop.

An experimental straight-through variation of the Model 60 compared to a Model 40. Field performance wasn't equal to that of the production Model 60, so it didn't make the team.

With a Model B tractor and a Model 40 All-Crop harvester the small farmer could control his own harvest and forego the drudgery of the threshing ring. With only a 45-in. width-of-cut it was slow but still much faster and easier than the alternative.

The Model 44 was ready to take the Model 40's place in 1946. It looked like a shrunken Model 60. It was not produced — by that time farmers required greater harvesting capacity.

The Model 40 used rubber-on-rubber shelling just like its big brother. Its 2,700 square inches of separating area was 72 percent of the 60. The Model 40 was capable of successfully harvesting over one hundred crops just like the 60.

The "turtle back" styling of the separator housing, formed on automotive body-panel dies, marked a first in the industry. The grain bin held a whopping 11 bushels, so combining a good crop of wheat meant three or four stops to empty the bin for each acre. Daily capacity might reach 10 to 12 acres, not record-breaking but a lot better than following the threshing rig for a month. With a price of only $345, most small farmers could afford one but it would have to be cash — their used binders weren't bringing much in trade.

Two well-restored Model 40 combines were displayed at the "Gathering of the Orange" at Freeport, Ill., in 1993. Like other All-Crop's, the 40 was available as both bagger and grain-bin models and both are shown here. The bagger model was built in very small numbers so this one is a rare example.

From 1938 through 1940 the LaPorte plant built 15,100 Model 40 All-Crop harvesters, 9,000 in 1940 alone. Model 40 production ceased as World War II, with its material shortages, made it more prudent to build the Model 60 with its greater field capacity.

A Bigger All-Crop For Bigger Tractors The Model 90 All-Crop Harvester

As A-C placed larger tractors on the market, it was only natural to match them with a larger All-Crop harvester. The 60 was geared to two-three-plow tractor power, but the WD45 and D17 tractors were easily one or two plows larger and capable of powering more harvesting capacity. LaPorte combine engineering had already hatched the SP100 in 1953 and followed with the Super 100. Using the same technology and many common parts and components, they designed the Model 90 pull-type combine and it went into production in 1957 with a run of 500 units.

The Model 90 could cut a 90-inch swath with an 84-inch cutterbar and 6-inch gather. Later a 24-inch extension was offered to permit harvesting three rows of soybeans with extra header length for down beans. The auger platform brought the crop in to the 42-inch feeder housing and an undershot raddle delivered the crop to the 42-inch wide, 15-inch diameter cylinder. Like the 60, the 90 had rubber-on-rubber shelling with both cylinder bars and concave bars covered with rubber, providing more gentle handling of delicate seed crops and soybeans.

The Model 90 had a grain bypass (called the bustle-back by company people) built into the back wall of the separator, immediately behind the cylinder which would permit threshed grain to travel directly to the grain raddle, bypassing the rack. Straw was retarded

A D17 diesel with a Model 90 All-Crop harvester working in small grain. The 90 cut a 7 1/2-ft swath in grain or took three rows of soybeans when equipped with the optional 2-ft header extension. The two-row cornhead added even more versatility.

by a toothed comb, causing it to fall directly to the straw rack. Later versions of the 90 had an air-blast dome built into the top of the separator housing behind the cylinder to optimize this initial separation. The 38- by 120-inch (4,560 in.^{2}) step-up rack was similar in design and function to the SP100. Total cleaning area of the three sieves was 4,012 square inches. The grain bin held 30 bushels, and could be unloaded in about 90 seconds, either standing or on-the-go.

The Model 90 Peanut Special was a bagger machine. It had a peg-tooth cylinder and concave along with dirt removal panels in strategic locations throughout the separator. A cornhead was available for the Model 90. The design was similar to the one for the Models 60, 66, and 72. However, only 200 were built.

A total of 7,540 Model 90 combines were built from 1957 through 1960. When the Model 72 came out in 1959, most dealers and customers turned to this proven machine that was about $500 lower-priced. The Model 90 market performance could not compare to either the older Big Bin 66 or the new 72, so combine production and the sales effort were concentrated on the Model 72 as a single entry in the pull-type combine segment. A straight-through combine to replace the Model 90 was designed and a prototype built in 1959, but by then it was evident that the pull-type combine was about to be history.

Cornheads for All-Crop Harvesters

Cornheads were available for the 66, 72, and 90 All-Crop harvesters. All were two-row models that shared the snapping rolls and other components with the A-C two-row corn harvester. The 66 cornhead was adaptable to older 60's as far back as Serial No. 86337, which was the start of the quick adjustable cylinder speed. A total of 1,150 cornheads were built: 850, Model 66 cornheads; 100, Model 72 cornheads; and 200, Model 90 cornheads.

The Model 90 had a step-up straw rack and air-blast dome with by-pass separation. The air-blast dome with combs slowed the straw but allowed the grain to enter the by-pass and go directly to the separator raddle.

Cornhead attachments for the Model 60, 66, 72, and 90 All-Crop harvesters brought field shelling to the family farm. However the family farm was turning to self-propelled combines and only 1,150 units were built.

Self-propelled harvesting came to the family farm with the arrival of the SP100 in 1953. With the same crop versatility as the pull-type All-Crop's, it could successfully harvest over 100 different crops and was popular for harvesting large acreages of high-value small-seed grasses and legumes.

The All-Crop Goes Self-Propelled Models SP100 and Super 100

Allis-Chalmers, like Gleaner and others, saw the opportunity in the self-propelled combines of the future and began a development program in 1947 to combine all the desirable characteristics of the All-Crop harvester with the convenience and productivity of a self-propelled unit. This required substantial upsizing of the combine design itself and the design of the propulsion mechanism. Although there were common small parts, there were no major components in common with the pull-type Big Bin 66 then in production.

The SP100 came into the market in the last half of 1953 and about 750 combines were built that year. The header had a 108-inch sickle with a 6-inch gather to cut 114 inches. A 3-foot extension increased the width of cut to 144 inches, but this was intended to be used only in headed grain crops, and would limit performance in down and lodged crops. The auger conveyor delivered the cut crop to a 48-inch conveyor which in turn fed the 48-inch cylinder. True to its ancestry, the SP100 had rubber cylinder bars and concave bars to continue the All-Crop's fine reputation in delicate seed crops. The 38-inch rack was originally 108 inches long (4,104 in.^{2} of separating area), but later specs show the length as 116.5 inches (4,427 in.^{2}). Total cleaning area was 4,012 square inches. By comparison the Gleaner Model A had 6,000 square inches of total separating area and 2,889 square inches of cleaning area.

The grain bin held 35 bushels, but could be increased to 45 with an extension. The grain bin was directly over the steering wheels, which no doubt made the optional power steering most welcome when it eventually became available.

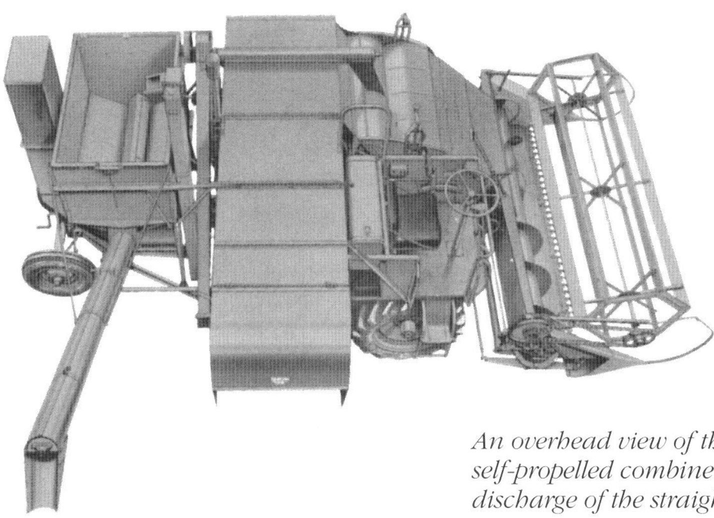

An overhead view of the SP100 shows the unusual layout for a self-propelled combine. It worked but operators preferred the rear discharge of the straight-through machine.

Charley Scranton retired as LaPorte chief engineer in 1957, but stayed on as a consultant through 1959. During that time he designed two straight-through combines. One was this pull-type which might have replaced the Model 90 if the market for pull-type combines remained strong. An unusual feature was the transport arrangement. With the header lifted to full height, the hitch could swing over so that the combine was directly behind the tractor.

The Super 100 gained larger tires, improved steering, retractable-finger auger, improved instrument panel, and more convenient controls. Only 1,000 were built.

The early SP100's were powered by the West Allis four-cylinder 201-cubic inch engine best known in the WD tractor. When the WD45 tractor came out a few months later, the new four-cylinder 226-cubic inch Power-Crater quickly found its way into the SP100. The traction drive was a variable V-belt coupled to a two-forward, one-reverse transmission. The variable V-belt traction drive was manually controlled until an optional hydraulic traction-drive control was offered.

A low speed auxiliary transmission was available to reduce the slowest working speed from about 1.3 mph to about 0.5 mph for use in high-yielding crops or poor operating conditions. In 1957 the Super 100 (Serial No. 5601 and up) replaced the SP100 and introduced many desirable new features. These included larger drive and steering tires for increased

Although A-C already owned Gleaner, some consideration was given to building small self-propelled combines at LaPorte and larger models at Independence. Scranton's 1959 straight-through machine was a candidate if that program had materialized. About the same size as the Super 100, it was called the Model 110. It was powered by the West Allis four-cylinder 226-in.^{3} engine. Further study concluded that self-propelled combine production should be concentrated at Independence.

flotation, folding unloading auger, standard retractable finger auger, higher air intake screen for a new larger capacity radiator, and a long list of product and performance improvements. The steering wheel was changed to an inclined position from the bus-like flat steering wheel of the SP100. An optional rear straw spreader-chopper chopped the straw and spread it behind the combine was available.

Although the Super 100 was offered through 1958, it must have been obvious to A-C management that the high-capacity self-propelled combine of the future would be a straight-through machine, since they had acquired Gleaner-Baldwin in February 1955. Although some consideration was given to building small self-propelled combines in LaPorte, a decision was ultimately reached to concentrate all self-propelled combine production in Independence.

All-Crop Harvester Specifications

	Pull Type						Self-Propelled	
	40	60	66	B.Bin 66	72	90	SP100	Super 100
Years built	1938-40	1935-53	1953-54	1954-59	1959-69	1957-60	1953-57	1957-58
Width cut (in.)	45	66	72	72	78	90-114*	114-150†	114-150†
Cutterbar (in.)	40	60	66	66	72	84-108*	108-144†	108-144†
Platform	Canvas	Canvas	Canvas	Canvas	Auger	Auger	Auger	Auger
Bin cap. (bu.)	11	18	18	25	25	30	35-45	35-45
Cylinder	Rubber bar‡							
Width (in.)	36	60	60	60	60	42	48	48
Diameter	15	15	15	15	15	18	18	18
Separator	Unit rack‡							
Width (in.)	36	32	32	32	32	38	38	38
Length (in.)	75	117	120	120	120	120	116.5	116.5
Sep. area (in.^{2})	2,700	3,744	3,840	3,840	3,840	4,560	4,427	4,427
Cleaning area (in.^{2})	INA§	2,820	2,820	2,820	2,820	4,012	4,012	4,012
Drive	PTO	PTO/AE ǁ	PTO/AE ǁ	PTO/AE ǁ	PTO/AE ǁ	PTO/AE ǁ	---	---
Engine	DNA#	A-C	A-C	A-C	A-C	A-C	A-C	A-C
Cylinders	---	4	4	4	4	4	4	4
Displacement (in.^{3})	---	125	125	149	149	149	201**	226
Traction drive							Variable V-belt	Variable V-belt
Transmission							2 for.,1 rev.	2 for.,1 rev.
Weight (lb)	1,700	2,650	3,350	3,650	4,280	4,653	6,750	7,160

* 24-inch extension for three rows of soybeans.
† Optional 36-inch header extension for short straw and lighter crops.
‡ Applies to all models in a given category.
§ Information not available.
ǁ PTO or auxiliary engine.
Does not apply.
** 1954 and later — 226 in.^{3}

All-Crop Harvester Prices

Year	Combine	List Price ($)
1935	Model 60	595
1947	Model 60	909
1951	Model 60	1,220
1956	66 Big Bin	1,375
1960	66 Big Bin	1,676
1960	Model 72	1,995
1960	Model 90	2,539
1960	90 Peanut special	2,769
1960	66 cornhead	1,592
1960	90 cornhead	1,653

Specifications of the High-Speed Combine

Width of cut (in.)	60
Cylinder width (in.)	60
Cylinder type	8 rubber bars
Concave	Rubber-faced
Rack	Vibrating type
Separating area (in.^{2})	3,830
Cleaning area (in.^{2})	1,980
Cleaning shoe	Two adj. sieves and scalping sieve
Bin capacity (bu)	18
Power required	2 plow tractor
Canvasses	Two, rubber-faced
Range of cut (in.)	2 to 22
Weight (lb)	2,600
Width	10 ft, 8 in.
Height, top of deck	6 ft, 3 in.
top of elevator	8 ft, 4 in.

Serial Numbers of All-Crop Harvester Models 40, 60, 66, 72, 90, SP100, and Super 100

Year	60*	60*	40		
1935	6848-7397	---	---		
1936	7398-15050	---	---		
1937	15051-23800	23829-26695	---		
1938	23801-23825†	26696-42015	101-200		
1939	---	42016-49831	201-6200		
1940	---	49837-58836‡	6201-15200		
1941	---	60105-71391			
1942	---	71392-86336			
1943	91169-95388	86337-91162§			
1944	95389-104449	---	---		
1945	104450-120473	---	---		
1946	120474-128763	---	---		
1947	128764-149588	---	---		
1948	149589-168563	60A	---		
1949	168564-183923	A101-4679	---		
1950	---	A4680-32584	---		
1951	---	A32585-60244	---		
1952	SP100	A60245-75100	66		
1953	SP101-859	---	B101-20379		
1954	---	---	B20380-33419 II		
1955	---	---	B33420-51410		
1956	-5600	---	B51411-59718		90
1957	SP5601-	---	B59719-66686		101-600
1958	-6600	---	B66687-70371	72	601-4411
1959	---	---	B70372-	101-724	4412-7100
1960	---	---	---	725-2001	7101-7640
1961	---	---	---	2002-2932	---
1962	---	---	---	2933-4632	---
1963	---	---	---	4633-5887	---
1964	---	---	---	5888-7011	---
1965	---	---	---	7012-7511	---
1966	---	---	---	7512-	---
1967	---	---	---	----	---
1968	---	---	---	9001-	---
1969	---	---	---	9093-	---

* Model 60 serial numbers not used: 41829-41831, 49832-49836, 91163-91168.
† Serial No. 23801 — Start of separate clean grain elevator and unloading auger.
‡ Serial No. 58587-58836 — Includes 250 Model 60 Hillside combines.
§ Serial No. 86337 — Start of quick adjustable cylinder speed.
II Serial No. 30601 — Start of Big Bin 66 (25 bushel bin).

Corn, Cotton, and Special Harvesters

The two-row corn harvester would fit the WC, WD, and WD45 tractors. Total production was 43,551 units.

Where Corn was King — Corn Harvesters

Although by the late 1930s Allis Chalmers was increasingly well-represented all over the United States and Canada, it was primarily a Corn Belt company. That's where most of the dealers were, where its equipment best fit the agriculture, and where it had enjoyed most of its success. The All-Crop and WC had convinced many skeptical farmers that the company could meet most of their needs as a reliable machinery supplier. Just one significant limitation — A-C could not harvest its customers' most important crop, corn.

This was not much of a problem for Corn Belt dealers. A substantial number of them had the contract to sell the well-known and respected New Idea pull-type corn picker, but that did nothing for the company itself. The sales potential for the right corn harvester in corn country was no doubt even greater than that of the combine and A-C needed a share of that market to be recognized as a full-line farm equipment company. The company had learned some important facts with the successful WC tractor and All-Crop harvester — such as keep it simple, lightweight, low-cost, and geared to the family farm. Adherence to that philosophy was evident in the two new corn harvesters which were rolled out at LaPorte in June 1940. Both were mounted units, a two-row for the WC and a one-row for the Model B. The former was a keeper, the latter wasn't.

The concept of combining the husking rolls with the snapping rolls eliminated hundreds of parts, as did the mounted design which dispensed with a large heavy frame, wheels, and a long PTO line. That took care of simplicity, weight, and cost, and made it a natural for the WC owner on a family farm.

Prototypes of the WC two-row and B one-row mounted corn harvesters at LaPorte in June 1940. The two-row went into production in 1941 and the one-row was abandoned.

Several unusual design elements placed the two-row corn harvester in a category by itself. First, it was suspended below the rear axle of the tractor so that the operator was not completely surrounded by noise and moving parts. Mounting was also quite easy once the tractor was prepared with proper wheel tread setting and various mounting brackets. The corn-picking components were just as simple. A free-wheeling lifter-spider wheel on each side of the row helped guide the cornstalks into the snapping rolls. A single gathering chain for each row worked primarily because the snapping rolls were close to the ground and relatively flat. Stripper plates kept the butts of the ears out of the snapping rolls and prevented shelling.

For husking, one roll was all rubber while the other was a steel roll fitted with rubber plugs and steel husking pegs. The writer knows from several years of personal experience that the system was much more effective when the cornstalks were slightly damp, and could lead to great frustration when the fodder was extremely dry and brittle. In those dry years the corn harvester was prone to put fodder in the wagon, something a cattle feeder might tolerate but was not acceptable to hog farmers. To help remove loose trash the harvester had a fan, driven by the tractor's belt pulley, over each husking roll assembly. A pair of stalk-ejector rolls at the rear of each row helped get rid of stalks that didn't go through the snapping rolls.

The A-C sweet corn harvester was a version of the two-row mounted corn harvester with special adaptations for that crop. The most obvious was the topper which was a pair of rotary cutters that topped the stalks to get rid of as much fodder as possible. These were driven from the tractor belt pulley, as were the fans for the picker. Another major modification was in the snapping rolls, which had sharpened knives to cut the cornstalk with its ear rather than snapping the ear. This prevented damaging the ear since the stalk was cut off before the ear could reach the snapping rolls. Sweet corn harvester production reached 409 units, but many were converted to standard units at the factory.

The two-row mounted corn harvester went into production in 1941 with 1,000 units coming off the line. None were built, however, in 1942. Production resumed in 1943, but was again interrupted in 1946. All-out production resumed in 1947 with 5,000 harvesters, followed by 5,500, 7,800, 6,000, 5,000, 3,000, and 3,000 in the post-war years of 1948 through 1953.

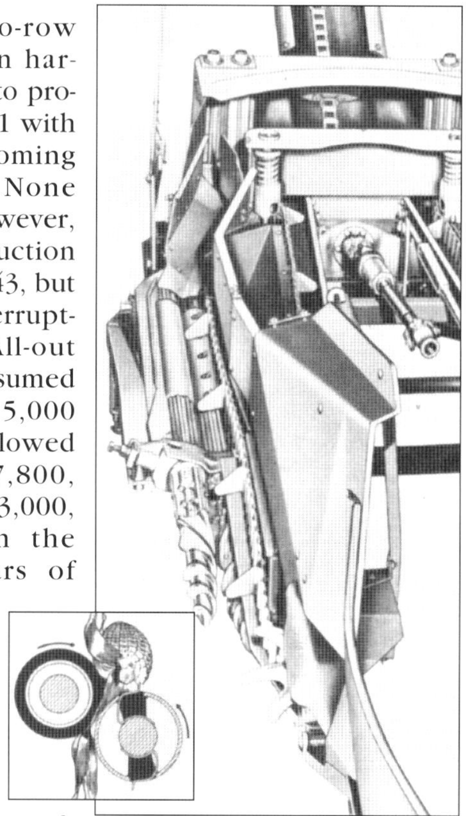

The combined snapping and husking rolls reduced the number of parts and the low snapping-roll angle helped pick-up down corn.

Including sweet corn harvesters, nearly 44,000 units were built in 11 years of production. Really quite impressive numbers for a unit that could be mounted on only two tractors. It certainly was not in the All-Crop league, but was good for its limited potential. The two-row harvester was priced at $745 in 1947 and rose to $982 in 1951.

The sweet corn harvester had stalk-toppers and special knives on the snapping rolls to cut stalks off before the ears entered the snapping rolls. Only 409 units were built and many of them were converted to conventional pickers.

The demise of the Model B one-row didn't stop work on a single-row machine. A one-row mounted picker was built for the Model C, but it was not produced either. The solution was a one-row pull-type picker adaptable to any small tractor. It was essentially one-half of the two-row with wheels, hitch, and PTO shaft. The 1951 price was $786.

Some important improvements appeared on the Model 33 two-row corn harvester in 1955 with the addition of rotary feeders and quick-adjustable stripper plates. A four-bladed rotary feeder over the husking section helped move corn through the machine and forced trash to pass through the husking rolls. The quick-adjustable stripper plates could be moved in or out

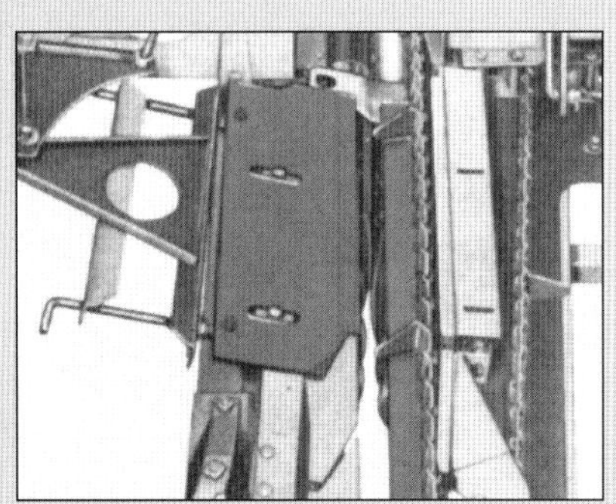

A one-row mounted picker for the Model C seemed like good idea for the small farmer but it was never produced.

The Models 33 two-row and 35 one-row pickers had rubber-on-rubber husking, rotary feeder (right) and on-the-go adjustable stripper plates (above) for a significant improvement in field performance.

The third attempt at developing a one-row picker produced the one-row pull-type picker. No doubt it was the best bet because it worked with any tractor.

with a lever near the operator, and allowed matching openings to ear size and stalk conditions. A-C's experience with rubber-on-rubber threshing in the All-Crop no doubt prompted the use of rubber-on-rubber husking. Twenty-six "sure-grip" rubber husking pegs per row meshed with matching indentions in the opposite rubber roll for better cleaning of the ears.

The No. 33 sold for $1,265 in 1956.

These features made a dramatic improvement in the picker's performance, especially in dry fodder conditions. By 1956 these improvements were applied to the one-row pull unit in the Model 35. At the same time the 34 two-row and 36 one-row snappers were added to the line. Snappers had smooth rolls in place of the rubber-on-rubber huskers, but otherwise were exactly the same as the 33 and 35. The 35 one-row was priced at $860 in 1956.

The No. 17 (shown) and the No. 14 pickers were modified No. 33 designs for the D17, D14, and D15 tractors.

When the D14 and D17 tractors came down the assembly line in 1957, both were in need of a corn harvester. The Nos. 14 and 17 became available in 1958 to satisfy this requirement. They were adaptations of the No. 33, changed to mount on the new models. The most obvious change was a single fan on the elevator replacing the twin fans over the husking rolls of the 33. Snapper versions of the 14 and 17 pickers were not available. All six

The 190 two-row mounted picker was a much-upgraded version of the 170 picker and could be mounted on the D19, D17, WD45, and WD tractors. It was the last picker in the A-C line.

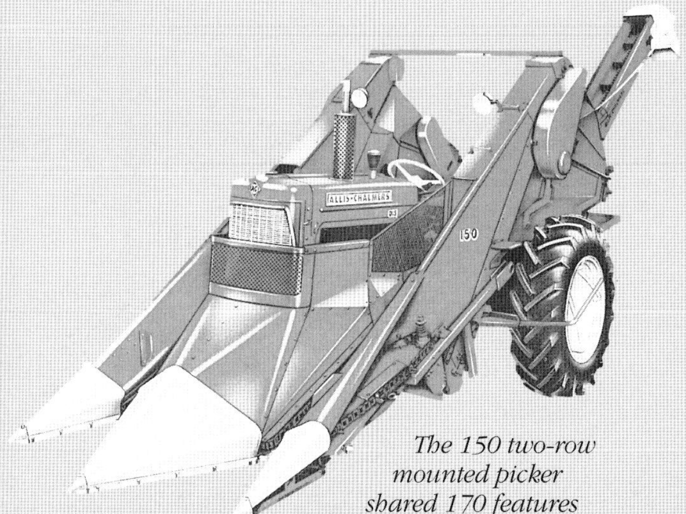

The 150 two-row mounted picker shared 170 features and fit the D15 Series II, D15, and D14 tractors.

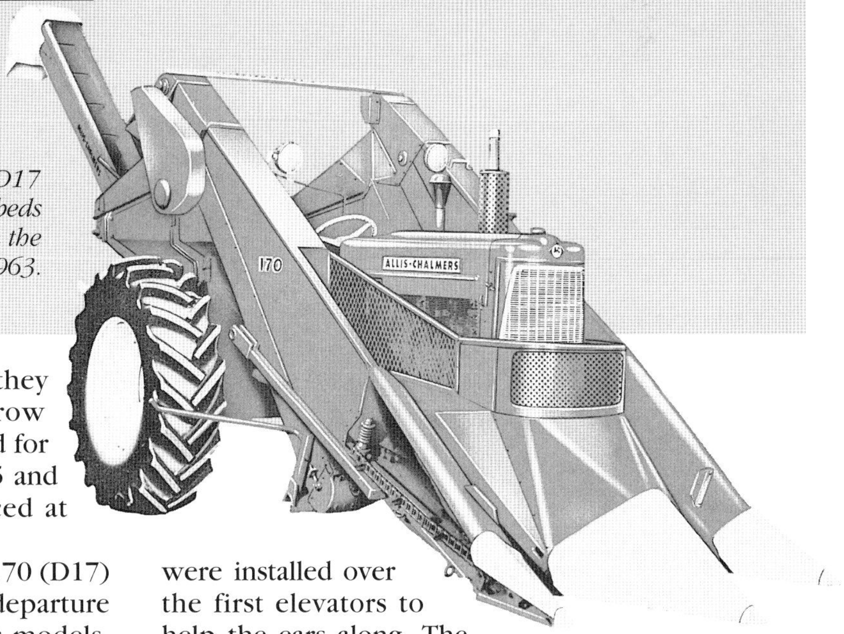

The 170 two-row mounted picker for the D17 was an all-new design with full husking beds and long snapping rolls. One of them won the National Corn Picking contest in 1963.

models were continued until 1961 when they were replaced by the 150 and 170 two-row mounted pickers. The Models 14 and 17 sold for $1,775, while the 33 and 34 listed at $1,475 and the 35 and 36 one-row pickers were priced at $1,022.

The new 150 (for the D14 and D15) and 170 (D17) two-row mounted pickers were a radical departure from A-C's design philosophy of the prior models. From gathering snouts to the end of the wagon elevator, the new pickers had the increased capacity necessary to harvest high-yielding new hybrids that were revolutionizing the corn culture. More aggressive snapping rolls with replaceable hardened blades handled high populations of tough stalks. Two gathering chains per row improved feeding and delivery to the first elevator, which carried ears to husking beds with four rubber-on-rubber rolls each. Floating conveyors over each husking bed assured that ears kept moving. Shelled corn savers under each husking bed put shelled corn in the wagon, not on the ground. A pair of stalk-ejector rolls at the top of the first elevator ejected any broken stalks and trash that made it that far. A large fan with dual outlets over the elevator hopper blew loose trash out. In 1962, rotary feeders were installed over the first elevators to help the ears along. The performance of these corn pickers was proven when Arlyn Zee, of Albany, Wisconsin, won the 1963 National Corn-Picking Contest with his 170 picker in some of the driest conditions ever seen in a national competition.

With the introduction of the D19, the new corn picker design was adapted to it with the Model 190 picker. Besides the rotary feeders, the 190 gained a third gathering chain on the outside of each row. A larger elevator hopper and relocated cleaning fan completed the upgrading. Later snapping rolls were redesigned to provide 30 percent more snapping action and dual rotary feeders over the first elevator helped reduce bunching and plugging there.

The 190 picker was adapted to the D19, D17, WD45, and WD tractors. Production continued through 1966 when the line was dropped. The 190 was never adapted to field shelling, unlike many other mounted pickers. Narrow rows and field shelling doomed mounted corn pickers to the salvage yards — seed corn production being one exception. The Gleaner combine with its high-capacity cornhead would be A-C's answer to field shelling. Records show that was the right answer.

The Flight-Line Elevator

With balers, combines, and corn pickers, Allis-Chalmers had the equipment the farmers required to get their crops out of the field. All that was needed to put them into storage and complete the system was a good portable elevator. The company filled this gap in the line by purchasing 500 elevators from the Mt. Hawley Company in Peoria, Illinois, in 1955. These were 17-inch units available in lengths from 26 to 50 feet. The sheet metal was galvanized for a long, non-rusting life. Engine, electric motor, and tractor power take-off drives were available.

In 1957 the Independence plant manufactured 500 elevators, similar to the earlier version except for width, which was expanded from 17 to 20 1/2 inches. Available lengths remained 26 to 50 feet, and a full line of drives, feeders, and downspouts were offered. No more were built after the 1957 production run. The flight-line elevators remained in the line until inventory was exhausted. They were listed as a current item in the 1960 price list. A 50-foot elevator with an 11-foot tilting feeder, downspout, and PTO drive had a 1960 list price of $922.

One batch of 500 of the 17-in. Flightline elevators was purchased from the Mt. Hawley Co. of Peoria, Ill., in 1955.

The 20 1/2-in. Flightline elevator was built at the Independence plant in 1957. Only one run of 500 was manufactured.

Cotton Pickers, Strippers, and Harvesters

One of agriculture's most labor-intensive jobs went mechanical in the late 1940s. As the labor pool that hand-picked cotton began to desert the fields for the industrialized cities during World War II, there was a loud call from cotton producers for help in harvesting their cotton. The engineers of several farm equipment companies responded with machines that could mimic human fingers well enough to pick the ripe, opened cotton bolls and put them in a basket, while leaving the green, unopened bolls for a later harvest. The first mechanical picker must have been an incredible sight for its observers. Considering the complexity of the job to be done, any success at all was a major victory for agricultural engineering. It wasn't easy and it wasn't cheap, but cotton could be picked by machine with one operator and his two-row picker replacing dozens of hand laborers.

Allis-Chalmers began development of the cotton picker at LaPorte in 1944. Cotton picker activity was transferred to the newly acquired Gadsden, Alabama, plant in 1948 and production began the following year with 25 units. Cotton pickers were based on two overall concepts, the drum-type and unit-type. The drum-type picker had vertical rows of spindles around the circumference of a single-axis drum, with a drum on each side of each row. The unit-type had many vertical rows of spindles on a conveyor arrange-

Experimental cotton pickers at work in Mississippi in 1947. Development began in 1944 and first production followed in 1949.

ment which travelled around two axes. The Allis-Chalmers pickers were the unit design. Both types utilized rotating spindles to grasp the cotton fibers. Of course cotton picker sales people spent considerable time discussing the merits and short-comings of the two concepts, each proclaiming superiority for his brand in vociferous terms and with great loyalty. Let it be said that both, when properly maintained and operated, could pick cotton pretty well.

Each type had an enormous number of moving parts, primarily because of the number of picking spindles and their driving components. The unit-type pickers had far more spindles than the drum-type, therefore they were probably somewhat more difficult to maintain. This was offset by the fact that the greater number of spindles allowed each spindle to remain in contact with the plant for a longer period of time, thus increasing its chances of grasping the fibers and picking a boll.

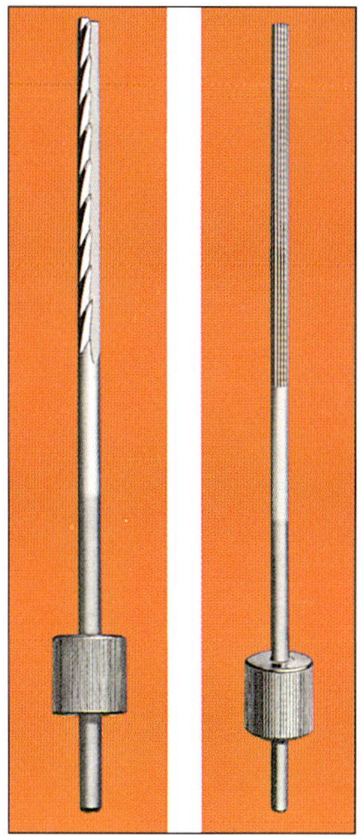

The Tupelo recessed-barb spindle (left) was more aggressive for clean picking. The original A-C knurled spindle picked cotton without "picker twist."

The standard unit machines had 1,280 spindles per row, 16 in each of 80 slats, 2,560 in a two-row picker. The high unit (1963 and later) for taller, higher-yielding cotton, had 1,760 spindles per row, 22 in each of 80 slats, 3,520 in a two-row picker. Two types of spindles were available on a picker. The original A-C design was a cadmium-plated knurled spindle which could be turned either direction. The later recessed-barb chrome-plated type, also known as the Tupelo spindle, was made in right- and left-turning versions. A later version of the Tupelo design had both right and left barbs cut into the same spindle so it could be used in either position. Later pickers used only the Tupelo spindles, which were more aggressive and more likely to grab bolls. Both were 8 inches long and had a knurled traction roller near the end. They were supported in two oil-impregnated bronze bearings. No other lubrication was used near the picker spindles because any oil on the fibers would substantially reduce the grade at the cotton gin.

The picking cycle began when the spindle was moistened with water from an on-board supply tank. The moisture helped the spindle grasp the cotton fiber. Each slat of 16 spindles (22 on a high unit) was

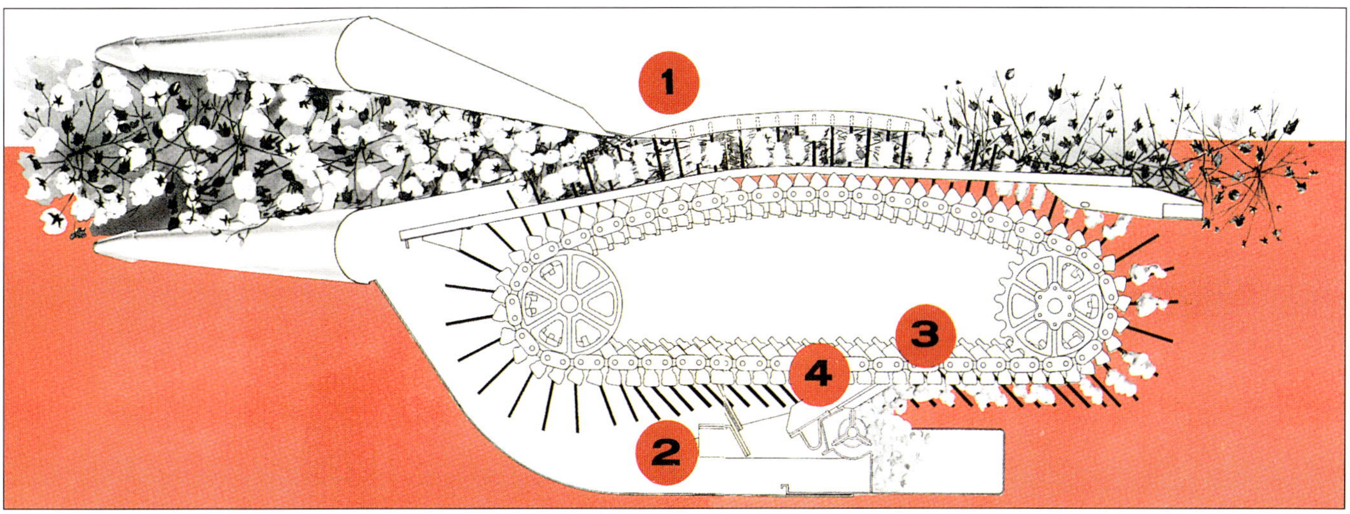

The cotton-picking cycle: 1. Plants are crowded into the picker unit's narrow throat. 2. Spindles are moistened with water, then revolve around on the carrier to enter the crop. Once in the crop they rotate to grasp and wind the fibers, pulling them from the boll. 3. The cotton is doffed from the spindles. 4. The rotating doffers complete the removal of the cotton from the spindles. A flow of air then delivers cotton to the basket.

The first production two-row picker with the basket raised for unloading. Hydraulics raised the left side of the basket, then the bottom conveyor elevated the cotton to the wagon.

The one-row cotton picker mounted on a Model CA brought mechanical picking to the smallest farm. It would also mount on WD and WD45 tractors. The picker had 1,024 spindles and the basket carried 750 lb (one-half bale) of seed cotton.

The 1955 two-row picker was mounted on a WD45 tractor. It had air delivery to the basket.

carried around to a point where the spindles could enter the crop, which had been compressed into a narrow width by the crowders in the inside wall of the unit. The rearward travel of the spindle was synchronized with the forward motion of the picker so that there was no relative movement between the spindle and the plant once the spindle had entered the plant. At that time the roller on the spindle contacted a drive track, causing it to rotate and grasp the fibers of a boll.

As the slat rounded the main unit drive sprocket, the spindles were removed from the plant zone and stopped rotating. They then entered the stripping area where they passed between stationary adjustable strippers which slid the bolls off the spindles. A three-blade rubber beater wheel aided the stripping action. The bolls then fell into a warm jet air stream which carried them to the basket. The thousands of steel fingers then went tirelessly back to work to gather more cotton.

The first production A-C cotton picker was a two-row model built on a modified Model WD tractor which ran backward. The picker could be removed and converted back to a tractor with a special package. The first 25 machines, built in 1949, were followed by 100 in 1950. By 1952, production reached over 1,200 machines.

The one-row picker which appeared in 1952 could be mounted on WD, CA, and later, WD45 tractors with a single-front wheel or dual-front end. Picking speed was the tractor's first gear, which was 2.0 to 2.4 mph depending on the model. The early one-row pickers had 1,024 spindles, 16 spindles in each of 64 slats. Later one-row pickers had 1,280 spindles, 16 spindles on 80 slats. The cotton was blown to the basket, which held about one-half bale of seed cotton. A basket extension increased capacity by 25 percent. The basket was dumped hydraulically into the waiting wagon. Production quantities were: 1952, 40; 1953, 500; and 1954, 750. A one-row picker in 1956 sold for $2,995.

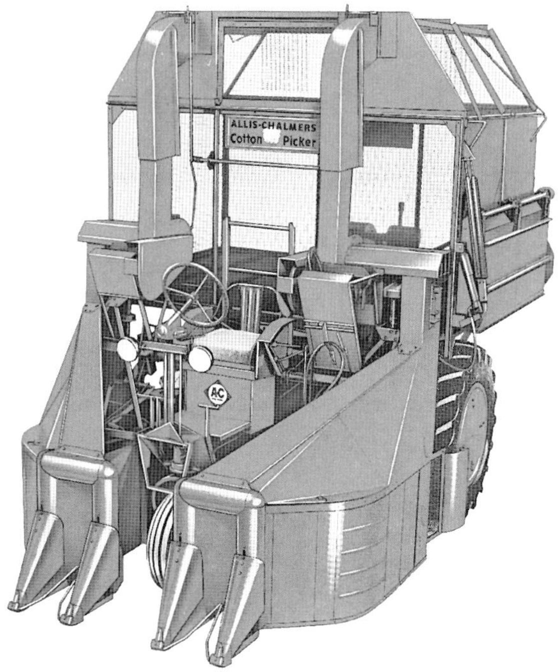

The 1959 Model 500 two-row picker was mounted on a D17 gas or diesel tractor which could be converted back to a standard tractor for the rest of the year. The Tupelo spindle improved the 500's field performance.

In 1955 the two-row picker had air delivery to the basket, with a fan on each side of the operator. By now the two-row was powered by the WD45 tractor with its 226-cubic inch Power-Crater engine. Reduced two-row picker production in 1955 limited the number built on the WD45. Picking speeds were about 2.2 and 3.4 mph, or first and second gear. The 188.6-cubic foot basket held about 1,200 pounds of seed cotton and was emptied by hydraulically lifting the left side to wagon height, then the basket bottom conveyor transferred the cotton to the wagon.

The 500 Two-Row Cotton Picker

The 1958 two-row picker was built on the D17 gas tractor. In 1959 the 500 two-row picker with new shielding over the units was announced. A total of 80 gas and 20 diesels were built that year. Production climbed to 210 gasoline and 65 diesels in 1960. The 500 had a choice of A-C knurled or Tupelo recessed-barb spindles, but the A-C spindles were soon displaced by the Tupelo spindle's superior performance. Basket capacity was increased to 259.6 feet and carried up to 1,400 pounds of seed cotton. New front shielding kept trash off the units and rear shielding protected the engine from tall stalks. The 1960 list price of a Model 500 two-row picker was $13,488. The optional diesel added $895, and an adjustable front axle tractor conversion sold for $710.

The 616 and 622 Pickers

As taller, higher-yielding cotton was grown on irrigated land, the 500 with 16 spindle height (1,280 total spindles per row) was not tall enough. The 22-spindle high unit (1,760 spindles per row) came out in 1960 and 1961 in the Model 622 two-row picker, along with the regular 16-spindle low-unit Model 616. The 616 had 2,560 spindles, while the 622 boasted 3,520 spindles. Both models were built around the D17 gas and diesel tractors. Field experience in heavy irrigated cotton soon revealed that the 622's needed more power, so the 262-cubic inch gas and 262-cubic inch turbocharged diesel engines (D19 power) were made available for the larger picker. Converting these pickers to tractors was not recommended, since that would mean putting D19 power through the D17 power train — most likely an invitation to disaster if the tractor were also overweighted.

The 616 and 622 pickers had larger 390-cubic foot baskets with hydraulic dumping capable of clearing a 10-foot cotton wagon box. Picking speeds were 1.96 mph in first-gear low range and 2.78 mph in first-gear high range. Changing from low- to high-range picking required changing the picker unit drive from a 15-tooth to a 21-tooth sprocket to synchronize unit speed with the increased ground speed. The 622 high-unit two-row with a gasoline engine had a list price of $14,880.

The 616 low-unit and 622 high-unit two-row pickers had hydraulically-dumped baskets which sped up the unloading process. This basket cleared the side of a 10-ft wagon.

This 622 and the 616 two-row pickers were mounted on D17 tractors, with the 262-in.^{3} gas and turbocharged diesel engines optional on the 622. Converting pickers with these optional engines back to tractors was not recommended.

The 116 and 122 One-Row Pickers

A pilot run of the 116 low-unit and 122 high-unit mounted one-row pickers occurred in 1962. While the earlier one-row picker had 64 slats and 1,024 spindles, the new one-row machines were both 80-slat units with 1,280 spindles in the 116 and 1,760 in the 122. Both were mounted on single-front wheel D17 tractors. The operator's station was mounted above the tractor's hood and the high-dumping basket was located over the rear axle of the tractor. The basket held 1,200 to 1,400 pounds of seed cotton. Two experimental one-row pickers were built on D19 tractors in 1963 and 1964, but there was no production. The 116 and 122 mounted pickers remained in production while the 616 and 622 two-row pickers were replaced by the new Models 801 and 802 in 1966.

The 116 and 122 one-row pickers were mounted on a D17 which ran forward. The operator had a lofty perch where he could see everything.

The 801 and 802 Cotton Pickers

The Models 801 low-unit and 802 high-unit Sample Guardian picker mechanisms were essentially the same as the 616 and 622. The 801 and 802 pickers were equipped with the 262-cubic inch gas and 262-cubic inch turbocharged diesel engines as standard equipment. Although they were mounted on a D17 chassis like prior models, they could not be converted back to tractors. The radiator was moved to a cleaner location, just in back of the operator and there were two fans, one for the radiator and the other in the normal position to deliver a flow of air over the engine. Most other specifications remained unchanged. The 801 and 802 were the last of the A-C cotton pickers and were phased out as the self-propelled stripper-type cotton harvesters, the 707 and 707XTB, took over.

The Models 801 and 802 two-row pickers were mounted on a highly-modified D17 chassis with a 262-in.^{3} gas or turbocharged diesel engine. It could not be converted to a tractor because radiator was located behind the operator, not in the normal tractor location. Picker components were similar to the 611 and 622.

Tractor-Mounted Cotton Strippers

The big two-row cotton pickers and their one-row sidekicks were well adapted to rank-growing high-yielding delta and southwestern cotton. In the high plains country, they were too big, too expensive, and far more machine than needed to harvest the storm-proof cotton grown there. High plains cotton was harvested with a simple, usually tractor-mounted but sometimes pull type, machine that took all the bolls at one pass.

A-C dealers needed a cotton harvester to get their share of the equipment market, but the existing pickers were not the answer. The Central Texas Iron Works of Waco, Texas, was building a two-row mounted stripper which would fit the D17 tractor. Rather than design and tool a stripper, the company elected to market the Central Texas unit.

The principle of the stripper was extremely simple; two long smooth metal rolls spaced close together and rotating away from each other stripped the bolls off the stalk as the outfit passed along the rows. Immediately under the smooth rolls was another pair of rubber-covered stalk-ejector rolls turning toward each other to pull the stalks through the stripper rolls. The cotton fell into augers turning in slotted conveyor troughs on each side of the stripper rolls, and was elevated to a pair of fans for delivery into a trailed wagon. The slotted auger troughs allowed much of the dirt and trash to fall to the ground. Green bolls were caught by boll boxes for later ginning when they ripened and opened.

The A-C cotton stripper was announced in early 1959. The first model was even more simple than that described above. It didn't have the stalk ejector rolls, boll boxes, or twin delivery fans. Cotton was carried to the wagon by a combination conveyor-fan system. Without the stalk ejector rolls, stalks with shallow roots or root rot would pull out of the soil and plug the machine.

In 1960 the new Model 2S stripper replaced the earlier model, and had the stalk ejection system which improved performance. Stalks were pulled through the stripper rolls before they were uprooted from the soil, so plugging was reduced if not eliminated. The twin fans were screened so that they provided a flow of air for delivery but cotton could not come in contact with the blades. The 2S stripper was easily mounted on WD, WD45, D17, and D19 tractors. Tractor lift arms carried the rear of the machine and lift cables supported the front. Fenders, which had to be removed on the early version, were now left on the tractor for operator safety. Wheel shields for use in bushy cotton were optional.

The early cotton stripper had a list price of $1,542 with optional wheel shields and was available only to the Amarillo, Dallas, and Oklahoma City branches. When the 2S came out in 1960, the price went up slightly to $1,567. Although the 2S cotton stripper filled an important market niche, it proved to be only an interim offering.

The 2S cotton stripper was a low-cost way to harvest high-plains storm-proof cotton. It mounted on D17, WD, and WD45 tractors.

Self-Propelled Cotton Harvesters The 707, 707XTB, and 760XTB

As new varieties of storm-proof cotton were developed, the need for improved strippers became apparent. In response to this change in the market, LaPorte engineering designed a pair of self-propelled strippers new from the ground up. Instead of relying on a tractor for the basis, these machines were built more like a combine with a separate frame which tied all components together into a complete unit. If fact, most drive-train components were common with the Models K and F combines.

Two models, the 707 and the 707XTB, were produced. The only difference was that the XTB was equipped with a burr-extractor. Both units were powered by the ubiquitous West Allis four-cylinder 226-cubic inch engine that developed 62.5 horsepower. A variable V-belt drove a three-speed transmission to provide infinite speed selection from 0.7 to 13.0 mph. Hydrostatic power steering was standard equipment.

The cotton, with the 707, was stripped from the stalks by brush-type rolls, with each roll having three rows of high-tensile nylon brushes rotating away from each other. Cotton was conveyed by augers from the stripping area to a cross auger which fed an undershot elevator. This in turn carried the cotton to twin fans which blew it into the 390-cubic foot basket. Green bolls were sidetracked to a 10- cubic foot boll box for later retrieval. All auger troughs and elevator bottoms were slotted to remove trash.

The 707XTB stripped the cotton in the same manner as the 707, but was also equipped with a Continental Gin burr extractor, which worked much like a small cotton gin to remove as much as two-thirds of the burrs and trash that remained after pre-cleaning. The first stage was air separation which carried open bolls to the 36-inch extractor, but allowed the heavier green unopened bolls to fall to the boll box. Open bolls were blown into the wad breaker which loosened burrs and distributed the material more evenly. The cotton was then blown down onto the top saw drum. An adjustable flow baffle set the material to the saw teeth on the drum. The saw teeth caught the fiber and pulled it under a series of adjustable-clearance grid-bars. The grid-bars knocked loose the burrs and trash which were slung off and onto the bottom saw drum. Any remaining cotton

The 707 self-propelled cotton stripper was built on a combine chassis and powered by the West Allis four-cylinder 226-in.^{3} engine. It had a variable V-belt drive for a wide range of stripping speeds.

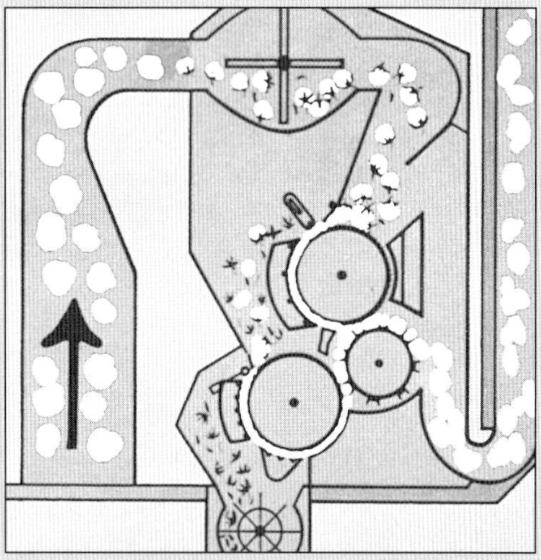

The Continental Gin burr extractor had a system of toothed saws which grasped the cotton fibers to pull them from the trash and burrs which dropped to the ground. This improved the cotton grade and cut transportation costs by reducing trips to the gin.

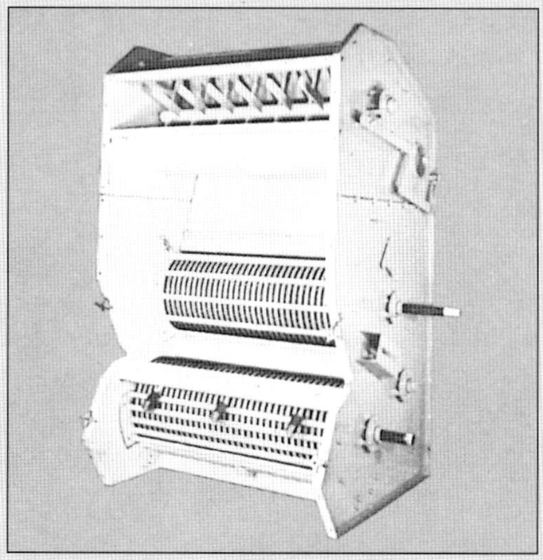

The burr extractor was essentially a small cotton gin.

The 707XTB was similar to the 707 except that it had a Continental Gin burr-extractor which removed much of the burrs and trash and produced a higher grade of cotton at the gin.

The 760XTB was similar to the 707XTB except that it had a Chevrolet six-cylinder 250-in.^{3} gas engine and optional cab.

was removed from the trash by the reclaiming saw. A counter-rotating doffer swept the cotton from the saw drums and back into the airstream for delivery to the basket. Burrs and trash were dropped to the ground.

The advantages of the burr extractor were cleaner lint with far more trash removed, fewer trips to the gin, and reduced ginning costs. The burr extractor set the Allis-Chalmers cotton stripper apart from the rest and it quickly acquired a good reputation in the marketplace. The XTB was so well accepted that the standard 707 was soon discontinued.

The 760XTB brought two significant changes in 1970. First the Chevrolet 250-cubic inch six-cylinder engine replaced the venerable West Allis four-cylinder 226 cubic inch, just as it had on the Model K Gleaner. The Chevy engine was rated at 78 horsepower. Only the XTB burr extractor version was offered and a cab was optional. The 520-cubic foot basket could clear an 11-foot trailer when dumping. The second big change was a new multi-row stripper head for narrow row cotton. The multi-row head was 154 inches wide, but was soon joined by a 76-inch size. Flexible V-shaped fingers set very close together combed the bolls off the stalks, and an adjustable reel moved the cotton back to the cross auger. A serrated stalk-walker below the fingers pulled stalks through the fingers for positive stripping and non-stop operation.

The multi-row head enabled the cotton strippers to harvest narrow-row cotton. It came in 154- and 76-in. widths. An 84-in. version came later.

The 860XTB and 860 Cotton Harvesters

The 760XTB was followed by the 860XTB in early 1973. Functional components including the burr extractor were unchanged from the 760XTB, except that the cab was now standard equipment. The stripper cab was an adaptation of the Gleaner combine cab with pressurized ventilation and heater. A large curved and tinted windshield provided good visibility of the header and hydrostatic steering and an adjustable steering wheel was standard equipment. The 860XTB had the same Chevrolet 250-cubic inch gas engine and power train as the 760XTB. The 76-inch and 154-inch multi-row heads were used.

In the 860 cotton harvester the Chevy six was replaced by the Allis-Chalmers 2800, 301-cubic inch direct-injection naturally-aspirated engine at 84 horsepower. The variable V-belt drive coupled with a three-speed transmission provided stripping speeds from 0.9 to 6.0 mph and road speeds up to 16 mph.

A 12-inch extension was available to increase basket capacity about 17 percent to 610 cubic feet. The 154-inch multi-row was the only header available for the 860 and automatic header-height control was optional. It improved efficiency and reduced operator fatigue.

The 880 — Last of a Long Line

As cotton culture changed to incorporate skip-row planting, the Model 880 cotton harvester was introduced in 1980 to handle the special harvesting requirements. A new four-row brush head could be configured to accommodate eight different combinations of narrow-row and skip-row planting patterns. The 154-inch multi-row (or broadcast) head was still offered and electro-hydraulic tilt control for more precise header operation was optional.

The 880 was powered by an A-C 433I four-cylinder turbo-intercooled diesel of 200 cubic inches with 95 horsepower. The same engine was in the 6080 tractor. A new hydrostatic drive with three-speed transmission had speeds of 0 to 17.6 mph. The 880 had the proven Continental Gin burr extractor and green-boll separator to maintain the fine reputation of A-C cotton harvesters.

The Model 860 had the A-C 301 in six-cylinder diesel engine. Since all self-propelled strippers had the burr extractor at this point, the XTB suffix was dropped.

The cab and Chevy six were standard equipment on the 860XTB. Only the burr-extractor model was offered.

The 880 harvester could harvest narrow-row cotton with the multi-row head or two-and-one skip-row and four-row solid-planted cotton with the four-row brush head.

This high-dumping 880 cotton harvester was equipped to harvest two-and-one skip-row cotton. The 880 had the A-C 433I 200 in four-cylinder diesel engine.

The 300 and 400 lifter-loaders lifted two, three, or four rows of beets and elevated them into a truck running alongside. Over 20-ft² of renk bed area separated the beets from the dirt and trash.

The 880 had a base price of $51,110. Completely equipped with a multi-row head, the price came to $65,242, but with the four-row brush head, the price rose to $70,989. The 880 was the last in the line as production of all cotton harvesters was phased out at the end of 1982.

Man Does Not Live by Bread Alone — Sugar Beet and Potato Equipment

During the mid-1960s when Allis-Chalmers had determined it would be a full-line farm equipment company, they acquired a line of sugar beet harvesters and potato harvesters, along with their complimentary machines. The line was manufactured by the Oppel Harvester Company, of Boise, Idaho. Oppel was well established in these specialty machines and enjoyed a reputation of building high quality products. A-C gained a proven line of machinery to sell and Oppel gained the distribution through the Allis-Chalmers dealer organization in areas where they had not been represented.

Sugar Beet Harvesters

Beet harvesters included the Models 300 and 400 lifter-loaders, the Big Husky 424 and 430 harvesters, and the Models 150, 250, and 350 tank-type harvesters. Depending on row widths, the 300 lifter-loader could handle two or three rows, and the 400 could dig three or four rows. After lifting, the beets passed over a renk bed to remove some of the dirt, then the beets were elevated directly into a truck. Power take-off horsepower requirements were approximately 40 and 50, respectively.

The Big Husky 424 beet harvester handled three 20- to 36-inch rows or up to four 24-inch rows. The 430 dug three 20- to 40-inch rows or up to four 30-inch rows. Although the Big Huskys loaded trucks on the go, they had a 1,000-pound storage capacity so stopping at row ends to empty the machine was not necessary. Hydraulic row finders kept the harvester on the row to reduce beet damage. Capacity was up to 2 tons per minute.

The 150, 250, and 350 beet harvesters were the top of the line with automatic row finders, a wide rear elevator for greater cleaning capacity, and high-capacity storage tanks for non-stop harvesting. The lifter wheels were 28 inches in diameter and were offered in three styles to match soil conditions.

Power toppers were available to eliminate the need for a separate topping operation. Rubber flails, running 400 rpm, knocked the beets from the lifter wheels into the renk bed for cleaning. The 150 could

The Big Husky harvesters were three- or four-row machines and had 1,000 lb of on-board storage which allowed making turns at the row ends without stopping to empty the renk bed.

The 350 beet harvester could dig three rows of beets and carry 7,500 lb in the tank. The 250 two-row harvester was similar to the 350 while the 150 one-row unit had a tank capacity of 4,000 lb.

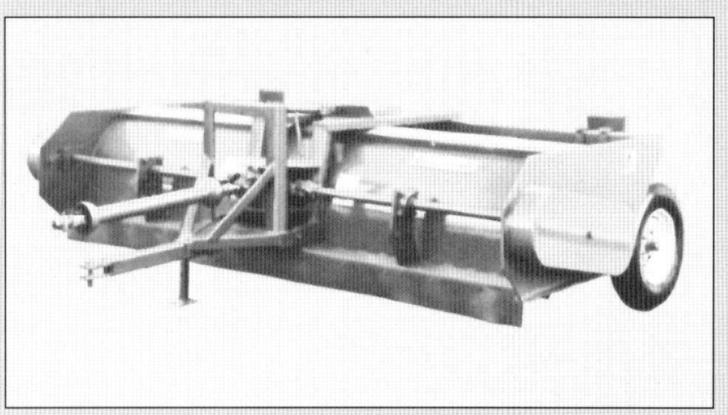

The Roto-Shredder was used to shred crop residue, devine potatoes, and top sugar beets. It was available in 84-, 128-, and 144-in. sizes.

A One-Ninety tractor and a 128-in. defoliator was removing leaves and topping beets ahead of the harvester.

A One-Ninety XT with a power mulcher preparing a fine seedbed for potatoes after the land was plowed. The mulchers were built in five sizes from 68 to 148 in.

lift one row, had a 22-inch rear elevator, and 4,000-pound capacity tank. The 250 was a two-row machine with a 26-inch rear elevator and 7,500-pound tank. The big 350 could dig two or three rows, had a wide 38-inch rear elevator, and a 7,500-pound tank. All three models had 26-inch unloading elevators and were equipped with tandem wheels for good flotation in soft field conditions.

The beet defoliator (also called a beet topper) came in two sizes: 90-inch with 36 flails and 128-inch with 56 flails. The purpose of the defoliator was to remove the leaves as completely as possible with minimum damage to the beets. If the topper chopped off the tops of the beets, profit was left in the field. Two horizontal shafts with flails removed the tops. The front shaft with steel flails got most of the greens, then the second shaft with rubber flails finished the job. The Roto-Shredder was a versatile machine, capable of shredding corn or cotton stalks, devining potatoes, or with rubber flails, topping

The potato windrower dug two rows of potatoes and conveyed them to two adjacent undug rows. The following potato harvester could then harvest four rows, nearly doubling its productivity.

The 230 narrow-row and 240 wide-row potato harvesters required an 80-PTO hp tractor for power. There was a choice of five digger points to adapt the machines to a wide range of soil and rock conditions. There was space for up to eight workers on the sorting platform for high field capacity.

sugar beets. It was offered in 80-, 84-, 128-, and 144-inch widths. An optional scalping attachment could scalp and polish beets in one pass.

The soil mulcher was used to prepare a fine seedbed after plowing. Power take-off driven, it was available in either three-point hitch-mounted or pull models. Five models were offered with 68- to 148-inch widths of cut.

Potato Harvesters

The Model 240 potato windrower was used to dig two rows of potatoes and windrow them on two adjacent undug rows, allowing the potato harvester to handle four rows, thereby increasing productivity by about one-third. The digging mechanism was the same as the larger potato harvester. Wheel tread was adjustable to match row widths and digging depth was hydraulically controlled from the tractor.

The potato harvesters were available in two row widths: the 230 for two 28- to 32-inch rows, and the 240 for two 34- to 40-inch rows. Except for width, the two were similar. The 160-cubic inch engine was available for auxiliary power.

There was a choice of five digger blades to match wet or dry, sandy or tight soils, and rocky or clean conditions. The straight-through deviner chain delivered vines out the rear of the harvester with minimum wrapping. Trash, rocks, and clods stayed on the high side of the cross conveyor while potatoes rolled gently to the low side. The sorting elevator was slightly inclined to keep potatoes from bunching or rolling back and the divider board helped separate trash from potatoes. The sorting table was hydraulically driven with speed adjustable from 0 to 180 feet per minute to prevent overcrowding and piling. Up to eight people could man the sorting table if needed. The 34-inch rubber-covered unloading elevator was hydraulically adjusted to control drop height and reduce bruising. Tandem wheels on walking-beam axles provided flotation and a smoother ride over ridges. The beet and potato harvester line was marketed by Allis-Chalmers from early 1966 through about 1970.

Cotton Picker Production

	Two-Row	622	616	One-Row	122	116
1949	25	---	---	---	---	---
1950	100	---	---	---	---	---
1951	600	---	---	---	---	---
1952	1,200	---	---	40	---	---
1953	443	---	---	500	---	---
1954	0	---	---	750	---	---
1955	92	---	---	0	---	---
1956	0	---	---	0	---	---
1957	0	---	---	0	---	---
1958	50*	---	---	50	---	---
1959	100†	---	---	27	---	---
1960	225†	50	---	0	---	---
1961	---	201	150	---	---	---
1962	---	178	112	---	2	10
1963	---	51	146	---	23	143‡
1964	---	81	150	---	40	146‡
1965	---	69	30	---	0	0
1966 (Replaced by 802 and 801, INA)						

* Built on D17 but not Model 500.
† Model 500 production.
‡ Includes one D19 one-row experimental unit each year.

A-C Corn Picker Production

Two-row mounted corn harvester	43,551
Sweet corn harvester	409
Model 33 and 34 two-row harvester	12,400
One row pull-type picker, all models	6,500
Model 14 two-row	200
Model 17 two-row	2,150
Model 150 two-row	250
Model 170 two-row	750
Model 190 two-row	1,000
Total	67,210

Lawn and Garden Tractors

During the late 1950s a new market rather suddenly appeared in the farm equipment business. Lawn mowers which had seats and steering wheels began to look like the big tractors. With a lot more pizazz than a riding mower, they caught on in a hurry.

Farm equipment companies were quick to realize that this was a parallel business to the big farm lines, and their search began for lawn and garden products with an emphasis on the tractors. Some companies developed products from scratch, while others turned to established manufacturers for private brands. Allis-Chalmers was in the latter category. Simplicity Manufacturing Company, just 40 miles north in Port Washington, Wisconsin, was well-established in the business, dating back to the truck gardeners' walk-behind garden tractors. They agreed to sell their 7-horsepower tractor and its line of attachments to A-C. In late 1961 the Model B-1 joined the line. It was exactly what the dealers and company personnel wanted.

All A-C center-mounted mowers were attached to the front axle so they followed the front wheels over uneven terrain which helped to eliminate scalping.

The B-1 had a 7 1/4-horsepower Briggs and Stratton one-cylinder engine and a three-speed transmission with speeds of 2.0, 3.75, and 6.0 mph. An optional low-speed drive reduced each speed 50 percent. Electric starter was standard equipment and live power outlets were located at the front, center, and rear for attachments. There was a complete line of implements and attachments, including a 42-inch rotary mower, 60-inch three-gang mower, 10-inch moldboard plow, cultivator, disk harrow, 42-inch

The B-1 was A-C's first garden tractor, announced in late 1961 for spring 1962 sales. It had a 42-in. mower and a 7 1/4-hp Briggs and Stratton engine with a three-speed transmission.

The Big Ten came in 1965 when the engine displacement was increased 1 in.^{3} Otherwise it was similar to the B-10.

dozer blade, 42-inch center-mounted grader blade, 32-inch tiller, snow thrower, and a sicklebar mower. Simplicity center-mounted mowers were attached to the tractor's front axle so that the mower always followed the front wheels over ridges and dips to eliminate scalping. This feature was incorporated into all the rider mowers and tractors, and remains an advantage to the present day.

Exactly like their farm tractor counterparts, bigger was better. By 1963 the B-1 was phasing out and being replaced by the B-10, a 9-horsepower tractor also powered by a Briggs engine. For one season, 1965, the little yellow tractor was known as the Big Ten, when the Briggs engine displacement was increased from 23 to 24 cubic inches. (After the first B-10, the last two digits of the model number indicate engine horsepower, unless otherwise stated.) It reverted to B-10 when the B-12 joined the team in late 1966. Except for the 12-horsepower engine and slighter larger tires on the B-12, the tractors were alike. Both manual and electric starting were offered. The line of attachments had grown both in number and size to match the increased tractor power.

Simplicity Purchased by Allis-Chalmers

The company purchased Simplicity in 1965. It was operated as a wholly-owned subsidiary with the name unchanged. Shortly after the buyout, the acquisition was challenged by the federal government on the basis that the two distribution channels, Simplicity and Allis-Chalmers, constituted too great a

The B10 and B12 were the garden tractor team in 1966. Briggs engines and three-speed transmissions still did the work.

The B112 (above) and the B110 were upgraded models based on the prior B-10 and B-12 tractors and continued the yellow color scheme.

share of the market and limited competition. Faced with the possible divestiture of Simplicity, the company built a new factory in Lexington, South Carolina, to assure a supply of products. The Lexington plant opened in 1968.

Lexington Plant Products

The first tractors built in the new plant were the B207, B110, and B112, all of which were based on prior models and continued the yellow color scheme. The B207 had a vertical-crankshaft engine while the others had horizontal-crankshaft engines, all from Briggs and Stratton.

In 1970 the B208, B210, and B212 were upgraded from the previous models and for the first time a hydrostatic drive model, the HB212, was offered. The B210 and B212 got a new variable-speed drive which, coupled with their three-speed transmissions, gave speeds from 0.8 to 6.6 mph. The HB212 hydro had speeds from 0 to 7.2 mph.

By 1971 a new color scheme — cream hood, orange body, and cream wheels — had been adopted. The line consisted of the following tractors: the 206, 207, 207E (electric start), and the 208, all powered by vertical-crankshaft Briggs and Stratton engines. The 200's all had belt drive to three-speed transmissions. They were continued as the Homesteader 6, 7, and 8 for 1972 with similar specifications. Only the Homesteader 8 remained in 1974.

A new line of larger tractors designed in the Lexington plant was also introduced in 1971. They included the 310, 310D, 312, 312D, 312H, 314, 314D, and 314H. Styling was all new. All models had electric starter, the "D" tractors had electric lift and

The 1971 Homesteader line consisted of 6, 7, and 8-hp lawn tractors and sported the new orange and cream color scheme.

The 416 hydrostatic-drive tractor was the top of the Lexington-built 400 line. With a 48-in. mower and vacuum collector it manicured big lawns.

power take-off, while the "H" tractors added hydrostatic drive.

The 300 Series used Kohler single-cylinder horizontal-crankshaft engines. Gear-drive models had a variable drive with seven ranges in each of three gears for speeds from 0.8 to 5.7 mph. The hydro models had speeds of 0 to 8.1 forward and 0 to 3.8 mph reverse. The 300 Series was replaced by the 400 Series in early 1973. Three sizes were offered: the 410, 414, and 416. The 410 had a standard three-forward, one-reverse transmission with a new four-forward, four-reverse shuttle drive optional. The shuttle was standard on the 414 and 416 with hydrostatic drive as the option.

The 300 Series of 1971 were the first to have Kohler engines along with the new colors. The gear-drive tractors had a seven-speed variable drive with three ranges for a total of 21 forward speeds.

The legal dispute was settled and Allis-Chalmers retained Simplicity. This led to closing the Lexington plant in 1973 and the return of all production to Port Washington the following year.

The 616, 620, and 720 Tractors

Simplicity developed a larger hydrostatic-drive tractor which A-C began marketing as the Model 616 in 1972. The 616 was somewhat larger than the typical garden tractor of that time, and was much like a small farm tractor. Styled like a One-Ninety farm tractor, it had a big tractor image. The 616 was powered by an Onan 16.5-horsepower two-cylinder engine which was smooth and quiet. With the three-speed transmission, the hydrostatic drive provided speeds of 0 to 10.4 mph forward and 0 to 6.2 mph in reverse. The 616 had a category 0 three-point hitch and power take-off as standard equipment, and a complete line of implements was available.

In 1973 the engine power was increased to 19.5 horsepower and the Model 616 became the 620. Black decals matched the farm tractors. Except for power, specifications remained unchanged. Two years later, that tractor became the Model 720, with engine power remaining at 19.5 horsepower. Sheet metal was not changed but new decals gave it a fresh look. The 720 was a fully-equipped unit with three-point hitch and power take-off, but a new 720

The 720 with a 650-lb capacity all-hydraulic loader handled the heavy work on small farms.

More a small farm tractor than a garden tractor, the 16.5-hp Model 616 hydrostatic drive tractor had a complete line of implements and accessories to handle all kinds of jobs.

The 1973 Model 620 had 19.5 engine hp. The Onan twin-cylinder engine was smooth and quiet. Hydrostatic drive with three-speed transmission provided speeds from 0 to over 10 mph.

Special without those features was offered for gang mowers, loaders, and other operations not needing these features.

A complete line of implements for the 720 included a 60-inch rotary mower, 57-inch sicklebar mower, 52-inch snow thrower, 60-inch grader and dozer blades, and a loader with 650-pound capacity. Companion equipment from other manufacturers included a backhoe, landscaping implements, and other special purpose tools.

Simplicity-Built Products

When the Lexington plant closed in 1973, the tractors built there were discontinued in favor of a line which was common with the Simplicity products. Two series were in the 1974 line, the 700 and the 600.

The 700 Series consisted of four models all powered by Kohler engines: the 710 three-speed, 712 shuttle, 712 hydro and 716 hydro. Power was transmitted by a shaft from the engine to a bevel-gear box, then by belt to the transmission. Styling was a miniature version of a 175 or 185 farm tractor and two small optional lights were arranged vertically on each side of the grill. The lights were moved inside the grill in 1976. The 710 six-speed replaced the three-speed and a 716 six-speed was added. In 1978 a new 718 hydro became the flagship of the line, then in 1979 the 900 Series replaced the 700's.

The 1974, 608 and 610 garden tractors had vertical-crankshaft Briggs engines and three-speed transmissions. Attachments included the 36- and 42-inch mowers, a 30-inch tiller, and a 32-inch snow thrower. Lawn tractors, by definition, are limited to non-ground-engaging attachments while garden tractors, which have more rugged power trains, may be equipped with tillers, plows, and cultivators, etc.

The 1974 700 Series was manufactured by Simplicity and had five sizes from 10 to 18 hp with gear, shuttle, and hydrostatic drive.

The 608 (above) and 610 garden tractors had the same frame design as the Lexington 200 Series tractors. The front frame and axle were a single component and the frame swiveled just ahead of the rear wheels.

The 800 Series Garden Tractors

For 1976, the 808GT and 810GT (garden tractors) replaced the 608 and 610, although specifications and attachments were similar. The 811GT supplanted the 810GT in 1978, and in 1979 they became the 810GT II and 811GT II with heavier transmissions. The same year the 816GT, with a new Briggs twin-cylinder engine and 48-inch mower, came out. The 818GT with a twin-cylinder Briggs engine, 42-or 48-inch mower, and 42-inch snow thrower topped the GT line in 1980. Based on the GT tractors, the T811 and T816 replaced them in 1981. The big change was the Vari-Shift drive which provided variable speeds in each of the three-forward and one-reverse gears. Speeds ranged from 0.6 to 5.5 mph forward and 2.4 to 4.4 mph in reverse. Briggs and Stratton vertical-crankshaft engines still provided the motivation.

600 Series Lawn Tractors

In 1975 the 608LT lawn tractor became the first of a lower-priced line for users who had no need for tillers or other ground-engaging attachments. This unit was joined in 1977 by the 611LT. These were built on lighter chassis than the earlier 608 and 610 garden tractors. They had Briggs vertical-shaft engines and three-forward, one-reverse transmissions. In 1979 they received five-forward, one-reverse transmissions and became the 608LTD and 611LTD tractors. By 1982 the 611H and 616H extended the 600 Series. Both had hydrostatic drive and the 616H had a two-cylinder engine and was built on the heavy frame so it could carry a 48-inch mower. Soon after, the 811 and 816 Specials were added. They were also built on the heavy frame and had five-forward, one-reverse gear drive.

The 900 Series

The new 900 Series replaced the 700 line in 1979. Models included the 910 six-speed, 912 hydro, 914 shuttle, 916 hydro, and 917 hydro. Kohler engines turned the wheels through drive trains very similar to the earlier 700 series. Except for the new two-cylinder 917 hydro, all had single-cylinder engines. In 1981 the 912 six-speed replaced the 910 and the 919

The T816 Vari-Shift with a 10-in. plow worked like the big ones. Vari-Shift was a variable V-belt drive with speeds from 0.9 to 6.5 mph. The T816 had a two-cylinder Briggs and Stratton engine.

The 611LTD lawn tractor with a 36-in. mower deck and vacuum collector groomed a suburban lawn to near perfection.

The 919 twin-cylinder hydrostatic-drive tractor was the largest in the 900 Series. It had plenty of power for a 48-in. mower or a 38-in. tiller.

twin-cylinder hydro was added. For 1982, the 914 traded its shuttle for a hydrostatic drive. During this period there was increasing interest in diesel power, particularly among commercial users. The 920, powered by a 18.5-horsepower two-cylinder Briggs and Stratton-Lombardini diesel engine, was offered for this market. With hydrostatic drive the 920 had 0 to 7 mph forward and 0 to 4 mph reverse speeds. Attachments included a 48-inch mower, 42-inch snow thrower, and 38-inch heavy-duty tiller, among many others.

In 1985 the line consisted of 14 lawn and garden tractors: the 608LTD, 611LTD, 611H, 611 Special, 616 Special, and 616H lawn tractors, the T811 and T816 Vari-Drive garden tractors, and the 912H, 914H, 916H, 917H, 919H, and 920H hydrostatic garden tractors. All of these tractors became Deutz-Allis products after May 1985.

Besides the lawn and garden tractors and their attachments, Allis-Chalmers offered a broad spectrum of outdoor power and leisure products. Among them were riding mowers, walk-behind mowers, snow-blowers, tillers, chain saws, and many smaller items. Leisure products included a line of off-road motor bikes and scooters, golf carts, and small commercial vehicles. Most of the leisure products met with limited market success and were soon dropped from the line.

The Terra-Tiger

The most interesting of this group of special products was the amphibious vehicle known as the Terra-Tiger. Equipped with six 11×20 low pressure (2 psi) tires and a watertight fiberglass body, it was at home both on water and land. The early models were powered by a JLO L252 two-stroke cycle engine of 15.07 cubic inches that developed 10.1 horsepower. Later versions had a JLO L295E with 17.8 cubic inches and 17.75 horsepower. A small torque-converter transmitted power to the six wheels. Controls were simple: a T-bar for steering, throttle, and brake.

The Terra-Tiger could carry two people and had 11 cubic feet of storage for equipment, with a total load capacity of 600 pounds. The more powerful model was rated at up to 30 mph on land, 2 mph in water. It was claimed to climb up to 45 degree inclines.

The Terra-Tiger was a fun machine and it found many uses. Farmers, hunters, fishermen, and outdoorsmen used it for work and play everywhere. Machines like the Terra-Tiger were the forerunners of the present day three- and four-wheel all-terrain vehicles.

The Terra-Tiger was a fun-machine. Up to 30 mph on land, 2 mph in water, and capable of climbing 45 degree slopes, it could take two people just about anywhere they might want to go.

The Beginning of the End

The farm equipment industry blazed out of the "Soaring Seventies" right into the "Excruciating Eighties," one of the most severe recessions to hit American agriculture in many decades. Farmers stopped buying. Equipment companies continued to build machinery in light of optimistic forecasts for a short recession. Inventories filled the pipeline and overflowed when the forecasts were dead wrong. The market dropped to levels below everyone's wildest imagination and stayed there. Huge discounts to move the bloated supply of big ticket items such as tractors and combines began to bleed the financial resources of every company. When the bleeding became a life-threatening hemorrhage some companies, including Allis-Chalmers, were unable to survive in their original form. A-C was unable to outlast the economic drought and began to spin off assets in an effort to maintain core enterprises. One of the first to be sold was the Simplicity Manufacturing Company but they continued to supply merchandise to A-C.

The Agricultural Equipment Division was reorganized as a separate entity, the Allis-Chalmers Agricultural Equipment Company, in preparation for its dispersal. Discussions were held with four major equipment companies to find a mutually beneficial combination. One of these was Klockner-Humboldt-Deutz (KHD) of Cologne, West Germany.

When KHD bought the company in May 1985, they did not buy the tractor plant, having no intention of building tractors in the United States. They did, however, purchase 6000, 8000, and 4W305 production from May until December 6, 1985. To show that the new firm, Deutz-Allis, intended to honor all warranties, etc., Deutz-Allis identification was placed over the Allis-Chalmers hood decals and grill emblems. Production tractors and all field inventory tractors except the Model 6140 received the new identification.

The last tractors came off the West Allis assembly line on December 6, 1985. The last 8000 Series tractor was an 8070 FWA Serial No. 3354. The last tractor was a 6070 Serial No. 1972.

December 6, 1985

There was no way to describe the atmosphere at the West Allis tractor plant on December 6, 1985. Hundreds of people, some of whom had spent all their working lives building orange tractors, were about to walk out gate No. 7 for the last time. It was a sad day for anyone who had been a part of A-C tractor history.

The last 8000 Series tractor was an 8070 FWA Serial No. 3354. This tractor was eventually shipped to Saudi Arabia. The last tractor off the line was a Model 6070 Serial No. 1972 and it found a home in Racine County, Wisconsin. As of mid-1993 the land where the tractor plant stood is bare ground, awaiting development.

Life Goes On . . .

Although the West Allis tractor plant exists only in the memories of hundreds of dedicated people and in books like this, the story did not end for other elements of the Allis-Chalmers Agricultural Equipment Company.

LaPorte was closed and many of the implements were manufactured for Deutz-Allis by the Vermeer Mfg. Co. of Pella, Iowa. These included the chisel plows and Min-Til, disk harrows, field cultivators, and a new planter line which had been developed by LaPorte. Most of those products have since been discontinued.

Deutz-Allis continued building the famous Gleaner combines in the Independence plant, and in 1989 produced a line of large tractors from 130 to 193 PTO horsepower. These tractors did not share any components with the former large A-C tractors.

Deutz-Allis was taken over by the Agco Corp., a domestically-owned company, in 1990. Since then their product lines have grown extensively and the familiar orange paint has returned to a full line of Agco-Allis tractors. The best news for owners of A-C equipment is that the Agco Corporation is committed to on-going parts and service availability for the orange machines. With immense facilities in Batavia, Illinois; Coldwater, Ohio; and Independence, Missouri, for manufacturing, parts inventories, and service capability, Agco can assure A-C owners that their equipment will have long, efficient, and productive lives.

Vorkers took each others' pictures at Allis-Chalmers. One worker captured Nevers Pittman (left) nd Eddie Beadle shaking hands as the last tractor rolled past their work station.

Last tractor rolls off line at A-C

By PATRICK LEE
Journal labor reporter

West Allis — As Hubert Patterson left the Allis-Chalmers Corp. tractor plant for the last time Friday, his voice was calm but there was pain in his eyes — and bitterness.

He was losing his job.

"The whole future looks down," said Patterson, 40. "When you build your future around something, you want to send your kids to college, to get a better education than you, then how would you feel? It's unexplainable."

Patterson found himself out on the street after almost 20 years of service when the Allis-Chalmers tractor plant shut down permanently Friday afternoon. He was one of about 100 remaining production workers laid off when the plant stopped production, as announced in April.

The closing brought to 800 the number of workers laid off from Allis-Chalmers, the union said. A skeleton crew of repairmen and cleanup workers will remain at the plant for the next couple of weeks to wrap things up.

The closing also marked the end of most production at the Allis-Chalmers West Allis complex, and the end of an era for the company that had once been the state's largest employer.

"You know what it is, it's a sad story, that's what," said Bob Myszewski, 35, a painter who has worked at A-C for 12 years.

Roger Mocherman, general manager of the tractor plant, said in a prepared statement:

"The end of tractor production at West Allis is an event we have planned for since the company made its decision to get out of the agricultural equipment business some months ago when we sold that business to Kloeckner-Humboldt-Deutz [a West German company]."

"As we view the continued deterioration of the agricultural economy today, it was the right business decision for Allis-Chalmers," he added. "Yet as manager of the tractor plant, I think I share with my

Please see **Plant,** Page 22

The tractor plant closing was covered by the Milwaukee Journal of December 7, 1985.

West Allis Tractors Production (1914-1931)

	10-18	6-12, 6-12 Cane	15-30, 18-30	20-35	25-40	12-20, 15-25
1914	1	---	---	---	---	---
1915	INA*	---	---	---	---	---
1916	INA*	---	---	---	---	---
1917	INA*	---	---	---	---	---
1918	INA*	---	---	---	---	---
1919	71	303	96	---	---	---
1920	31	467	694	---	---	3
1921	45	72	104	---	---	7
1922	38	249	---	181	---	201
1923	4	211	---	334	---	431
1924	---	99	---	315	---	229
1925	---	8	---	635	---	352
1926	---	---	---	682	---	358
1927	---	---	---	1,814	---	36
1928	---	---	---	4,867	---	---
1929	---	---	---	4,339	---	---
1930	---	---	---	934	657	---
1931	---	---	---	---	128	---
Total	190†	1,409	894	14,101	785‡	1,617

* Information not available.
† Incomplete, records for 1915 through 1918 not available.
‡ Continued as E in production records, but is same tractor.

West Allis Tractor Production (1947-1958)

	WD	WD45	WD45D	CA	G
1947	---	---	---	---	6
1948	9,236	---	---	---	10,955
1949	26,200	---	---	10	12,219
1950	36,870	---	---	305	826
1951	32,861	---	---	10,218	1,263
1952	26,043	---	---	11,644	1,228
1953	15,170	13,871	---	9,181	1,539
1954	---	30,244	350	1,484	1,000
1955	---	23,725	3,319	4,296	940
1956	---	9,931	2,397	1,415	---
1957	---	6,259	423	347	---
1958	---	---	---	537	---
Total	146,380	84,030	6,489	39,437	29,976

West Allis Hundred Series Production (1964-1981)

	170	175	180	185	190	190XT	200	210	220	440*
1964	---	---	---	---	1,502	---	---	---	---	---
1965	---	---	---	---	1,384	4,293	---	---	---	---
1966	---	---	---	---	568	3,837	---	---	---	---
1967	1,700	---	1,700	---	564	4,221	---	---	---	---
1968	2,673	---	3,157	---	347	2,587	---	---	---	---
1969	995	---	3,141	---	115	2,552	---	---	934	---
1970	623	478	1,320	941	74	3,158	---	103	530†	---
1971	693	159	1,162	977	76	1,771	---	961	161	---
1972	1,097	104	731	840	73	---	2,342	407	240	138
1973	1,512	367	539	1,202	---	---	2,507	---	---	199
1974	---	1,146	---	1,582	---	---	3,006	---	---	300
1975	---	499	---	1,804	---	---	2,271	---	---	302
1976	---	844	---	1,574	---	---	---	---	---	---
1977	---	656	---	1,553	---	---	---	---	---	---
1979	---	685	---	1,513	---	---	---	---	---	---
1980	---	504	---	975	---	---	---	---	---	---
1981	---	---	---	321	---	---	---	---	---	---
Total	9,293	5,442	11,750	13,282	4,703	22,419	10,126	1,471	1,865	939

* Purchased from Steiger.
† Includes 100 with front-wheel assist.

West Allis Tractor Production (1929-1957)										
	U	UC,cane	UI	WC	E	A	B,IB	WF	RC	C
1929	1,962*		---	---	---	---	---	---	---	---
1930	4,579*	36†	---	---	---	---	---	---	---	---
1931	707*	1,058†	---	---		---	---	---	---	---
1932	156	131	---	---	53	---	---	---	---	---
1933	267	63	---	25	38	---	---	---	---	---
1934	387	258	---	3,096	248	---	---	---	---	---
1935	1,474	423	---	10,743	272	---	---	---	---	---
1936	1,338	740	---	17,912	28	25	---	---	---	---
1937	2,032	1,047	62	29,208	---	578	90	285	---	---
1938	738	753	81	14,441	---	310	11,699	1,047	3	---
1939	329	189	40	16,321	---	168	21,707	555	4,385	---
1940	453	230	92	12,017	---	114	16,189	408	1,040	104
1941	271	80	80	10,972	---	19	7,036	405	88	12,284
1942	265	---	135	8,765	---	11	4,619	300	---	6,384
1943	332	---	2	4,164	---	---	1,001	---	---	5,126
1944	18	30	---	6,957	---	---	4,708	188	---	6,787
1945	---	---	1	13,466	---	---	2,054	318	---	5,683
1946	---	---	---	4,791	---	---	1,109	238	---	2,790
1947	307	208	1	17,349	---	---	5,692	363	---	12,347
1948	968	250	---	8,109	---	---	10,509	1,389	---	16,766
1949	1,078	118	---	---	---	---	10,374	1,817	---	12,236
1950	615	162	---	---	---	---	9,458	999	---	3,513
1951	337	133	---	---	---	---	4,486	38	---	---
1952	396	204	---	---	---	---	3,987	---	---	---
1953	---	75	---	---	---	---	2,169	---	---	---
1954	---	---	---	---	---	---	735	---	---	---
1955	---	---	---	---	---	---	2,016	---	---	---
1956	---	---	---	---	---	---	794	---	---	---
1957	---	---	---	---	---	---	351	---	---	---
Total	19,009	6,188	494	178,336	639	1,225	120,783	8,350	5,516	84,020

* Called United in production records.

† Called Cultivator in production records.

West Allis D-Series Production (1957-1969)											
	D10	D12	D14	D15	D15D	D17	D17D	D19	D19D	D21	D21II
1957	---	---	8,425	---	---	2,710	517	---	---	---	---
1958	---	---	5,526	---	---	9,725	2,539	---	---	---	---
1959	933	734	6,129	---	---	8,927	2,125	---	---	---	---
1960	752	864	2,212	683	216	3,276	1,267	---	---	---	---
1961	478	150	---	3,915	654	3,970	994	64	1	---	---
1962	1,009	637	---	1,614	87	3,777	1,050	3,927	2,330	---	---
1963	874	411	---	3,709	197	4,664	915	2,043	901	414	---
1964	377	324	---	2,621	138	4,658	921	921	394	664	---
1965	286	315	---	1,124	559	2,850	564	---	---	50	194
1966	281	284	---	2,132	112	4,395	869	---	---	---	421
1967	191	229	---	1,392	73	2,646	523	---	---	---	922
1968	123	122	---	284	15	---	---	---	---	---	681
1969	---	---	---	---	---	---	---	---	---	---	111
Total	5,304	4,070	22,292	17,474	2,051	51,598	12,284	6,955	3,626	1,128	2,329

West Allis 7000 Series Production (1973-1981)

	7000	7000*	7030	7050	7010	7020	7040	7045	7060	7080	7580	8550
1973	---	---	1,583	683	---	---	---	---	---	---	---	---
1974	---	---	1,813	1,619	---	---	286	---	289	3	---	---
1975	639	---	---	---	---	---	2,830	---	1,456	566	---	---
1976	3,385	---	---	---	---	---	2,711	---	1,833	480	284	---
1977	1,346	---	---	---	---	289	1,453	218	1,061	430	320	75
1978	393	950	---	---	---	1,438	---	930	793	267	216	265
1979	---	551	---	---	918	1,109	---	1,245	906	379	267	210
1980	---	---	---	---	880	868	---	827	749	305	232	170
1981	---	---	---	---	638	511	---	688	709	273	291	253
Total	5,763	1,501	3,396	2,302	2,436	4,215	7,280	3,908	7,796	2,703	1,610	973

* 7000 with wet clutch.

West Allis 6000-8000-4W Series Production (1980-1985)

	6060	6070	6080	8010	8030	8050	8070	4W220	4W305
1980	296	---	151	---	---	---	---	---	---
1981	1,166	---	1,850	19	8	15	3	---	---
1982	1,431	---	1,565	692	1,084	908	426	78	108
1983	678	---	1,213	554	608	572	660	64	64
1984	456	608	1,073	343	445	691	813	31	162
1985	---	364	846	224	183	150	452	---	75
Total	4,027	972	6,698	1,832	2,328	2,336	2,354	173	409

Imported Tractors

	ED40	160	6040	5040	5045	5050	5020	5030	5015	6140
1963-64	450	---	---	---	---	---	---	---	---	---
1969-70	---	INA*	---	---	---	---	---	---	---	---
1971	---	995	---	---	---	---	---	---	---	---
1972	---	587	---	---	---	---	---	---	---	---
1973	---	900	---	---	---	---	---	---	---	---
1974	---	---	499	---	---	---	---	---	---	---
1975	---	---	---	411	---	---	---	---	---	---
1976	---	---	---	1,630	---	405	---	---	---	---
1977	---	---	---	1,772	---	1,505	1,218	---	---	---
1978	---	---	---	INA*	---	INA*	871	1,005	---	---
1979	---	---	---	INA*	---	INA*	1,024	250	---	---
1980	---	---	---	INA*	---	1,023	1,025	721	---	---
1981	---	---	---	---	599	526	1,244	544	---	---
1982	---	---	---	---	---	520	1,354	546	728	446
1983	---	---	---	---	---	73	346	148	1,550	402
1984	---	---	---	---	---	---	483	155	955	INA*
1985	---	---	---	---	---	---	INA*	INA*	INA*	INA*

* Information not available.